T0382925

NEW ERGONOMICS PERSPECTIVE

SELECTED PAPERS OF THE 10TH PAN-PACIFIC CONFERENCE ON ERGONOMICS, TOKYO, JAPAN, 25–28 AUGUST 2014

New Ergonomics Perspective

Editor

Sakae Yamamoto

Department Management Science, Tokyo University of Science, Tokyo, Japan

Co-Editors

Masahiro Shibuya

Division of Management Systems Engineering, Tokyo Metropolitan University, Tokyo, Japan

Hiroyuki Izumi

Institute of Industrial Ecological Sciences, University of Occupational and Environmental Health (UOEH), Kitakyushu, Japan

Yuh-Chuan Shih

Department of Logistics Management, National Defense University, Taipei, Taiwan

Chiuhsiang Joe Lin

Department of Industrial Management, National Taiwan University of Science and Technology, Taipei, Taiwan

Hyeon-Kyo Lim

Department of Safety Engineering, Chungbuk National University, Cheongju, Korea

CRC Press
Taylor & Francis Group
Boca Raton London New York Leiden

CRC Press is an imprint of the
Taylor & Francis Group, an **informa** business

A BALKEMA BOOK

CRC Press/Balkema is an imprint of the Taylor & Francis Group, an informa business

© 2015 Taylor & Francis Group, London, UK

Typeset by V Publishing Solutions Pvt Ltd., Chennai, India
Printed and bound in Great Britain by CPI Group (UK) Ltd, Croydon, CR0 4YY

Published by: CRC Press/Balkema
            P.O. Box 11320, 2301 EH Leiden, The Netherlands
            e-mail: Pub.NL@taylorandfrancis.com
            www.crcpress.com – www.taylorandfrancis.com

ISBN: 978-1-138-02751-0 (Hbk)
ISBN: 978-1-315-71436-3 (eBook PDF)

Table of contents

5   *Human computer interaction*

6 Current issue and ergonomics approach

New Ergonomics Perspective – Yamamoto (Ed.)
© *2015 Taylor & Francis Group, London, ISBN 978-1-138-02751-0*

Preface

The 10th PPCOE2014 was held at Tokyo Metropolitan University (Tokyo, Japan) from 25 to 28 August 2014. About 80 researchers from eight countries participated and 80 papers were presented. This book is composed of a selection of papers chosen by the PPCOE2014 Program Committee. You may be aware that this PPCOE (Pan Pacific Conference on Ergonomics) 2014 conference celebrates the 10th occasion that the meeting is held in a time span of 20 years. Our society, and the research fields covered, have changed considerably in these past 20 years. Firstly, caused by the development of ICT (Information and Communication Technology), i.e. the internet and various Social Networking Systems (SNS). These technology developments have changed the way of working as well as the production systems. The second reason is a social situation, caused by the transition from an aged society to an aging society. That means that research on how to educate the younger generation for communicating the know-how of older workers becomes necessary. Researchers must investigate support methods and the environment for the elderly further, and development and evaluation of the assistive technology for elderly people becomes necessary. These changes have an impact on our research scopes and themes. This 10th conference is a milestone as our research fields are developing now. If we understand this fact, we will know that it is essential to re-evaluate our research aims and scope. This meeting will provide a platform and a place to consider how to change to go into this new direction.

I would like to thank Emeritus Prof. Masaharu Kumashiro, the president of PPCOE, who gave me the opportunity to convene this conference. Also, I would like to acknowledge the help of Professor Masahiro Shibuya as secretary general of the PPCOE 2014 conference. Finally, a word of thanks to all participants.

Prof. Sakae Yamamoto, Ph.D.
Chair of the Pan Pacific Conference on Ergonomics (PPCOE) 2014
The president of the Society for Occupational Safety, Health and Ergonomics (SOSHE)

New Ergonomics Perspective – Yamamoto (Ed.)
© *2015 Taylor & Francis Group, London, ISBN 978-1-138-02751-0*

Message from the President: Storyteller of PPCOE history

Masaharu Kumashiro
Association for Preventive Medicine of Japan, Japan

Established 24 years ago, the PPCOE is now at a turning point, transitioning from a stage of growth to maturity. I would like to take this opportunity afforded me by the 10th PPCOE 2014 Conference Chair to share with you the history, and reveal the spirit and intent of PPCOE.

1 IT ALL BEGAN IN BEIJING IN OCTOBER 1988

In October 1988, the International Conference on Ergonomics, Occupational Safety and Health and the Environment was jointly organized by the Chinese Society of Metals and the Darling Downs Institute of Advanced Education in Australia. This conference, held at the Beijing International Hotel, can be said to be the beginning of international ergonomics interaction in China.

I was invited to be the keynote speaker at that conference. One night during the conference, I was visited by Professor Lu Demao and his colleagues from the Wuhan-based Safety and Environmental Protection Research Institute (SEPRI), a pioneer in ergonomics studies in China which began ergonomics research in the 1980s. Professor Lu Demao wanted to discuss ways to obtain support for the development of ergonomics in China. My idea was to propose establishing a regular ergonomics forum between Japan and China and to create a network of ergonomists between our two countries. Since I am a person who prefers to take action, right there we decided to hold the first Japan-China conference in the summer of 1990.

Upon returning to Japan, I sought the cooperation of colleagues in Korea and Taiwan. With the strong support of Prof. Kyong S. Park, Chair of the Ergonomics Society of Korea, and Mr. C.C. Lin of the Taiwan Council of Labor, good progress was made on laying the groundwork for the conference.

In June of 1989, however, the Tiananmen Square Incident erupted, necessitating that we either delay the Japan-China Ergonomics Conference or change its name. I absolutely did not want to change the date of conference, so gave consideration to two new names. One was the Asian-Pacific Conference on Occupational Ergonomics and the other was Pan-Pacific Conference on Occupational Ergonomics. Later, I was occasionally asked about the philosophical background behind the PPCOE name, but to tell you the truth, it was just a name given at the spur of the moment. The philosophy behind the conference, however, was to incorporate the culture and spirit of Asia into ergonomics, a concept that was borne in the West during the first half of the 20th century.

2 THE BIRTH OF PPCOE

With less than two years of preparation, the 1st PPCOE was held in Kitakyushu in 1990, attended by 292 people from 18 countries including the former Soviet Union. Unfortunately, only six people attended from China. This was understandable given the economic and political conditions in China at the time.

Thinking back on it, attending was a "Who's Who" list of international ergonomists—nostalgic names such as IEA President Prof. Ilkka Kuorinka, IEA Secretary General Prof. Hal Hendrick, Prof. Nigel Corlett and Prof. Brian Shackel, all there to celebrate the birth of the

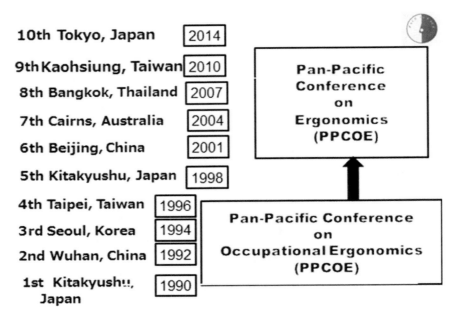

Figure 1.   History of PPCOE from 1990 to 2014.

PPCOE. We also had many exciting papers, such as a paper by Prof. Munipov from the Soviet Union on the analysis of human factors in the Chernobyl nuclear power plant disaster.

3   GROWTH

The first 10 years was a whirlwind of activity, something like the way we remember our childhood. Although the PPCOE is an organization, it was not firmly established to grow beyond being a volunteer activity. Nevertheless, by 2007, it had grown to an organization that included representatives from 10 countries with the addition of volunteers from Hawaii, Malaysia, and Vietnam.

Up until 1998, the PPCOE had been held every two years as a conference. Then, following that year, the "C" of PPCOE changed to stand for not only "Conference," but also for "Council." In other words, the PPCOE has both a regularly held conference and an ongoing council.

With each conference held in one of the countries, more ergonomists from a wider range of locations participated. For this reason, it was decided to again change the name of the organization. So PPCOE came to stand for Pan-Pacific Council on Ergonomics.

Figure 1 shows the countries and dates of the PPCOE that have been held. The conference was held every two years up to 1998. After that it has been held every three years.

The long awaited Asian Journal of Ergonomics was launched in September 2000. This was made possible by funding from the PPCOE Executive Council members so as to foster the development of ergonomists in the Asian-Pacific region.

4   FOUNDATION OF THE ERGONOMICS SOCIETY IN PPCOE COUNTRIES

Japan has the oldest ergonomics history in Asia. The Japan Ergonomics Society was established in December 1964 and joined the IEA in 1965.

By the way, Japan early on adopted human-centric engineering developed in the United States, which was introduced to the psychology community in 1919. Let's look at the year of

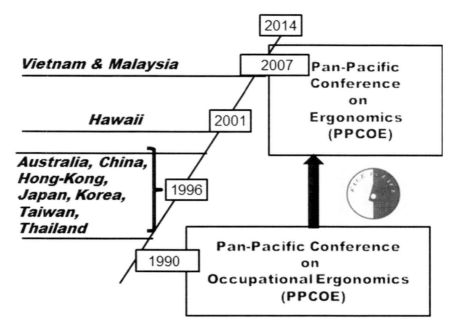

Figure 2.   PPCOE member nations and years of affiliation.

establishment for the ergonomics societies of the countries that were involved in the establishment of PPCOE. They are Australia in 1964, Korea in 1982, China in 1989, Taiwan in 1993 and Thailand in 2001. Hong Kong held an ergonomics advocacy symposium by PPCOE members in 1999. Using this as a stepping stone, the Hong Kong Ergonomics Society was born in 2001. An ergonomics advocacy symposium was also held in Vietnam in 2005. This made it possible to have Vietnam join us.

5   MOTTO OF PPCOE

PPCOE pursues ergonomics as a practical science. First of all, a data bank of success stories is to be provided through the PPCOE website and be open to access and use by anyone. And finally, the success stories of each country will be scientifically analyzed and made universally applicable. New theories may also arise from these success stories. Then we will aim to share the results with the world.

6   THE REBIRTH OF PPCOE

This was a quick summary of PPCOE's past to the present. Now, PPCOE activities are entering a new stage after 24 years of history. Fortunately, participants of the PPCOE Final in Tokyo are eager to take a first step on the road to its rebirth, the establishment of a new, mature PPCOE.

7   LOOKING FORWARD TO THE NEW PPCOE

As volunteers, we often come up against issues of financing that result in slow organizational growth and development. One way of countering this is to increase the number of

PPCOE member countries, creating a new PPCOE organization that fits with local conditions throughout the Asia Pacific region. Given this, the first priority of a resurgent PPCOE is to become a federation of volunteer ergonomists that is truly representative of the region's nations, and to this end, we should encourage participation through an established ergonomics research organization.

8 JOURNAL

Next is the launch of a new journal. The Asian Journal of Ergonomics was launched with contributions from PPCOE executive council members, but financial conditions forced its termination. We need to take this opportunity to plan the launch of a new journal, and to increase the number of subscribers concurrently with creating a truly representative Asia Pacific organization. The development of ergonomics researchers and practitioners in the Asia Pacific region is a major duty of PPCOE, and having a new journal will encourage up-and-coming graduate students to contribute.

9 FOCUSING ON ERGONOMICS EDUCATION

Ergonomics is an interdisciplinary field. Further, it is difficult to teach in undergraduate courses. In recent years it seems as if ergonomics education in the universities has been stagnating. Ergonomics will certainly fall into decline if we do not nurture and work to support the education of the next generation young ergonomists. As part of this, PPCOE has the duty to facilitate standardized ergonomics education and training. We should work to have universities located in each member country open their doors to students from PPCOE member countries. Then, based on this standardization, students will be enabled to participate in the courses of instructors in other countries from whom they want to learn and to receive credit that will be recognized by their home universities. To accomplish this, an Internet television lecture system can be used to provide courses among the universities of PPCOE member countries. In addition, for training in businesses and elsewhere, PPCOE members will be dispatched as instructors. Trainees can receive certificates and companies will be presented with awards.

Now, let's talk about our dreams, outline them, and realize them. The PPCOE founding board members look forward to the active participation of everyone. I sincerely hope that you will support the development of a new PPCOE.

BIOGRAPHY

Dr. Masaharu Kumashiro is President of Pan-Pacific Council on Occupational Ergonomics (PPCOE), and he was a Professor of the Department of Ergonomics at Institute of Industrial Ecological Sciences in the University of Occupational and Environmental Health (UOEH), Japan. Since April in 2012, he is emeritus professor of UOEH, Japan and is the president of the Association for Preventive Medicine of Japan. He also has consulting jobs in Mitsubishi Research Institute as a Visiting Research Fellow. He works at Fukuoka Occupational Health Support Center as an occupational health consultant, and he is a fellow at National Institute of Occupational Safety and Health, Japan. His specialties are; Aging and Work including "work ability" & "employability", Occupational Stress, Job Re-Design, work-related MSD and Developing Supporting System and Human Computer Interaction. He is a Fellow member of the Institute of Ergonomics & Human Factors in UK, and JES Certified Professional Ergonomist. In addition, He was Chair of the ICOH (International Commission on Occupational Health) Scientific Committee for Aging and Work until the end of March in 2012, and was President of the Society for Occupational Safety, Health and Ergonomics (SOSHE) until the end of March in 2013.

1   *Aging and occupational safety*

New Ergonomics Perspective – Yamamoto (Ed.)
© 2015 Taylor & Francis Group, London, ISBN 978-1-138-02751-0

Integrating Kansei Engineering method and Kano model for service of nursing home in Taoyuan, Taiwan

Chinmei Chou, Cindy Sutanto & Chi-Kuang Chen
Department of Industrial Engineering and Management, Yuan Ze University, Chung-Li, Taiwan

ABSTRACT: The purpose of this study is using Kansei Engineering method and Kano model to improve the service design in nursing home in Taiwan. Previous studies found that Taiwan is facing the rapid increase of the ageing population and lack of resources for taking care of elderly. Nursing home has become one of solutions to overcome these issues, but the current condition indicates some nursing homes can not satisfy the customer desires, especially for elderly who are more sensitive. Kansei Engineering method has ability to translate the customer's feeling into design specification and minimize the subjective interpretation of emotion or Kansei and Kano model has ability to take out unspoken customer desires and provide a unique way by distinguishing the impact of different customer needs on total customer satisfaction in the early stage of product or service development to get the higher degree of effectiveness and efficiency. The integration between Kansei Engineering method and Kano model is used to analyze the relationship between service attributes performance and customer emotional response by categorizing the service attributes based on importance priority. Thus, the result may generate the proper result to improve the service design of the nursing home based on the elderly desires.

Keywords: Kansei Engineering; Kano model; service design; nursing home; elderly

1 INTRODUCTION

Nowadays the ageing population in Taiwan continues to increase annually and has become trend[1][2]. As predicted by the Council for Economic Planning and Development (CEPD), the increase in proportion of older people will change Taiwan from ageing society at present to an aged society in 2017[3][4] and to a super aged society in 2025[5]. The consequence of rapid increase of ageing population is the raising of many healthcare issues including the long-term care of disabled people[6]. The modernization of the global economy also will affect to family structure changing[4] and declining birth rates[7][8]. As the result, the available resources for taking care of elderly at home are gradually reduced and these changes present a considerable challenge to meet the needs of Taiwan's ageing society[9].

Nursing home is a multi-residence housing facility that provide service package to elderly and it plays important role in a long-term care system because it is not merely temporary, but a part of human lives who mostly old, sick, and easily hurt[8]. The elderly feeling should be involved in the daily life in order to make them impress how their lives change, how important they are to the sociality, and how the nursing home can help them to achieve their desires[5] the service quality of nursing home is worthy of concern in long-term care facilities[10][11] because service is not just what customer get, but it is how the customer get and feel about the service[12].

Nagamachi introduced Kansei Engineering (KE) as a technology that translates the customer's feeling into design specification and minimize the subjective interpretation of emotion or Kansei. KE is also an application of quality management which begins with customer expression word or 'Kansei' word, with the purpose is to maximize customer satisfaction[13]. However, the relationship between product or service quality and customer response is not always linear. Nonlinear view point can be seen in service characteristics which produce satisfaction, are not the same as those which produce dissatisfaction[14]. The Kano model has ability to take out unspoken customer desires[15] and provide a unique way by distinguishing the impact of different customer needs on total customer satisfaction in the early stage of product or service development to get the higher degree of effectiveness and efficiency[16][17].

2 RESEARCH METHODS

2.1 *Subjects*

Study has been conducted in nursing homes in Taoyuan County during November 2013– March 2014. Total of 42 elderly from 4 nursing homes in Taoyuan which selected randomly were interviewed for about 30–50 minutes for each. The elderly gender was split into 28 males and 14 females.

2.2 *Research methods questionnaires*

5 types of questionnaires were used: Kansei words, Expectation, Perception, and pair of Kano questionnaires. The first step is collecting Kansei words from different sources such as newspaper, magazine, internet, etc. in order to get the most complete semantic description possible[14][18]. The second step is reducing the number of words which are similar and have identical meaning to find high level of Kansei words[15]. The method used to reduce the number of words was an affinity diagram[19][20]. The third step is constructing questionnaire based on Servqual dimension (tangible, reliability, responsiveness, assurance, empathy) by collecting the service attributes through technical document, literature study, related Kansei studies, and service group[21]. 5-Likert scale was used to get customer responses.

Expectation and perception questionnaires were used to measure the gap score for each service attributes. Pair of Kano questionnaire (functional and dysfunctional) was used to measure whether the services are provided or not provided. The responses thus were combined and categorized based on Kano evaluation table (shown in Table 1)[22]. Frequency analysis was applied to evaluate and define Kano category for each service attributes regarding the highest frequency they had[23][24].

There are 6 Kano model categories[22]: 1) Attractive/A, the increase of service attributes quality will increase the consumer satisfaction, but the decrease of service attributes quality will not decrease the consumer satisfaction 2) One-dimensional/O, the service attributes quality is proportional to the customer satisfaction 3) Must-be/M, the increase of service attributes quality will not increase the consumer satisfaction, but the decrease of service attributes quality will decrease the consumer satisfaction 4) Indifferent/I, whether the service attributes quality increase or decrease will not affect to consumer satisfaction 5) Reverse/R, the service attributes quality is inversely proportional to the customer satisfaction

Table 1. Kano evaluation table.

		Dysfunctional				
		1	2	3	4	5
Functional	1	Q	A	A	A	O
	2	R	I	I	I	M
	3	R	I	I	I	M
	4	R	I	I	I	M
	5	R	R	R	R	Q

4

6) Questionable/Q, whether the service attributes quality increase or decrease, it has possibility to satisfy and/or to disappoint the consumer.

To integrate the Kansei and Kano model, regression analysis using SPSS 16.0 was conducted by linking the Kansei words response with the elderly perceived service attributes. Stepwise regression method was used in order to get the strongest correlated variables by removing some variables which had low contribution to each Kansei word. The method for analyzing the regression result is to check the negative gap which is the negative difference between perception and expectation mean value. The first step is to check whether the number of affected Kansei words among significant service attributes is the same for each negative gap. The action if it is the same is to prioritize service attributes with the most negative gap value, otherwise the service attributes with the higher number of Kansei words will be the priority. Whether the number of Kansei words is the same or not, the priority action from this result is continuous improvement to those service attributes selected. The second step is to check if the service attributes have positive gap and if the number of Kansei words among the significant service attributes is the same. If it is the same then the action is to priority the service attributes with the lowest gap value, otherwise the service attributes with the higher number of Kansei words will be the priority. Thus, this result will be chosen to be the main focus to propose service design improvement and to define which service attributes should be maintained.

3 RESULTS

There were 16 Kansei words used in this study: cleanliness, elegance, modern, brightness, careful, helpful, delightful, peaceful, spacious, quietness, responsiveness, delicious, convenience, relax. The results showed Kansei words 'cleanliness' and 'brightness' had the highest mean of evaluation score which is $\bar{x} = 3.50$ and Kansei words 'elegance' and 'delicious food' had the lowest mean of evaluation score which is $\bar{x} = 2.90$.

Total of 24 service attributes for each questionnaires were categorized into five Servqual dimensions. Validity and reliability test was conducted using SPSS 16.0 to evaluate the adequacy of relationship model between constructs and the measurement items of the research instrument. The result showed all of the questionnaires were reliable. Some of variables did not reach the standard of validity test. However, based on some literature review and peer evaluation, it was decided to keep those invalid variables on the next analysis in order to have better comprehension of the study.

The customer expectation and performance questionnaire were first analyzed. The result showed the elderly have high expectation where all the mean of each service attribute reached score above 4. The gap score which is the difference value between perception and expectation score showed the range of difference between how the services are experienced and expected by the elderly. The lower of the gap value, the wider the range of difference between the services are experienced and expected. The result showed 'suitable temperature at patient rooms' and 'feel safe and feel at home' had the lowest gap scores (−1.26 and −1.19).

The result of Kano categorization showed 14 items were on Attractive category, 6 items on One-dimensional category, 4 items on Indifferent category. Five of six service attributes in 'reliability' dimension had attractive (A) quality, which are appropriate employees response, medical treatment and doctor visiting are well scheduled, available and adequate patient family visiting time, the employees solve the elderly problem sincerely, all equipment (AC, TV, radio, light, etc.) work properly, and the rest was indifferent (I) quality which is all patient activities are well scheduled. This showed that the elderly expected the ability of employees to perform the promised service dependably, consistently, and accurately. The elderly did not expect their activities are well scheduled (this had the lowest score of expectation questionnaire, 4.19), therefore whether this attribute quality increase or decrease will not affect the elderly satisfaction (Indifferent category). The result of integration between Kansei and Kano model showed Kansei word 'delightful' had the most significant model ($R^2 = 0.441$,

p-value < 0.001) and was influenced by service attribute 'quick medical treatment response when patient need it' with highest coefficient (0.588). There were three services attributes with the highest number of Kansei words which are medical instrument & physical facilities are visually appealing, feel safe and feel at home, quick medical treatment response when patient need it. The service attribute 'medical instrument & physical facilities are visually appealing' was considered as the priority for improvement since it had five Kansei words related and the lowest gap score. The service attributes 'feel safe and feel at home' was neglected to be prioritization of improvement because it had negative coefficient relationship with particular Kansei words. Based on the research framework we already mentioned, the first prioritization would be the service attribute 'medical instrument & physical facilities are visually appealing' since it had the highest number of Kansei words related.

4 DISCUSSION

In this study, Servqual dimensions helped to measure the service from all the aspects. Secondly, it was applied to measure the gap score which is the difference between customer perception and customer expectation toward the service provided. Some researchers proved that measuring service quality based on gaps method is a richer approach because it is a multidimensional construct[25] and found that gap scores provide the actual service performance and useful to identify the weakness and the strength of the firms[26]. The result can help to determine where improvement should be targeted[27].

Service attributes that were categorized in Attractive category would be prioritized to be improve to increase and maximize customer overall satisfaction and fulfil customer emotional needs (Kansei) because little improvement of the service quality can increase significant level of customer satisfaction. Services attributes in One-dimensional and Must-be categories are unlikely to be associated with strong emotions such as in Attractive category[15]. They tend to be compulsory quality which is appropriate to be well provided in daily basis because the improvement of these service attributes will not give significant effect to increase customer satisfaction.

Although Kansei Engineering has been widely used in design product or service of several sectors, Jiao *et al.* (2006) found that the mapping relationship between Kansei words and design elements are often not clearly available[28]. Designers are often not aware of the underlying coupling and interrelationships between various design elements with regard to customers' affective satisfaction achievement. By using the stepwise regression, Kansei words (emotions) which impacted each service attribute were investigated, modeled, and evaluated. Each model explained that certain Kansei words gave influence for one or more service attributes. In reverse, a service attribute might have impact on one or more Kansei words. The higher number of Kansei words in a service attributes, the more important this service attributes was because it is more close-related with customer emotions.

5 CONCLUSION

This study already proved that integrating Kansei Engineering method and Kano model based on Servqual model could gave more significant result and more detail information, and how those three methods complemented to each other. In total, there were 10 service attributes to be prioritized (they had negative gap scores and were categorized in Attractive category) by the nursing homes. Based on Kansei Engineering method, one service attribute (feel safe and feel at home) was neglected of improvement prioritization since it had negative relationship (negative coefficient) with particular Kansei words. Service attribute 'medical instrument & physical facilities are visually appealing' became the first prioritization because it had the highest number of Kansei words.

REFERENCES

[1] Hsu, Wen-Ming, Ching-Yu Cheng, Jorn-Hon Liu, Su-Ying Tsai, and Pesus Chou. "Prevalence and causes of visual impairment in an elderly Chinese population in Taiwan: the Shihpai Eye Study." *Ophthalmology* 111, no. 1 (2004): 62–69.

[2] Chen, Chen-Ru, Wen-Hsi Lydia Hsu, Hua-Lin Tsai, and Chun-Wei Lu. "A study of employer background in care demand in southern Taiwan." In *Proceedings of the 11th WSEAS international conference on Applied Computer and Applied Computational Science*, pp. 134–138. World Scientific and Engineering Academy and Society (WSEAS), 2012.

[3] Bartlett, Helen P., and Shwu-chong Wu. "11 Ageing and aged care in Taiwan." *Ageing in the Asia-Pacific region: issues, policies and future trends* 2 (2000): 210.

[4] Kang, Shih-Chung. "Initiation of the Suan-Lien Living Lab–a Living Lab with an Elderly Welfare Focus." *International Journal of Automation and Smart Technology* 2, no. 3 (2012): 189–199.

[5] Chen, Chi-Kuang. "Graceful Ageing Service Model—The Basic and Application Research". Industrial Engineering and Management Yuan Ze University, 2013.

[6] Chang, Hsiao-Ting, Hsiu-Yun Lai, I-Hsuan Hwang, Mei-Man Ho, and Shinn-Jang Hwang. "Home healthcare services in Taiwan: a nationwide study among the older population." *BMC health services research* 10, no. 1 (2010): 274.

[7] Tien, Chi Hu, and Wen Chi Tsai. "Resource Integration Strategies for Elder Education Organizations: A Case Study in Taichung, Taiwan." *Online Submission* (2013).

[8] Yeh, Tsu-Ming, and Shun-Hsing Chen. "Integrating Refined Kano Model, Quality Function Deployment, and Grey Relational Analysis to Improve Service Quality of Nursing Homes." *Human Factors and Ergonomics in Manufacturing & Service Industries* 24, no. 2 (2014): 172–191.

[9] Bartlett, Helen P., and Shwu-chong Wu. "11 Ageing and aged care in Taiwan." *Ageing in the Asia-Pacific region: issues, policies and future trends* 2 (2000): 210.

[10] Gupta, Parul, and R.K. Srivastava. "Customer Satisfaction for Designing Attractive Qualities of Healthcare Service in India using Kano Model and Quality Function Deployment." *MIT Int J Mech Eng* 1, no. 2 (2011): 101–107.

[11] Yeh, Tsu-Ming. "Determining medical service improvement priority by integrating the refined Kano model, Quality function deployment and Fuzzy integrals." *African Journal of Business Management* 4, no. 12 (2010): 2534–2545.

[12] Peranginangin, Ezra, Kuo Hsiang Chen, and Meng-dar Shieh. "Toward Kansei Engineering Model in Service Design: Interaction For Experience In Virtual Learning Environment." In *Proceedings of the 9th International Conference on Electronic Business, Macau*. 2009.

[13] Nagamachi, Mitsuo, ed. *Kansei/affective engineering*. CRC Press, 2010.

[14] Llinares, Carmen, and Alvaro F. Page. "Kano's model in Kansei Engineering to evaluate subjective real estate consumer preferences." *International Journal of Industrial Ergonomics* 41, no. 3 (2011): 233–246.

[15] Hartono, Markus, and Tan Kay Chuan. "How the Kano model contributes to Kansei engineering in services." *Ergonomics* 54, no. 11 (2011): 987–1004.

[16] Raharjo, Hendry. "Dealing with Kano Model Dynamics: Strengthening the Quality Function Deployment as a Design for Six Sigma Tool." *Jurnal Teknik Industri* 9, no. 1 (2007): 15.

[17] Sireli, Yesim, Paul Kauffmann, and Erol Ozan. "Integration of Kano's model into QFD for multiple product design." *Engineering Management, IEEE Transactions on* 54, no. 2 (2007): 380–390.

[18] Huang, Yuexiang, Chun-Hsien Chen, and Li Pheng Khoo. "Kansei clustering for emotional design using a combined design structure matrix." *International Journal of Industrial Ergonomics* 42, no. 5 (2012): 416–427.

[19] Bergman, Bo, and Bengt Klefsjö. "Quality from customer needs to customer satisfaction". Studentlitteratur (2010).

[20] Terninko, John. *Step-by-step QFD: customer-driven product design*. CRC Press, 1997.

[21] Schütte, Simon. "Designing feelings into products: Integrating kansei engineering methodology in product development." (2002).

[22] Kano, Noriaki, Nobuhiko Seraku, Fumio Takahashi, and Shinichi Tsuji. "Attractive quality and must-be quality." *The Journal of the Japanese Society for Quality Control* 14, no. 2 (1984): 39–48.

[23] Sauerwein, Elmar, Franz Bailom, Kurt Matzler, and Hans H. Hinterhuber. "The Kano model: How to delight your customers." In *International Working Seminar on Production Economics*, vol. 1 (1996): 313–327.

[24] Bayraktaroglu, Gül, and Özge Özgen. "Integrating the Kano model, AHP and planning matrix: QFD application in library services." *Library Management* 29, no. 4/5 (2008): 327–351.

[25] Parasuraman, A., Valarie A. Zeithaml, and Leonard L. Berry. "Servqual." *Journal of retailing* 64, no. 1 (1988): 12–37.

[26] Tseng, Shih-Chang, and Shiu-Wan Hung. "A framework identifying the gaps between customers' expectations and their perceptions in green products." *Journal of Cleaner Production* 59 (2013): 174–184.

[27] Asubonteng, Patrick, Karl J. McCleary, and John E. Swan. "SERVQUAL revisited: a critical review of service quality." *Journal of Services marketing* 10, no. 6 (1996): 62–81.

[28] Jiao, Jianxin Roger, Yiyang Zhang, and Martin Helander. "A Kansei mining system for affective design." *Expert Systems with Applications* 30, no. 4 (2006): 658–673.

New Ergonomics Perspective – Yamamoto (Ed.)
© 2015 Taylor & Francis Group, London, ISBN 978-1-138-02751-0

REST-based Ambient Intelligent, pervasive Care Services Platform (RACS) in the cloud for aging-in-place

Chuan-Jun Su

Department of Industrial Engineering and Management, Yuan Ze University, Taiwan

ABSTRACT: Several powerful trends are contributing to a "graying" of much of the world's population, especially in economically-developed countries. This inexorable shift in demographics is already starting to place significant economic burdens on those still in the workforce and on society as a whole. To mitigate the negative effects of rapidly ageing populations, societies must act early to plan for the welfare, medical care and residential arrangements of their senior citizens, and for the manpower and associated training needed to execute these plans. A key aspect of these efforts is developments of information technology to automatically provide personal care services with comprehensive real-time data on the charges. This article describes the development of a "RESTful driven Ambient Intelligent Care Services Platform" (RACS) for aging-in-place in the Cloud, which creates an environment of Ambient Intelligence through the use of Ontology, Restful Web Services, Sensor Networks and Mobile Agent technologies. The RACS presented in this paper is capable of offering pervasive, accurate and contextually-aware personal care services. Architecturally the REST-implemented RACS leverages Restful web service and Cloud Computing to provide economic, scalable, and robust healthcare services over the Internet.

Keywords: Pervasive care services; aging-in-place; ambient intelligence; RESTful web services; cloud computing

1 RESEARCH BACKGROUND AND MOTIVATION

Rapid advances in medicine and technology are leading to marked increases in human longevity. Combined with declining birth rates, societies around the world are now facing challenges associated with rapidly aging populations. A recent report from Taiwan's Department of Health shows that nearly 90% of senior citizens in Taiwan live with one chronic disease, while over half live with three or more chronic diseases. Many of these people are incapable of living independently, and the requirements for care will only continue to increase. These issues are raising broader concerns about the overall care and quality of life of the elderly.

Home Care is conceived as the integration of medical, social and familiar resources addressed to the same goal: enabling the integral care of the elders in their own residence [1]. It emphasizes the concept of the "aging-in-place" which refers to living where care-receivers have lived for many years, or to living in a non-healthcare environment, and using products, services and conveniences to enable care-receivers to not have to move as circumstances change [2]. The practice of the aging-in-place would significantly have the advantages of underpinning to elders' feelings of dignity, quality of life and independence. However, it is very difficult in Home Care achievements. The care-receivers are widely distributed over the city.

The Home Care services delivering would not only take the care-givers (e.g. Home Care assistants) should work in an open environment but also make the resources, in care-receiver volving the care-givers labor, medical resources and so on, hard to plan. It directly makes the care-givers' workload and cost of the care increasing [3][4].

The goal of health services provision is to improve health outcomes in the population and to respond to people's expectations, while reducing inequalities in both health and responsiveness. The health care needs of the population should be met with the best possible quantity and quality of services produced at minimum costs. The use of information technology is already contributing in significant ways to enhancing healthcare delivery and to improving the quality of life. However, deployments of information technology have only scratched the surface of possibilities for the potential influence of computer and information science and engineering on the quality and cost-effectiveness of healthcare [5].

In this article we present a RESTful driven Ambient-Intelligent Care Services Platform (RACS) based on the concept of Ambient-Intelligent (AmI) using Ontology, Restful Web Services, Sensor Networks and Mobile Agent technology. It aims to raise the quality of Home Care, make Home Care services more proactive, and reduce the overall cost in the provision of care services. Ontology that provides the formal descriptions of the care activities is used in the process of creating the personalized care plan for each care-receiver. The RESTful Web Services provide the various applications which are relevant to the care activities needs. Sensor Networks allows easily collecting the environment information involving the temperature, humidity, person's location and so on, while Mobile Agents furnish contextually-aware, timely and accurate information for the provision of quality care, with care and treatment data recorded automatically. We expect RACS to raise the convenience, contextual awareness and accuracy of pervasive personal Home Care services.

Section 2 of this paper presents research on AmI in the context of solutions used for care services, and discusses the issues about how to generate the personalized services from personal profiles. Section 3 presents the system design and architecture of the RACS. Section 4 covers usage scenarios and implementation. Finally, Section 5 provides conclusions and suggestions for future research.

2   AMBIENT INTELLIGENCE (AmI)

The early developments in AmI took place at Philips electronics founded and headquartered in the Netherlands. The board of directors commissioned a series of internal workshops to investigate different scenarios that would transform the high-volume consumer electronics industry from its then "fragmented with features" world into a world in 2020 where user-friendly devices would support ubiquitous information, communication and entertainment [6].

AmI refers to a vision of the future in which people are empowered by an electronic environment that is aware of their presence, and is sensitive and responsive to their needs. It aims at improving quality of life by creating the desired atmosphere and functionality via intelligent, personalized interconnected systems and services [7]. AmI implies a seamless environment of computing, advanced networking technology and specific interfaces. It is aware of the special characteristics of human presence and personality, addresses human needs and is capable of responding intelligently to spoken or gestured commands, and can even engage in intelligent dialogue with the user [8]. A typical context of an ambient intelligence environment is a Home environment [9]. AmI is built on the concept of contextual-awareness, which raises possibilities of developing a new generation of interactive applications and systems. However, context influences, and often fundamentally changes, interactive systems [10].

The term contextual-awareness was coined in ubiquitous computing as being "the ability of a mobile user's applications to discover and react to changes in the environment they are situated in" by Schilit and Theimer [11]. Perhaps due to technical limitations, the actual contextual information that Schilit and Theimer used in their ActiveMap Service was limited to location only. They suggested that future contextually-aware applications should include much more than locational information, a suggestion that numerous researchers have attempted to build upon Barkhuus's research [12]. The Context Broker Architecture (CoBrA) is an Agent-based architecture for supporting context-aware systems in smart spaces (e.g., intelligent meeting rooms, smart homes, smart vehicles, etc.) [13]. Central to this architecture is an intelligent agent called context broker that maintains a shared model of context on the behalf of a community of agents, services, and devices in the space and provides privacy protections for the users in the space by enforcing the policy rules that they define.

3 AMBIENT INTELLIGENCE APPLICATIONS IN CARE SERVICES

Contextual awareness is a concept that has been described for some time, but technologies (e.g. wireless technologies, mobile tools, sensors, wearable instruments, intelligent artifacts, handheld devices, etc.) are only now becoming available to support the development of applications. Such technologies could help health care professionals to manage their tasks while increasing the quality of patient care [14]. Su and Chen proposed an "Intelligent Community Care System (ICCS)" by applying RFID (Radio-Frequency Identification) and Mobile Agent technologies to enable care givers and communities to offer pervasive, accurate and contextually-aware care services [15]. RFID allows care-givers to easily locate the care-receivers, while Mobile Agents furnish contextually-aware, timely and accurate information for the provision of quality care, with care and treatment data recorded automatically. Fraile et al. presented a hybrid Multi-Agent architecture, named HoCa, for the control and supervision of dependent environments [16]. The HoCa architecture provided the basic idea in incorporating the alert management system based on SMS and MMS technologies and context control system based on Java Card and RFID technologies. With the past researches (e.g. mentioned above), the basic idea referring to the infrastructure of developing a care services system platform equipped AmI mechanism was performed. The RACS proposed in this paper would be built and integrated the novel ideas, (e.g. cloud computing, RESTful web services, etc.), to provide the more robust care services on-top of the past contributions.

4 PROFILES AND PERSONALIZED SERVICES

The user profile is the data instance of a user model that is applied to adaptive interactive systems. It also acts as the key component which is adopting to provide the personalized services or information in the personalization system [17]. Golemati M. et al. showed that the generation of the user profile which is available in semantically inferring the personalized services, from user information collection, profiles construction to profiles representation [18]. Eslami MZ et al. proposed an effective service tailoring process and architecture to personalize homecare services according to the individual care-receiver's needs. In the proposed approach, the tailoring process is divided into six steps which present how to configure the personalized services from the user profile [19]. Yang CL et al. developed a Personalized Service Recommendation System (PSRS) in a home-care environment. The PSRS has capability of providing proper services based on the user's preferences and habits which are recorded in the user profile. Through the user profile, the system will be able to automatically launch to safety alert, recommendable services and healthcare services in the house [20].

The feasible solution, which presents how to generate the personalized services from user profile, can be summarized into two processes: generating the goal activities from the user profile and defining the services which can perform the activities. In generating the goal activities, the ontology, named "goal ontology", is created to present the relationship between the user profile and goal activities. In defining the services, the ontology, named "task ontology", is created to present the relationship between the goal activities and the services. This solution has broadly used in many researches [18] [19] [20]. In this paper, this solution is also cited to support the personalized services configuration.

5 RESEARCH METHODOLOGY

This paper aims to establish a RESTful driven AmI Care Services Platform (RACS) for Personalized Aging in Place in the Cloud based on the concept of AmI using Ontology, Sensor Network, Restful Web Services and Mobile Agent technology, as shown in Fig. 1. We designed the RACS to be capable of identifying the personalized needs of the elders and delivering the corresponding contextualized care services automatically. Functionally, the RACS is able to infer the appropriate care plan which defines the needed care activities through the elders' profiles and relevant external web services to deliver the care services with practical real-time information. Whenever the RACS can't receive the response from the care-recipient during the care services delivering, the RACS would notify the care-givers. The system would also keep track of the care services performed. The system would allow caregivers to administrate and modify the care activities and services, and would also allow the caregiver to immediately access contextually-relevant data on their tablet PC or PDA to answer questions from the care-recipient.

In the design of the RACS, a large or complex task is divided into several modules, each dealing with a variety of tasks. The RACS can be divided into four major modules as shown in Fig. 1. The Repositories Layer which acts the core in the RACS is responsible for not only storing the personal profiles and the web services description but also processing the personalized care plan based on calculating the personal profiles, care ontology and service ontology. The RESTful Web Services provide the various applications which are relevant to the care activities needs such as reminding service, weather information service and so on. The Sensor Network is constructed to detect the real-time environment information and

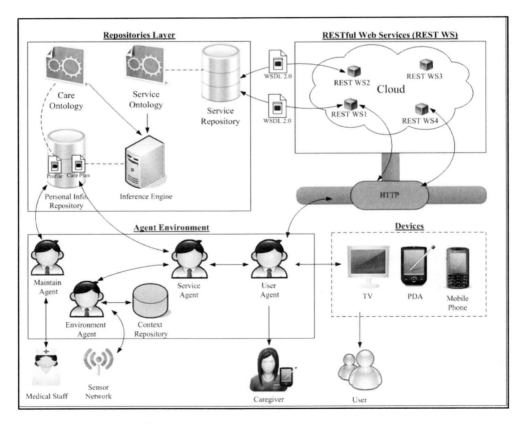

Figure 1.   The RACS architecture.

context data such as the temperature, humidity, location and so on. The Agent Environment is performed to communicate between the users and applications. The agents would perform the care services which are defined in the care plan to the care-recipient by dispatching the RESTful Web Services with the real-time context data through the HTTP protocol.

6 IMPLEMENTATION

The RACS was built on Microsoft Windows Server 2003 Standard Edition, with the Java™ SE Development Kit 1.6 installed. Microsoft SQL Server™ 2008 Express Edition was used as the data-base server. A middle layer called Java Database Connectivity 3.0 was installed to access the database server. TopBraid™ Composer—Maestro Edition (TBC-ME) was used to develop the RACS because the Eclipse-based TBC-ME is not only a visual ontology editor but also serves as a knowledge-base framework capable of integrating inference engines through hybrid inference-chaining. The built-in capability of running multiple inference engines shortens the development cycle of RACS. The Agent Environment in RACS is implemented in FIPA (Foundation for Intelligent Physical Agents)—compliant agent framework JADE (Java Agent Development Environment).

Usage Scenario: The scenario illustrates the usage of RACS: "John Wang is 72 years old. Despite his age and having various medical conditions, he prefers to remain living in his own home. John has a hemorrhagic stroke with mild claudication and hypertension, he needs to go to the hospital to rehabilitate three times a week and take the medicine that controlling blood pressure twice daily. However, John was difficult to walk to the hospital because claudication, and for taking medicine, he often forgot that. It is a real torment to him. Therefore, the caregivers may use RACS to provide personalized healthcare plan for solving John's problems."

In order to derive a personalized healthcare plan which fulfills John's desirable goal "remain living in his own home", a professional firstly submit John's profile to RACS via the GUI provided by Maintenance Agent. The profile will be subsequently converted into an ontology specification and stored into the personal information repository. The inference engine will then takes action to generate a personalized care plan based on the Care ontology and Service ontology.

John has a clinical state "Hemorrhagic Stroke", the inference engine in this example generates a relevant care service "Medical Appointment Service" based on the Care Ontology, which will be included in John's care plan. Similarly, the clinical state "Hypertension" will be inferred by the inference engine. The generated care services "Reminder Service" and "Weather Service" will also be included in John's care plan.

If the generated care plan is approved by John's care-giver, the tasks required to complete the services in the plan will be further inferred based on the Service ontology. In this example, the "Medical Appointment Service" may consist of "Appointment Making" and "Round-trip Pickup" (John suffers from claudication) activities to aid John in making appointment. A "Voice Reminder" activity using an event-driven calendar activity may be generated for the "Reminder Service" for taking medicine. If a hearing problem is recorded in John's profile, such other options as text-message or video-message will be generated. Weather has great impact on blood pressure. Because John suffers from Hypertention, in addition to the "Reminder Service" for routine medication, he should also be aware of the weather change. Similar to the "Reminder Service", the "Weather Service" provides the information regarding weather change especially sudden drop of temperature and reminds John to closely monitor his blood pressure.

Once the services and related tasks are inferred, a care plan is generated and stored in the personal information repository as shown in Fig. 2(a). The web services associated with the tasks can be subscribed for plan execution. The generated care plan may be adjusted periodically by his physician according to his health conditions or the outcome of plan execution. For example, if John needs pick-up service only on Monday, Wednesday, and Thursday at 9:30 A.M. every week. This can be accomplished by a GUI built with the service as illustrated in Fig. 2(b). The web services are implemented in Java under REST architectural style and deployed in a cloud setting. The services can therefore be reliably accessed from anywhere by using virtually any Internet-enabled device as shown in Fig. 2(c).

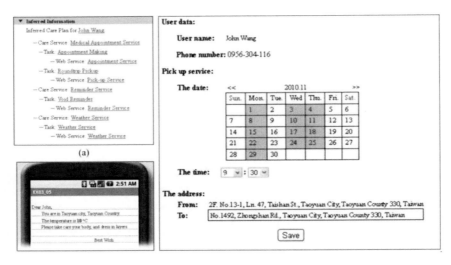

Figure 2.   Care plan generated for the example scenario.

7   CONCLUDING REMARKS

In this paper, we present the design and development of RACS based on seamless integration of such enabling technologies as Ontology, Sensor Network, Web Services and the FIPA-compliant agent framework JADE to provide personalized and context-aware health care services. The RACS employs ontologies to improve the integration and orchestration of all the technologies involved such as web services, context-aware healthcare services, sensor network, and software agents. Rules of inference are formulated, which utilize the structural information in ontologies to derive personalized, context-aware healthcare plan for elders. Architecturally the FIPA-compliant agent infrastructure and Cloud-based web services make the RACS scalable and interoperable. The RACS contributes a comprehensive platform toward offering intelligent care services pervasively for achieving the ultimate goal of aging-in-place.

REFERENCES

[1] Valls A, Gibert K, Sánchez D, Batet M (2010) Using ontologies for structuring organizational knowledge in Home Care assistance. International Journal of Medical Informatics 79:370–387.

[2] Aging in Place, http://www.seniorresource.com/ageinpl.htm.

[3] Bajo J, Fraile JA, Pérez-Lancho B, Corchado JM (2010) The THOMAS architecture in Home Care scenarios: A case study. Expert Systems with Applications 37:3986–3999.

[4] Gaddam A, Mukhopadhyay SC, Gupta GS (2010) Smart Home for Elderly Using Optimized Number of Wireless Sensors. In: Mukhopadhyay SC, Leung H (eds) Advances in Wireless Sensors and Sensor Networks. Springer-Verlag Berlin, Heidelberg, pp 307–328.

[5] S. Graham, D. Estrin,E. Horvitz, I. Kohane, and E. Mynatt, Information Technology Research Challenges for Healthcare: From Discovery to Delivery, "Discovery and Innovation in Health IT" workshop, San Francisco, October 29, 2009.

[6] Kikhia B (2008) Acceptance of Ambient Intelligence (AmI) in Supporting Elderly people and people with Dementia, in Department of Business Administration and Social Sciences. Master thesis, Lulea University of Technology: Sweden.

[7] Loenen EJ (2003) On the Role of Graspable Objects in the Ambient Intelligence Paradigm. Smart Objects Conference SoC 2003, Grenoble.

[8] Mendes M, Suomi R, Passos C (2004) Digital Communities in a Networked Society. Baker & Taylor Books.

[9] Bieliková M, Krajcovic T (2001) Ambient Intelligence within a Home Environment. ERCIM News. http://www.ercim.org/publication/Ercim_News/enw47/bielikova.html. Accessed October 2001.

[10] Schmidt A (2005) Interactive Context-Aware Systems Interacting with Ambient Intelligence. In: Riva G, Vatalaro F, Davide F, Alcañiz M (eds) Ambient Intelligence. IOS Press, pp 159–178.

[11] Schilit BN, Theimer MM (1994) Disseminating Active Map Information to Mobile Hosts. IEEE Network 8:22–32.

[12] Barkhuus L (2003) Context information vs. sensor information: a model for categorizing context in context-aware mobile computing. In: Symposium on Collaborative Technologies and Systems, San Diego, CA, pp 127–133.

[13] Chen H (2004) An Intelligent Broker Architecture for Pervasive Context-Aware Systems. UMBC eBiquity Publications. http://ebiquity.umbc.edu/paper/html/id/212/An-Intelligent-Broker-Architecture-for-Pervasive-Context-Aware-Systems. Accessed 14 December 2004.

[14] Bricon-Souf N, Newman CR (2007) Context awareness in health care: A review. International Journal of Medical Informatics 76:2–12.

[15] Su CJ, Chen BJ (2010) An Intelligent Community Care System Using Network Sensors and Mobile Agent Technology. In: Vincent G. Duffy (ed) Advances in Human Factors and Ergonomics in Healthcare. CRC Press, Boca Raton, pp 558–568.

[16] Fraile JA, Bajo J, Lancho BP, Sanz E (2009) HoCa Home Care Multi-agent Architecture. In: Corchado JM, Rodríguez S, Llinas J, Molina JM (eds) International Symposium on Distributed Computing and Artificial Intelligence 2008 (DCAI 2008). Springer-Verlag Berlin, Heidelberg, pp 52–61.

[17] Gauch S, Speretta M, Chandramouli A, Micarelli A (2007) User profiles for personalized information access. In: Brusilovsky P, Kobsa A, Nejdl W (eds) The Adaptive Web. Springer-Verlag Berlin, Heidelberg, pp 54–89.

[18] Golemati M, Katifori A, Vassilakis C, Lepouras G, Halatsis C (2007) Creating an Ontology for the User Profile: Method and Applications. Proceedings of the First International Conference on Research Challenges in Information Science. Ouarzazate, Morocco. pp 407–412.

[19] Eslami MZ, Zarghami A, Sapkota B, Sinderen M (2010) Service Tailoring: Towards Personalized Homecare Services. Proceedings of the 5th International Workshop on Architectures, Concepts and Technologies for Service Oriented Computing (ACT4SOC). Athens-Greece.

[20] Chang YK, Yang CL, Chang CP, Chu CP (2009) A Personalized Service Recommendation System in a Home-care Environment. Proceedings of the 15th International Conference on Distributed Multimedia Systems (DMS 2009). San Francisco.

New Ergonomics Perspective – Yamamoto (Ed.)
© *2015 Taylor & Francis Group, London, ISBN 978-1-138-02751-0*

Age effects on reach to grasp movement

Min-Chi Chiu

Department of Industrial Engineering and Management, National Chin Yi University of Technology, Taichung, Taiwan

Hsin-Chieh Wu & Nien-Ting Tsai

Department of Industrial Engineering and Management, Chaoyang University of Technology, Taichung, Taiwan

ABSTRACT: Human interact with circumstance, the reach to grasp movement is indispensable element of daily living. With aging, the physiological and neuromuscular decline changes the gross and fine motor function for elderly. Therefore, this study aims to realize the object weight and location on reach to grasp movement between young and older adults.

A total of 12 healthy subjects, 6 young and 6 older adults were recruited to participate in this study. The average age for young and old adults were 23.5 (SD = 0.5) and 71.2 (SD = 2.9) years old, respectively. The target is a cylinder with 3 cm-diameter and 12 cm-height. Nine locations (3 distances: 80% arm length/AL, 100% AL and 120% AL with 3 orientations: midline, 60 degree right and 60 degree left from midline) and three weights (100, 600 and 1000 g) were be arranged as 27 trials. All subjects randomly complete 81 trials (27 × 3 times) continuously.

Measurements include kinematic analysis, electromyography of upper extremity and reaction time during a reach-to-grasp task. For kinematic data, the joints range of motion were recorded by an ultrasound-based movement analysis system (Zebris CMS-HS/Zebris Medical GmbH, Germany) including wrist flexion-extension, wrist radial-ulnar deviation, forearm pronation-supination, elbow flexion-extension, shoulder flexion-extension and shoulder adduction-abduction. The reaction time was calculated by the movement trajectory of hand. For the muscle activity, four muscle groups include Biceps, Triceps, Brachioradialis and Pronators Teres were measured by electromyography (Zebris CMS-HS/Zebris Medical GmbH, Germany).

The results of this study indicated that age has a significant influence on joint range of motion and muscle activity (p < 0.05). For elderly, it displays greater wrist flexion, less elbow flexion, forearm pronation and less shoulder flexion than young adults. Moreover, older adults required higher muscle activity in Biceps, Triceps, Brachioradialis and Pronators Teres during reach to grasp movement. Furthermore, object location affect the joint range of motion, object weight significantly impact the muscle activity of Brachioradialis for elderly. It's interesting to find that lighter object weight required longer reaction time for elderly. The findings of this study can provide useful information for clinical training, rehabilitation goal setting, relevant product design and barrier-free considerations.

Keywords: Aging; kinematics; object orientation; reach to grasp; electromyography

1   INTRODUCTION

Reach to grasp is a fundamental movement for people to interact with environment and to accomplish the activities of daily living. This functional action demands the comprehensively integrating with multiple neurological and sensorimotor systems. Jeannerod (1981, 1984) proposed that the act of reach to grasp roughly comprise reaching (moving the arm to a location near the target), grasping (adjusting hand posture during approaching target) and manipulation (accurately grasped and manipulated target) phases. In general, the two-digit and whole hand grasps, the peak aperture (the distance between thumb and fingers) occurs at approximately 60~70% of total movement time [3]. Supuk et al. (2011) further identified that reach-to-grasp movement composed three phases (hand acceleration, and deceleration and final closure of the fingers) [7]. Many studies declared that the grasping movement is highly complex which might related with many intrinsic (target size, shape, weight and material) and extrinsic (location, hand-object distance) factors [1, 4].

On other hand, aging is one of the irreversible factors to impact the human biomechanical function. Many functional capacities seems to remain stable until age 65 years, after age 75 years, it diminishes slowly, age differences are mostly apparent in prehension pattern frequency, hand strength, performance time and joint range of motion during reach-to-grasp movement [6]. Previous studies revealed that senior people presented lower values of maximum height of wrist, greater reaction times and more reaction durations than young adults during reaching an object [2, 5]. For grasping, age-related differences in finger coordination had been considered. Elderly were expected with higher safety margins and lower synergy indices. However, elderly showed lower safety margins for the thumb during rotational hand action [8].

Hence, this study aims to elucidate the effects of object weight and location (distance and orientation) on reach-to-grasp task between young and old adults. Relevant kinematic data including joint range of motion and electromyography involving muscle activity of muscles of upper extremity would be measured and discussed. The findings of this study can provide useful information to understanding the aging influences on reach-to-grasp movement.

2 METHODOLOGY

2.1 Subjects

A total of 12 healthy subjects, 6 young and 6 older adults were recruited to participate in this study. The average age for young and old adults were 23.5 (SD = 0.5) and 71.2 (SD = 2.9) years old, respectively. All subjects were free from neurological and musculoskeletal injuries and illness of upper extremity. All subjects signed a written consent form which was ethics approval obtained from the IRB of Chaoyang University of Technology, Human Factors Laboratory committee. The averaged height and weight for young and elderly were 165.0 cm (SD = 12.6); 55.8 Kg (SD = 11.2) and 158.0 cm (SD = 3.4); 72.5 (SD = 9.2) Kg, separately. Table 1 reveals the subjects' information. There were statistically different in averaged age, body height and weight (p < 0.05), however, it didn't has different in the arm length (p > 0.05).

2.2 Apparatus

An ultrasound based, three-dimension motion analysis system (Zebris CMS-HS/Zebris Medical GmbH, Germany) was used to collect relevant kinematic date of upper extremity (as Fig. 1). The Win Date (v.2.19.44) software which was provided by Zebris Medical GmbH had been used to record joint ROM of the shoulder, elbow, and wrist. One triangle

Table 1. The subjects' information.

Groups	Young	Elderly	P-values
Age	23.5 (0.5)	71.2 (2.9)	0.00*
Height (cm)	165.0 (12.6)	158.0 (3.4)	0.04*
Weight (Kg)	55.8 (11.2)	72.5 (9.2)	0.00*
Arm length (cm)	59.1 (5.2)	56.7 (2.0)	–

*Significant level at <0.05.

Figure 1. An ultrasound 3D motion analysis system (Picture resources Zebris Medical GmbH, Germany).

markers were placed on the dorsal surface of metacarpal; two triple markers were placed on the forearm and upper arm. Three single-point markers were placed on the lateral side of acromiohumeral, nail of thumb and index. Since a triple marker can be used to define one body segment. The joint angles were calculated by the relationship between two body segments to indentify the range of motion of wrist flexion-extension, wrist ulnar-radial deviation, forearm pronation-supination, elbow flexion-extension, shoulder flexion-extension and shoulder adduction-abduction. All data were collected with the sampling rate at 50 Hz.

An eight-channel electromyography which was built-in with motion analysis system (Zebris CMS-HS/Zebris Medical GmbH, Germany) was applied to measure the muscle activity of four muscle groups. Four pairs of disposable electrodes were placed on the muscle belly of each muscle group including: the biceps, triceps, Brachioradialis and Pronator Teres according to the guideline of Electromyography. All EMG data were collected and analyzed using the MyoresearchTM (Noraxon) software at 1000 Hz with the band pass filtered at 6–600 Hz. The Maximal Voluntary Contraction (MVC) of each of the four muscles was measured independently before each experiment. The EMGs amplitude was normalized to the MVC for obtaining the muscle electrical activity (EA%). The sampling rate was been set at 900 Hz.

2.3 Experimental design

The nested-factorial experimental design was adopted to discuss the effects of age, target weights, and locations on reach to grasp movement. The similar targets, a cylinder with diameter of 3 cm and height of 12 cm, have three different weights including 100 g, 600 g and 1000 g. Nine locations which are composed by 3 distances and 3 orientations were presented as target locations. The distance was defined by the subject's arm length (AL) which was measured from acromioclavicular joint to fist center of hand. Three distances were 80% AL, 100% AL and 120% AL, respectively. For orientation, the midline direction was refer the sagittal axis of right acromioclavicular joint, three orientations were midline, 60 degree to right and 60 degree to left from midline, separately. Moreover, nine locations were been signed as A, B, C, D, E, F, G, H and I (as Fig. 2). All subjects were asked to randomly complete 81 trials (27 × 3 times) continuously. The experimental design was showed as Table 2.

2.4 Date analysis

For data processing, Analysis of Variance (ANOVA) was conducted to assess the effects of age, target weight and location on the reach to grasp movement. Statistical analyses were performed using the SPSS version 14.0 statistical analysis software. Significant level was set at p > 0.5. Post hoc testing was conducted using Duncan's multiple range tests.

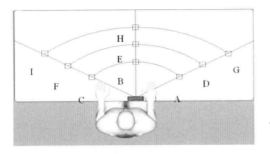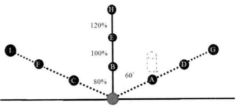

Figure 2. Target locations.

Table 2. The experimental design.

Variables	
Independent	Dependent
1. Weights (100 g, 600 g and 1000 g)	1. Kinematics date (joint range of motion) Wrist Flexion-Extension (WFE) Wrist Ulnar-Radial Deviation (WUR) Elbow Flexion-Extension (EFE) Forearm Pronation-Supination (FPS) Shoulder Flexion-Extension (SFE) Shoulder Adduction-abduction (SADD)
2. Locations (3 distances × 3 directions = 9 locations) Distances: 80% AL, 100% AL and 120% AL Directions: 60 degree right from midline, midline and 60 degree left from midline	2. Electromyography (%MVC) Biceps Triceps Brachioradialis Pronators Teres
3. Age (young and old adults)	

3 RESULTS

3.1 *Aging effect on the range of motion of upper extremity during reaching to grasping*

The range of motion of wrist, elbow and shoulder during reaching to grasping movement was presented as Table 3. For young and old adults, there were significantly angle differences in wrist flexion-extension, elbow flexion-extension, forearm pronation-supination and shoulder flexion-extension ($p < 0.05$). During reach to grasp a target, elderly have greater wrist flexion, but less elbow flexion, forearm pronation and lower shoulder flexion angle than young adults.

3.2 *Age effect on muscle activity of upper extremity during reaching to grasping*

For electromyography, the muscle activity including biceps, triceps, Brachioradialis and Pronators Teres during reach to grasping movement were be measured and calculated. As Table 4, there were significant differences in muscle activity of biceps, triceps, Brachioradialis and Pronators Teres between young and old adults ($p < 0.05$). Elderly required higher muscle activity in biceps, triceps, Brachioradialis and Pronators Teres than young adults to execute the reach to grasp movement.

3.3 *The effects of target weight and location on reaching to grasping*

The effects of target weight and location on reach to grasp movement were indicated as Table 5. For young adults, target weight has significant influence on muscle activity of

20

Table 3. The Range of Motion (ROM) between young and old adults during reaching and grasping.

ROM[a]	WFE[b]	WRU	EFE	FPS	SFE	SADD
Age						
Young	41.5 (8.8)	27.5 (4.6)	96.4 (22.2)	50.1 (7.4)	50.5 (12.9)	47.8 (17.3)
Elderly	47.3 (10.7)	27.7 (7.7)	82.3 (17.7)	47.1 (10.0)	47.3 (12.3)	47.6 (18.7)
P-value	0.00*	0.77	0.03*	0.04*	0.01*	0.90

[a] ROM: unit is degree.
[b] WFE = Wrist Flexion-Extension, WRU = Wrist Radial-Ulnar deviation, EFE = Elbow Flexion-Extension, FPS = Forearm Pronation-Supination, SFE = Shoulder Flexion-Extension, SADD = Shoulder Adduction-abduction.
* Significant level at <0.05.

Table 4. The muscle activity between young and old adults during reaching to grasping.

EA%	Biceps	Triceps	Brachioradialis	Pronators Teres
Age				
Young	0.17 (0.09)	0.06 (0.04)	0.20 (0.17)	0.27 (0.15)
Elderly	0.33 (0.23)	0.16 (0.13)	0.35 (0.15)	0.45 (0.17)
P-value	0.00*	0.00*	0.00*	0.00*

* Significant level at <0.05.

Table 5. The ANOVA results of target weight and location on reaching to grasp.

		Variables					
		Young adults			Elderly		
Measurements		Weight	Location	Weight × Location	Weight	Location	Weight × Location
ROM	WFE[a]	–	–	–	–	0.02*	–
(degree)	WRU	–	–	–	–	–	–
	EFE	–	0.00*	–	–	0.00*	–
	FPS	–	–	–	–	–	–
	SFE	–	0.00*	–	–	0.00*	–
	SADD	–	0.00*	–	–	0.00*	–
EA%	Biceps	0.00*	–	–	–	–	–
	Triceps	–	–	–	–	–	–
	Brachioradialis	0.00*	–	–	0.00*	–	–
	Pronators Teres	–	–	–	–	–	–

[a] WFE = Wrist Flexion-Extension, WRU = Wrist Radial-Ulnar deviation, EFE = Elbow Flexion-Extension, FPS = Forearm Pronation-Supination, SFE = Shoulder Flexion-Extension, SADD = Shoulder Adduction-abduction.
* Significant level at <0.05.

biceps and Brachioradialis ($p < 0.05$). The results of post hoc testing indicated that grasping weighty target require more muscle effort and present higher muscle activity in biceps and Brachioradialis (as Table 6). Target location has remarkable effect on the range of motion of elbow flexion, shoulder flexion, shoulder adduction ($p < 0.05$). In general, for reaching to grasping movement, the near distance (80% arm length) causes more elbow flexion and right orientation induces less shoulder flexion and adduction angle of joint range of motion (as Table 6).

For elderly, it was similar to young adults that target weight didn't have significant influence on range of motion and target weight only has an effect on muscle activity of Brachio-

Table 6. The results of Duncan's multiple range tests for target weight and location.

Measurements		Young adults		Elderly	
		Weight[b]	Location[c]	Weight	Location
ROM (degree)	WFE[a]	–	–	–	HFIBECGD < DA
	EFE	–	HG < IED < AFB < C	–	GH < DIE < FBC < A
	SFE	–	A < D < G < BCEHFI	–	A < DG < B < CEHF < HFI
	SADD	–	ABD < BDHG < DHGE < C < F < I	–	A < DGB < GBEH < C < F < I
EA%	Biceps	100 g < 600 g < 1 Kg	–	–	–
	Triceps	–	–	–	–
	Brachioradialis	100 g < 600 g < 1 Kg	–	100 g < 600 g < 1 Kg	–

[a] WFE = Wrist Flexion-Extension, WRU = Wrist Radial-Ulnar deviation, EFE = Elbow Flexion-Extension, FPS = Forearm Pronation-Supination, SFE = Shoulder Flexion-Extension, SADD = Shoulder Adduction-abduction.
[b] Weights: 100 g, 600 g and 1 Kg (1000 g).
[c] Locations: A, B, C, D, E, F, G, H, I.
* Significant level at <0.05.

radialis ($p < 0.05$). As Table 6, the post hot tasting reveals that heavy target weight induces higher muscle activity of Brachioradialis. Target location has statistical influence on the range of motion in the wrist flexion, elbow flexion, shoulder flexion and shoulder adduction ($p < 0.05$). For older adults, the reaching to grasping movement, the near distance and right orientation (location A and D) induces greater wrist flexion. Moreover, the near distance induces more elbow flexion and right orientation lead into less shoulder flexion and adduction angle of joint range of motion (as Table 6). However, there was no interaction effect between target weight and location on the range of motion and muscle activity of upper extremity.

4 DISCUSSION AND CONCLUSION

The results of this study indicated that age has a significant influence on joint range of motion and muscle activity during reach-to-grasp movement. For elderly, it displays greater wrist flexion, less elbow flexion, forearm pronation and less shoulder flexion than young adults. Moreover, older adults required higher muscle activity in Biceps, Triceps, Brachioradialis and Pronators Teres to conduct the reach to grasp movement.

Moreover, target weight has significantly effect on muscle activities of upper extremities. Grasping weighty target require more muscle effort and present higher muscle activity in biceps and Brachioradialis for young adults. For elderly, it was similar to young adults that target weight has an effect on muscle activity of Brachioradialis.

Furthermore, target location has remarkable effect on the range of motion of wrist flexion, elbow flexion, shoulder flexion, shoulder adduction. For young adults, the near distance (80% arm length) causes more elbow flexion and right orientation induces less shoulder flexion and adduction angle of joint range of motion during reaching to grasping. For old adults, target location has statistical influence on the range of motion in the wrist flexion, elbow flexion, shoulder flexion and shoulder adduction. The near distance and right orientation (location A and D) induces greater wrist flexion. Moreover, the near distance induces more elbow flexion and right orientation lead into less shoulder flexion and adduction angle of joint range of motion.

For reaction time, there didn't have significantly different between two age-groups. Furthermore, object location affect the joint range of motion, object weight significantly impact the muscle activity of Brachioradialis for elderly. These primary findings of this study can provide useful information for clinical training, rehabilitation goal setting, relevant product design and barrier-free considerations.

ACKNOWLEDGEMENTS

This study was funded by the National Science Council of Taiwan (grant no. NSC 99-2221-E-040-007-MY2). The authors especially acknowledge Humanbio Instrument, Inc (Taiwan) for their instrument and technical support.

REFERENCES

[1] Chiou RY-C, Wu DH, Tzeng OJ-L, Hung DL, Chang EC. Relative size of numerical magnitude induces a size-contrast effect on the grip scaling of reach-to-grasp movements. *Cortex* 2012;**48**:1043–1051.

[2] Geronimi M, Gorce P. Aging effect on movement of prehension with obstacle. *Journal of Biomechanics* 2007; **40**:S2.

[3] Jone L A, Lederman SJ. Human Hand Function. *Oxford University Press* 2006.

[4] Karok S, Newport R. The continous updating of grasp in response to dynamic changes in object size, hand size and distractor proximity. *Neruopsychologia* 2010;**48**:3891–3900.

[5] Poston B, Van Gemmert AWA, Barduson B, Stelmach GE. Movement structure in young and elderly adults during goal-directed movements of the left and right arm. *Brain and cognition* 2009; **69**:30–38.

[6] Shiffman LM. Effects of Aging on Adult hand function. *The American Journal of Occupational Therapy* 1992; **46**: 785–792.

[7] Supuk T, Bajd T, Kurillo G. Assessment of reach-to-grasp trajectoryes toward stationary objects. *Clincal biomechanics* 2011;**26**:811–8.

[8] Varadhan SKM, Zhang W, Zatsiorsky VM, Latash ML. Age effects on rotational hand action. *Human movement science* 2012;**31**:502–518.

New Ergonomics Perspective – Yamamoto (Ed.)
© 2015 Taylor & Francis Group, London, ISBN 978-1-138-02751-0

Development of walking ability evaluation method with accelerometer and gyrometer

Tasuku Ito
Waseda University, Tokorozawa, Saitama, Japan

Hidetaka Nozawa, Shigeya Okada, Minoru Hatakeyama & Norihiko Shiratori
MicroStone Corporation, Saku, Nagano, Japan

Akira Ichikawa
Saku Central Hospital, Saku, Nagano, Japan

Macky Kato
Waseda University, Tokorozawa, Saitama, Japan

ABSTRACT: Fall accident is one of the problems in aged society. And Joint disease is also severe, especially in the field of rehabilitation. They are related to the walking ability. Thus, the evaluation and training of walking ability are important to prevent fall accidents. Conventionally, it needs long term of experiments to evaluate the walking ability by motion capture systems, force plates, and so on. In addition, the conventional methods spend costs expensively. The evaluation should be obtained easily in order to reduce the subjects' burden. In this study, evaluation by the accelerometer and gyrometer was practiced in place of the conventional methods. Two 8ch acceleration and gyro sensors (RF10, MicroStone Corporation) were installed to a subject's sacrum and vertebra to measure his spine stick picture motion during his walking. Power spectrum, movement track and RMS were used for comparison with measurements of motion capture system (Vicon). All the results have shown the validity and reliability of the simple measurement method. It was revealed that the stick picture motion has the possibility to be applied to the walking ability indicator.

Keywords: Accelerometer; Gyrometer; Walking Ability; Motion Capture

1   INTRODUCTION

The proportion of the elderly people in Japan is increasing in the aged society in recent years. One of the most severe problems of the aged society is the unexpected accident in everyday life. Fall accident is one of them. Approximately 2,500 people had been injured with fall accidents at home in 2009 [1]. On the other hand, the number of elderly people who have joint disease is increasing in the field of rehabilitation. Most of them caused from fall accident and fracture of bones. One of the reasons of fall accident is the decline of walking ability. Thus the training and evaluation of walking ability are required to prevent the fall accidents. In the past, Ohtaki was attached acceleration sensors to thigh near the knee of the right leg, lower leg near the ankle and the center of vertebra to show they can analyze the walking features [2]. Hanawaka attached the accelerometer and gryometer to the toe of the participants. It was not easy to measure higher speed movements and the measurement of moving objects. The study has shown the possibility to measure these movements [3]. Also, Kato established the method to evaluate the standing ability with the accelerometers. As a result, it was revealed that the children's body movements were large. In addition, Elderly people's movements tend to incline forward [4]. In conventional method to evaluate the motion ability, motion capture system has been used frequently. However, many of them need long preparing time and expensive devices even though the experiment has spent just a short time. Especially, the subjects' burden cannot be ignored in the case of the elderly people. Thus, the simple evaluation method would be needed in the field of rehabilitation.

2 METHOD

The subject was a healthy 24 year old male athlete whose height is 158 cm and weight is 65 kg. Two RF10 (MicroStone Corporation), the device which records three dimensional acceleration and angular velocity, were attached to him as shown in Figure 1. The attached points were sacrum and vertebra. In addition, Motion capture system (Vicon) was installed around a treadmill to measure the movement of the attached points. Measurements of these devices were held simultaneously during he walked on the treadmill at 4 km/h. Sampling frequency was 200 Hz. The length of the synchronized dataset was approximately 9 seconds. The series of movements, which were calculated from the dataset of RF10 after filtering with 0.5 Hz HPF and 5.0 LPF, were compared with the series of movements measured by Vicon. Power spectrum, Coherence, motion tracks and RMS of movements were used.

3 RESULTS

3.1 *Comparing the features of the measurements*

3.1.1 *Power spectrum by FFT*
Figure 2 shows the comparison of power spectrums of x-, y- and z-axis of movements by FFT for the measurement results. Comparison of corresponding pairs, all the spectrums have same peaks. In brief, the peak of the x-axis, which means the subject's lateral direction, for sacrum and vertebra by the both measurements can be observed around 0.9 Hz. The peak of the y-axis, which means the subject's direction of walking, for the both points by the both measurements can be observed around 1.8 Hz, twice as the x-axis. In addition, a little peak can be observed around 0.5 Hz and 0.6 Hz on the y-axis. The peak of the z-axis, which means the subject's vertical direction, for the both points by the both measurements can be observed around 1.8 Hz as same as the y-axis.

3.1.2 *The movement tracks on the horizontal plane and vertical plane*
Figure 3 shows that the movement tracks for each attached points during the trial. Each of them has two kinds of coordinates. One is the x-y coordinate, which means the movement of the point observed from above the subject. Another is the x-z coordinate, which means the movement of the point observed from behind the subject. The corresponding movement

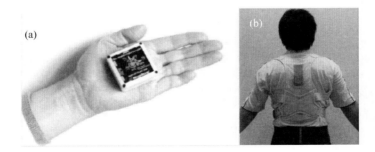

Figure 1.   (a) RF10 Accelerometer; (b) Attached position.

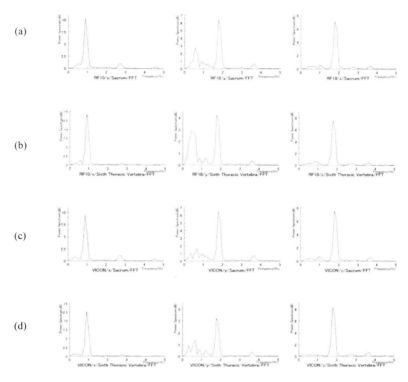

Figure 2. Parts of power spectrums of x-, y-, z-axis (a) sacrum RF10; (b) vertebra RF10; (c) sacrum Vicon; (d) vertebra Vicon.

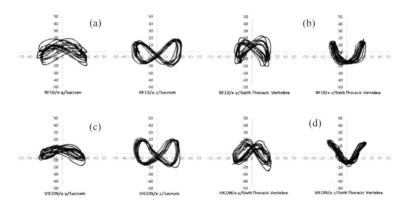

Figure 3. The movement tracks observed from the above direction (x-y) and behind direction (x-z); (a) sacrum RF10; (b) vertebra RF10; (c) sacrum Vicon; (d) vertebra Vicon.

tracks of each point are similar in the figure. Every point shows the different shapes, however, the shapes look figure of eight.

3.1.3 *Relationships between the RMS of RF10 and that of Vicon*

Figure 4 shows the relationships between the RMS of each measurement. The coefficient of determination at the sacrum was 0.87. The positive regression line can be observed. The inclination of the regression line for the sacrum was approximately 0.97. The coefficient of determination at the vertebra was 0.66. The positive regression line can be also observed. The inclination of the regression line for the vertebra was approximately 1.06.

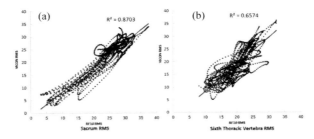

Figure 4.   Correlation coefficient between the movement track of the both measurements (a) sacrum; (b) vertebra.

4   DISCUSSION

4.1.1   *Power spectrums which mean the walking rhythm*
The peak of 0.9 Hz on the power spectrums of lateral direction is thought to mean the effect of the footstep. The center of balance tends to go right at the moment of right foot stroke, and it tends to go opposite side at the moment of left footstep. Thus the unit of the walking cycle observed in x-axis consists of two-foot steps. On the other hand, the peaks of y- and z-axis were observed around 1.8 Hz, because they're thought to have a different rhythm. In brief, the acceleration of both steps would make the subject go same direction. Thus, the peaks of the power spectrums have twice of x-axis. Almost the same features can be observed in the both measurements except the y-axis. The small peaks can be observed around 0.3 Hz and 0.6 Hz. The frequency of the peak is lower than the 0.9 Hz, which is the peak of x-axis. The reason is thought to be the drifts. That is because some drifts can be observed from the original fluctuation of the movement.

4.1.2   *The movement tracks which mean the walking stability*
As seen in the Figure 3, many times of foot stroke seem to have similar features, because the shape of eight can be observed clearly in each figure. Naturally, the tendency of sacrum point is different from those of the vertebra. Thus, the movement of the straight line between the two points would be regarded as the inclination of the upper body posture. I can be also regarded as the indicator of the subject's walking ability. Almost same features can be also observed in the movement tracks.

4.1.3   *RMS which means the volume of the movement*
RMS means the volume of the movement of the measurement points. Confirming the volume of RF10 movements would be required to reveal that it has stability as same as the conventional method, motion capture system of Vicon. The high coefficient of determination shows the strong relationship between two types of measurements. In addition, the volumes of RMS for each measurement are so close that the inclination of the regression line for both of the sacrum and the vertebra were almost 1.0. It means that the suggested method can measure the same dataset so that it can be used in place of the conventional one.

4.1.4   *The validity and reliability of the new evaluation system by RF10*
The results of comparing the both methods revealed that the suggested method would have enough validity and enough reliability as the conventional method. In brief, the purpose of this study is the development of walking ability evaluation method without expensive devices and hard burdens for the subjects. Vicon, the conventional method, is regarded as one of the

de facto standard measurement devices. Thus RF10, which indicated the same results in the same situation, can be also regarded as one of the certain devices to evaluate the walking ability.

In the future, it will be able to spread to the other fields, such as nursing, rehabilitation, and manufacturing, which have occupational physical problems.

REFERENCES

[1] Ministry of Health, Labour and Welfare. Demographic Statistics; 2009.
[2] Yasuaki O, Koichi S, Hikaru I. A Method for the Long-Term Gait Assessment Utilizig Accelerometers and Gyroscopes. The Japan Society of Mechanical Engineers. **67**, 655: 2001, p.192–198.
[3] Masuo H, Masaharu S, Naoyuki FUJIMOTO, Akira SATOH. Gait Measurement System and its Applications, Yokogawa Technical Report, **51**, 3; 2007, p.21–24.
[4] Macky K, Yoshie S, Takeshi S, Comparison of Postural Control Ability among Three Age Groups using Acceleration Measurement, Journal of Human Ergology, **40**, 1, 2; 2011, p.101–108.

New Ergonomics Perspective – Yamamoto (Ed.)
© *2015 Taylor & Francis Group, London, ISBN 978-1-138-02751-0*

Occupational safety issues at workplace among older fishermen in Malaysia

Rosnah Mohd Yusuff
Faculty of Engineering, Universiti Putra Malaysia, Selangor, Malaysia

Mohd Rizal Hussain & Nabilah Latif
Institute of Gerontology, Universiti Putra Malaysia, Selangor, Malaysia

Anita Abdul Rahman
Faculty of Medicine and Health Science, Universiti Putra Malaysia, Selangor, Malaysia

Sharifah Norazizan Syed Abd Rashid & Haslinda Abdullah
Faculty of Human Ecology, Universiti Putra Malaysia, Selangor, Malaysia

ABSTRACT: Sea fishing is a dangerous and risky occupation. Fisherman often faced hazardous conditions while at work. It has been reported that majority of hospitalized injuries among commercial fishermen are caused by machinery. Working on wet, slippery surface, handling heavy equipment, cables and ropes are familiar situations which are further aggravated by bad weather conditions. Falls were also reported as a major contributor to commercial fishing accidents. Therefore, a wide variety of workplace hazards confront the work force. The fisheries sector is an important sub-sector in Malaysia and plays a significant role in the national economy development. Fish is largely consumed by the population and the fishery industry is also a source of employment for those living in the coastal areas. A study was conducted to explore occupational safety issues at the workplace among older fishermen in Malaysia, since most of the fishermen are aged above 45. A total of 397 sea fishermen in the states of Terengganu, Kedah and Perak aged 45 years and above were involved in a face-to-face interview using developed questionnaire. The questionnaire was designed based on Focus Group Discussions conducted earlier with the stakeholders. This is to ensure that crucial issues related to safety of the fisherman were addressed. The questionnaire focused on the safety problems at work and what are the causes of these problems. Data was analyzed using Statistical Program for Social Science (SPSS). The respondents were mostly males and had primary school education. The finding showed that almost one-third of the respondents reported that they had experienced boat accidents mostly caused by bad weather. One-fifth of the respondents in this study have also experienced accidents/physical injury in their workplace, with animal bites and sting being identified as the main contributor. Falls was reported by almost half of the respondents and slippery floor, strong waves and tripping were identified as the major causes. By identifying the problems and factors faced by the fishermen, future improvements to make the workplace safer and comfortable, especially for the older fishermen can be designed. Appropriate personal protective equipment such as gloves, water proof jackets, appropriate signages, improve visibility and appropriate shoes need to be designed for the workers, suitable to be used in the sea and weather conditions of Malaysia.

Keywords: safety; older fishermen; Malaysia; workplace

1 INTRODUCTION

The fisheries sector (including marine capture fisheries, aquaculture, and inland fisheries) is an important sub-sector in Malaysia and plays a significant role in the national economy development. Based on statistic by Department of Fisheries Malaysia (DOF) in 2010, this sector contributed 2,014,534.84 metric tonnes of fish production, which valued at RM9, 495.28 million. Comparing this in 2009, it was increased to 8.86 percent and 10.02 percent both in quantity and in value. Furthermore, in the same year this sector was contributed 1.3 percent to national Gross Domestic Product (GDP). Apart from that, it is also a source of employment, foreign exchange and a source of cheap animal protein supply for population in the country. Fish constitutes 60–70% of the national animal protein intake, with per caput consumption of 47.8 kg per year. The rate of demand for fish as the main source of protein is expected to increase from the current annual consumption of 630 000 tonnes to over 1 579 800 tonne by 2010 (using an estimated population of 26 330 000 with a per caput consumption of 60 kg/year). In 1997, the fisheries sector provides employment for more than 79 000 fishermen [1].

However, many previous studies highlighted that sea fishing, which is a part of fisheries sector, is hazardous and risk occupation [2–5]. Finding also shows that majority of hospitalized injuries among Alaska commercial fishermen are causes by machinery [6]. As fisherman, working on wet, slippery, handling heavy equipment, cables and ropes are a familiar situation. Furthermore, inclement weather and the poisonous spines of some fish are considered natural [7]. Falls also were reported as a major contributor to commercial fishing fatalities among Alaska fisherman. The most common circumstances associated with falling overboard were working with fishing gear, being alone on deck, losing balance or slipping, heavy weather, gear entanglement, and alcohol [8]. This problem also evident internationally where a study reported that falls overboard in the U.S represented 25% of all fishing fatalities, compared to 27% in Norway, Denmark (30%), Ireland (20%) and Iceland (33%) [9].

Therefore, this paper was to explore occupational safety issues at workplace among older fishermen in Malaysia, where its finding might help as reference value in planning programs and strategies for intervention in reducing injuries and accidents among older fisherman towards safety environment in workplace.

2 MATERIALS AND METHODOLOGY

2.1 *Participants*

Data was obtained from a cross-sectional study entitled 'Workplace Safety and Health among Older Workers: Programs and Strategies for Intervention' conducted between the years 2009–2011 by the Institute of Gerontology, Universiti Putra Malaysia. A total of 397 sea fisherman aged 45 years and above from the states of Terengganu, Kedah and Perak in Peninsular Malaysia were involved in this study. Participants were identified through Persatuan Nelayan (Fisherman Welfare Association) from every districts/area in that states. Each district/area was provided 25–30 sea fishermen to be as respondents in this study.

2.2 *Instruments*

Respondents were interviewed face-to-face by Enumerator using developed questionnaires form. The questions were developed based on inputs and information given by fishermen from Focus Group Discussion (FGD). This FGD was chosen as a guideline in the qualitative method where complex themes can be selected to be the topic for discussions and analysis. Focus group discussion is a good approach to gather people from similar backgrounds or experiences to discuss a specific topic of interest. The strength of FGD relies on allowing the participants to agree or disagree with each other so that it provides an insight idea on how a group thinks about an issue, coordinate the range of opinion and ideas, and make judgement on the inconsistencies and variations that exists in a particular community in terms of beliefs and their experiences and practices [10–12].

In addition to details of demographic background and fishing background the questionnaire also including several aspects that related to safety issues at workplace such:

i. Safety problems at workplace: the respondents indicated type of problems that they were faced at workplace whether boat accident, accident/physical injury caused by machinery, equipment, building infrastructure and public facilities, chemical, animal or others, fall and etc.

ii. Factor causing safety problems at workplace: this consisted question on how safety problems are happened whether causing by shallow/narrow river confluence, beacon lamp not function/available, jetty/port is busy or congested, bad weather, not wearing/using personal protection equipment, slippery floor, tripping over equipment, or machine while working and etc.

2.3 *Analyses*

Data was analyzed using SPSS (Statistical Programme for Social Sciences). Level of significance used for the data was set at $p < 0.05$ (two-tailed). Descriptive analysis was performed to identify frequency, percentage, Mean and Standard Deviation on background demographic items (gender, race, age, level of education, marital status, household size, years fishing and monthly income) and safety issues at workplace.

3 RESULT

Table 1 details the demographic background of the respondents. It shows that most of respondents were male (99.7%), Malay (90.2%), average age was 56.43 years old (SD = 7.748), primary school education (65.2%), now married (90.4%) and average household and monthly income size were 5.4 (SD = 2.573) and RM884.15 (SD = RM588.781).

In response to the questions about safety issues at workplace that show in Table 2, it emerged that 27.7 percent of respondents reported having experience boat accident dur-

Table 1. Demographic background of respondents.

Items	N	%	Mean	SD
Gender				
i. Male	396	99.7		
ii. Female	1	0.3		
Race				
i. Malay	358	90.2		
ii. Chinese	39	9.8		
iii. Indian	–	–		
iv. Others	–	–		
Age	397	100	56.43	7.748
Level of education				
i. Never been school	25	6.3		
ii. Primary education	259	65.2		
iii. Lower secondary education	73	18.4		
iv. Upper secondary education	33	8.3		
v. Higher education	5	1.3		
vi. Others	2	0.5		
Marital status				
i. Never married	18	4.5		
ii. Now married	359	90.4		
iii. Divorced/separated	5	1.3		
iv. Widowed	15	3.8		
Household size			5.4	2.573
Monthly income			884.15	588.781

33

Table 2. Safety problems and factors causing safety problems faced by respondents at workplace.

Safety problems	N	%	Factors causing safety problems	N	%
Boat accident			Shallow/narrow river confluence		
i. Yes	110	27.7	i. Yes	12	3.0
ii. No	287	72.3	ii. No	385	97.0
			Busy and congested jetty/port		
			i. Yes *	13	3.3
			ii. No	384	96.7
			Beacon lamp not working/available		
			i. Yes	29	7.3
			ii. No	368	92.7
			Poor boat condition (damage, too old)		
			i. Yes	55	13.9
			ii. No	342	86.1
			Bad weather		
			i. Yes	61	15.4
			ii. No	336	84.6
			Confusion of similar colour signal light for all situations		
			i. Yes	5	1.3
			ii. No	392	98.7
Accident/physical injury caused by machinery			Not wearing/using personal protection equipment		
i. Yes	13	3.3	i. Yes	20	5.0
ii. No	384	96.7	ii. No	377	95.0
Accident/physical injury caused by equipment			No personal protection equipment available/provided		
i. Yes	17	4.3	i. Yes	17	4.3
ii. No	380	95.7	ii. No	380	95.7
Accident/physical injury caused by building/public facilities			Using equipment/machinery/ chemical without caution		
i. Yes	1	0.3	i. Yes	13	3.3
ii. No	396	99.7	ii. No	384	96.7
Accident/physical injury caused by chemical			Unsatisfactory/damage of machinery/equipment personal protection equipment/ building infrastructure and public facilities		
i. Yes	2	0.5	i. Yes	6	1.5
ii. No	395	99.5	ii. No	391	98.5
Accident/physical injury caused by animal			Lack of knowledge/exposure on aspects of safety at workplace		
i. Yes	60	15.1	i. Yes	12	3.0
ii. No	337	84.9	ii. No	385	97.0
			Animals bite or sting		
			i. Yes	46	11.6
			ii. No	351	88.4
Fall			Strong wave		
i. Yes	201	50.6	i. Yes	164	41.3
ii. No	196	49.4	ii. No	233	58.7
			Slippery floor		
			i. Yes	176	44.3
			ii. No	221	55.7
			Tripping over equipment/machine		
			i. Yes	82	20.7
			ii. No	315	79.3

ing working, which caused by bad weather (15.4%), poor boat condition (13.9%), beacon lamp not working/available (7.3%), busy and congested jetty/port (3.3%), shallow/narrow river confluence (3.0%) and confusion of similar color signal light for all situations (1.3%).

Finding shows that accident/physical injury is another issue reported that faced by respondent at workplace, which causes by animals (15.1%), equipment (4.3%), machinery (3.3%), chemical (0.5%) and building infrastructure or public facilities (0.3%). Several factors were identified contributing to those problems above, which mostly by animals bite and sting (11.6%), respondents were not wearing/using personal protection equipment (5.0%), no personal protection equipment available/provided (4.3%), using equipment/machinery/chemical without caution (3.3%), lack of knowledge/exposure on aspects of safety at workplace (3.0%) and unsatisfactory/damage of machinery/equipment/building infrastructure and public facilities (1.5%). Respondents also reported experience problem on fall (50.6%) during work, which mostly causes by slippery floor (44.3%), strong wave (41.3%) and tripping over equipment/machine (20.7%).

4 DISCUSSION

Even though the analysis only focused on problems and factors that contributed to safety issues at workplace among fisherman, finding is helpful in given us the scenario of work environment faced by fisherman in their daily work. Almost one-third of respondents reported that they were having experience in boat accident, which mostly causes by bad weather. It is not surprising because Malaysia is a country that has an equatorial climate, giving it a warm and wet weather due to its proximity to the equator. On an average, Malaysia receives about 6 hours of sunshine each day with cloud formations occasionally leading to rainfall. There are two monsoon winds that influence the rainfall at different intervals of the year. The Southwest Monsoon usually occurs between May till September, bringing rainfall to the western side of Peninsular Malaysia. On the other hand, the Northeast Monsoon starts from November and lasts till March, bringing heavy rainfall to areas on the east side of Peninsular Malaysia. As this monsoon wind is particularly strong, it often brings heavy rain to the west side of Peninsular Malaysia as well during this period. According to this matter, previous finding shows that weather condition also contributed of fishing boat incidents in Atlantic Canada [13].

Accident/physical injury is common problem among fisherman [14]. Therefore, it shows that almost one-fifth of respondents in this study were experienced this problem in their workplace, which animals bite and sting was identified main contributor to this problem followed by less awareness among respondents in using personal protection equipment when working. This finding is similar with other studies, where commercial fishermen are exposed to some hazards including marine animals as well as more widespread hazards such as mechanical equipment, fatigue and stress [15–20].

In this study, falls was reported have highest number of respondents, where was contributed almost half of respondents. Slippery floor was identified major contributor to this problem. For fishermen, work capacity is very largely determined by the condition of the legs; therefore, leg injury can be a serious matter in commercial fishing [21]. Therefore, it is important for every fisherman to avoid any hazards that can lead to risk of falling during working.

5 CONCLUSION

Safety is complimentary issue that needs to give fully attention by everyone, including at workplace. By identifying problems and factors that contributed to this issue, it's will help us in future to improve workplace to become more safe and comfortable for older workers.

ACKNOWLEDGEMENT

We would like to thanks all the participants, research assistants and researchers in involving the study.

REFERENCES

[1] FAO. Safety at sea as an integral part of fisheries management. FAO Fisheries Circular No. 966, Roma, 2001.

[2] Jin, D., Kite-Powell, H. & Talley, W. The safety of commercial fishing: Determinants of vessel total losses and injuries. *Journal of Safety Research* 2001;32:209–228.

[3] Morel, G., Amalberti, R. & Chauvin, C. Articulating the differences between safety and resilience: the decision-making process of professional sea-fishing skippers. Human Factors: *The Journal of the Human Factors and Ergonomics Society* 2008;50(1):1–16.

[4] Murray, M., Fitzpatrick, D. & O'Connell, C. Fishermens blues: Factors related to accidents and safety among Newfoundland fishermen. *Work and Stress: An International Journal of Work, Health & Organization* 1997;11(3):292–297.

[5] Lise, H.L., Henrik, L.H. & Olaf, C.J. Fatal occupational accidents in Danish fishing vessels 1989–2005. *International Journal of Injury Control and Safety Promotion* 2008;15(2):109–117.

[6] Jennifer, M. Lincoln & Chelsea, C. Woodward. Reducing commercial fishing deck hazards with engineering solutions for winch design. *Journal of Safety Research* 2008;39(2):231–235.

[7] Sprent, P. Taking risks: *The science of uncertainty*. London:Penguin, 1988.

[8] Devin, L. Lucas & Jennifer, M. Lincoln. Fatal falls overboard on commercial fishing vessels in Alaska. *American journal of industrial medicine* 2007;50:962–968.

[9] Abraham P.P. International comparison of occupational injuries among commercial fishers of selected northern countries and regions. In: Lincoln JM, Hudson DS, Conway GA, Pescatore R, editors. *Proceedings of the International Fishing Industry Safety and Health Conference*. Cincinnati: National Institute for Occupational Safety and Health 2002. pp 455–465. Available from http://www.cdc.gov/niosh/docs/ 2003–102/2003102pd.html.

[10] Krueger, R.A. a. C., M.A. *Focus groups: A practical guide for applied research*. Thousand Oaks, London, New Delhi: Sage Publications, Inc., 2009.

[11] Morgan, D.L. *Focus groups as qualitative research* 1997;16. Sage Publications, Inc.

[12] Stewart, D.W., Shamdasani, P.N., & Rook, D.W. *Focus groups: Theory and practice*: Sage Publications, Inc., 2007.

[13] Yue, W., Ronald, P. & Casey, H. The effect of weather factors on the severity of fishing boat accidents in Atlantic Canada. *Risk Management* 2005;7(3):21–40.

[14] Norrish, A.E. & Cryer, P.C. Work related injury in New Zealand commercial fishermen. *Br J Ind Med* 1990;47:726–732.

[15] Holland, Martin D. Trawler safety-final report of the committee of inquiry into trawler safety. London: HMSO, 1969.

[16] Schilling, R.S.F. Hazards of deep-sea fishing. *Br J Ind Med* 1971;28:27–35.

[17] Barss, P.G. Penetrating wounds caused by needle-fish in Oceania. *Med J Aust* 1985;143:617–622.

[18] Jeays, L.W. Safety for fisherman on trawlers. Sydney: National Occupational Health and Safety Commission, 1987.

[19] Dutkiewicz, J,, Jablonski, L. & Olenchock, S.A. Occupational biohazards: a review. *Am J Ind Med* 1988;14:605–623.

[20] Sutherland, K.M. & Flin, R.H. Stress at sea: a review of working conditions in the offshore oil and fishing industries. *Work and Stress* 1989;3:269–285.

[21] Olaf, C. Jensen. Non-fatal occupational fall and slip injuries among commercial fishermen analyzed by use of the NOMESCO injury registration system. *American Journal of Industrial Medicine* 2000;37:637–644.

New Ergonomics Perspective – Yamamoto (Ed.)
© 2015 Taylor & Francis Group, London, ISBN 978-1-138-02751-0

Basic studies of decomposition of event-related potentials

Ryoichi Otsuka & Kimihiro Yamanaka
Tokyo Metropolitan University, Hino, Japan

Hidetoshi Nakayasu
Kanagawa University, Yokohama, Japan

ABSTRACT: In this study, we aim to identify characteristic waveforms of perception by measuring brain waves, assuming that the processes involved in information processing are independent, and performing wave decomposition. The ability to use brainwave signals to identify characteristic waveforms for perception occurring during information processing would aid in elucidating the emergence of human error. Specifically, we measured brainwaves with respect to a task based on Rasmussen's operator model skill base. Since the measured brainwaves are event-related potentials, the waveforms were acquired by using the summation averaging method. To check whether a feature quantity exists in the waveform, we conducted peak identification and determined N1, the negative peak with the largest magnitude in the latency period between 80 and 150 ms, which is generally considered to be the perceptual component. Next, we conducted independent component analysis on N1 and demonstrated that it is possible to use brainwave decomposition to extract the components that correspond to each process occurring during information processing. In the independent component analysis performed in this study, the observed signal x was defined on the basis of the source signal s generated in the brain and of the blend ratio A; the restoration signal y was estimated with the restoration matrix W from x.

Keywords: Human information processing; Event-related potential (P300); Wave decomposition; ICA (Independent Component Analysis)

1 INTRODUCTION

At present, the waveform of Event-Related Potential (ERP) as obtained by electroencephalographic measurement equipment is considered to be a composite wave made up of components for perception, cognition, and judgment. Because ERP is deeply related to the cognitive process and its characteristic components are considered to have functional significance for cognition, it is known as a composite brainwave that occurs during perception and cognition[1]. The calculation of the characteristic components is conventionally attempted by principal component analysis and similar techniques, but there is not yet an understanding of the information handling process, and detailed waveform analysis has not been performed[2].

The present study aims to find a characteristic waveform of perception through waveform analysis by measuring brainwaves and analyzing the results to see whether the information handling process is independent of other process, which is checked by Independent Component Analysis (ICA). If the characteristic waveform of perception in the information handling process can be found from brainwave signals, it may be possible to elucidate the developmental process of human errors[3,4].

2 WAVEFORM COMPONENT IDENTIFICATION PROCEDURE

Since the objective of the present study is to extract the components of sensation and perception from ERP, brainwaves were measured in relation to a task based on the skill base of the Rasmussen operator (SRK) model[5]. To verify whether a characteristic value is present in measured ERP waveforms, peak identification was performed and N1 latency of 80–149 ms and N2 latency of 150–240 ms were identified, which are conventionally considered to be sensation and perception components. Furthermore, it was investigated whether extracting the components related to each process of information handling through waveform analysis was possible by carrying out ICA on the N1 and N2 data. The ICA used in this research was carried out by first defining an observation signal x from the original intracerebral signal s and the blend ratio A and then estimating the restoration signal y from a restoration matrix W for x. The details are described in Section 4.

3 ERP MEASUREMENT EXPERIMENT

In the experiment, goals were set according to the SRK model. In the SRK model, the degree of difficulty of information processing increases and processing time becomes longer as the task transitions from skill-based to knowledge-based.

The participants were 5 male university student aged 21–23 years. For electroencephalographic measurements, each participant was fitted with plate electrodes and wore headphones. The participant was then seated in a soundproofed room; figure 1 shows the snapshot of the experiment. Participant performed an oddball task by voice. In the oddball task, two kinds of auditory stimulation (2 kHz and 1 kHz) were presented in a random order and the participants counted the number of each type of stimulus. Simultaneous, ERP was measured by response via a switch at hand. ERP measurements were conducted by an evoked potential measuring system (MEB-9100, Nihon Kohden), with 30 readings taken. The ERP measurements were taken at points Cz and Pz, in accordance with the International 10–20 system[6].

N1 and N2 components were extracted by the peak identification method because the characteristic value for the ERP waveform obtained is verified after removing noise. Generally, the condition difference (high or low auditory stimulation) and position difference (electrode position) relating to N1 has the characteristic that Cz < Pz at peak amplitude during a low tone and Cz > Pz at peak amplitude during a high tone[7]. In the present study, analysis was performed under the assumption that the N2 component has the same characteristics.

4 ICA PROCEDURE

In the present study, a restoration matrix W is found such that the distribution of the two observation signals is orthogonal, as shown by Equations (1) and (2). At this point, the assumption $W = A^{-1}$ is necessary[8].

Figure 1.   Snap shot of experiment.

$$x = As \qquad (1)$$
$$y = Wx \qquad (2)$$

In the first step of ICA, the correlation matrix R_x is found from Equations (3) and (4), where x_1 is the observed value at Cz and x_2 is the observed value at Pz because the solution can be found from uncorrelated matrices if the restoration signals are also uncorrelated.

$$E\{x_i x_j\} = \frac{1}{T} \sum_{t=1}^{T} x_i(t)\, x_j(t) \qquad (3)$$

$$R_x = \begin{bmatrix} E\{x_1 x_1\} & E\{x_1 x_2\} \\ E\{x_2 x_1\} & E\{x_2 x_2\} \end{bmatrix} \qquad (4)$$

Next, an eigenvalue λ of W is found, the orthogonal matrix direct eigenvector K_x is determined, and whitening is performed. The whitening process converts the observation signal so that the mean stochastic variable is 0 and the dispersion is 1, which allows definition of properties stronger than being uncorrelated. Accordingly, the whitening matrix V that converts the correlation matrix R_x into an identity matrix is found (Equation (5)). Applying the whitening matrix V to the observation signal x results in a whitening matrix Z (Equation (6)). Here, D is the diagonalized matrix of eigenvalues.

Next, ICA is performed. Here, fast ICA is used, and the solution space after whitening is confined to a unitary space characterized by an angle θ, which minimizes the entropy (mean information content) in the characteristic component; the entropy can be found by using a non-linear function G.

$$V = D^{-1/2} K^T \qquad (5)$$
$$Z(t) = Vx(t) \qquad (6)$$

The information content possessed by the waveform is minimized by minimizing the entropy, and the characteristic component can be extracted in a form that is more nearly independent. In the present study, a two-dimensional mixing matrix is estimated, and so there is one degree of freedom. It is therefore acceptable to assume that there is only one angle parameter during the unitary transformation.

Concerning the procedure, the value y of the characteristic component of N1 is first found for the whitening matrix Z, which integrates the unitary matrix (Equation (7)). Next, the relation between the value of Equation (9) and the angle during the unitary conversion is found. The angle at which entropy becomes a minimum is chosen as the angle of u_1 and the axis of the Primary Independent Component (IC1). The Secondary Independent Component (IC2) is hypothesized to be independent and orthogonal to IC1, and the axis is made orthogonal with IC1 by Gram–Schmidt orthogonalization according to Equation (10).

$$U = [u_1, u_2] = \begin{pmatrix} \cos\theta & -\sin\theta \\ \sin\theta & \cos\theta \end{pmatrix} \qquad (7)$$

$$p_i = exp[-|y_i|] \qquad (8)$$

$$G(y_i) = -\log p(y) \qquad (9)$$

$$u_2' = u_2 - \frac{u_1^t u_2}{u_1^t u_1} u_1 \qquad (10)$$

Finally, the whitening matrix V found by the above procedure and the integral of the unitary matrix U become the estimated value of the restoration matrix W, and an estimate of the original signal can be found from Equation (12).

$$W = A^{-1} = UV \tag{11}$$

$$S_1 + S_2 = \begin{pmatrix} w_{11} \\ w_{21} \end{pmatrix} x_1 + \begin{pmatrix} w_{21} \\ w_{22} \end{pmatrix} x_2 \tag{12}$$

5 RESULTS AND DISCUSSION

Table 1 shows the results of performing a two-way analysis of variance, taking the negative maximum amplitude of the 80–149 ms latency, which can be considered to be the N1 component, as a characteristic value, and the auditory stimulation (high or low pitch) and measurement position (Cz or Pz) as factors. Table 2 shows the analogous results for the 150–240 ms latency, which can be considered to be the N2 component. As can be seen in the tables, it was found that measurement position (Cz or Pz) is an influential factor in both N1 and N2. Accordingly, figures 2 and 3 show the amplitudes (mean value and standard deviation) of N1 and N2 under each condition for all participants. The horizontal axes in the figures show the type of auditory stimulation and the vertical axes show the amplitude. Additionally, the two curves represent the two measurement positions. The low position inspection results with

Table 1.   ANOVA for the N1 component.

Source of variation	SS	df	MS	F	P value	F crit
Stimulation (high or lowpitch)	18751.69	1.00	18751.69	0.07	0.79	4.49
Electrode position difference (Cz or Pz)	2354352.20	1.00	2354352.20	8.88	0.01	4.49
Interaction	213748.49	1.00	213748.49	0.81	0.38	4.49
Within	4241774.70	16.00	265110.92			
Total	6828627.08	19.00				

Table 2.   ANOVA for the N2 component.

Source of variation	SS	df	MS	F	P value	F crit
Stimulation (high or lowpitch)	72222.17	1.00	72222.17	0.28	0.61	4.49
Electrode position difference (Cz or Pz)	2390342.65	1.00	2390342.65	9.15	0.01	4.49
Interaction	305502.12	1.00	305502.12	1.17	0.30	4.49
Within	4179747.78	16.00	261234.24			
Total	6947814.73	19.00				

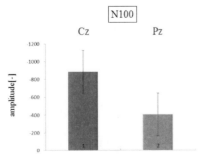

Figure 2.   N100 (high pitch stimulus).

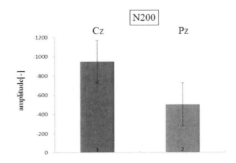

Figure 3.   N200 (high pitch stimulus).

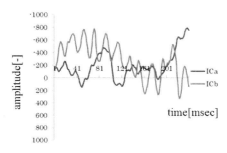

Figure 4.   Presumed primary signal with ICA.

Table 3.   Results of calculating the correlation.

Subject	Before ICA (observation signal)	After ICA (primary signal)
A	0.61	0.47
B	0.82	−0.47
C	0.84	0.02
D	0.71	−0.02
E	0.83	0.83

respect to measurement position are also shown in the figures. From the figures, Cz < Pz at the peak amplitude during a low tone and Cz > Pz at the peak amplitude during a high tone. According to these results, it can be said that at least the two components N1 and N2 are contained in the 1–240 ms latency of the measured ERP waveform.

As a result of the transformations performed during the ICA analysis procedure shown in Section 4, it could be confirmed that axes IC1 and IC2 are orthogonal after fast ICA processing.

From the estimates y_1 and y_2 of the original signals, an independently segregated restoration matrix that has N1 and N2 components as characteristics could not be obtained. However, a waveform that is probably biological noise present at the time of measurement was successfully separated.

Comparing S_1 and S_2 from figure 4, it can be observationally confirmed that the characteristics of N1 and N2 are contained in S_1, but characteristic components cannot be confirmed for S_2. This means that the independence of N1 and N2 was not achieved with respect to the two orthogonal independent components obtained in this experiment. The results of calculating the correlation coefficient of the two waveforms of the observed signal and the two waveforms of the estimated original signal are shown in table 3. Looking at the changes in correlation coefficient, it is found that the correlation coefficient became smaller after wave analysis for 4 of 5 participants. This result could mean that the waveforms of the two original signals have different sources. When the waveforms are segregated in this way, it may be possible to obtain a waveform that has independent components with large characteristics by repeated division of the waveform in a way that lowers the correlation with the observed signal.

6   CONCLUSION

The present study handled variables with a low number of dimensions, and so waveform analysis was performed by fast ICA. As a result, the correlation of the estimated original signal waveform was lower than the correlation of the observed signal for 4 of 5 participants. However, a completely independent analysis of N1 and N2 components was not achieved.

Carrying out techniques for post-whitening ICA processing with the aim of elucidating the relation between ERP and cognition process and the study of various ICA techniques to achieve this objective are left as future tasks[10].

REFERENCES

[1] H. Nittono, Event-Related Potential (ERPs) and cognitive activity a new perspective from engineering psychology, Japanese Association of Behavioral Science, Vol.42 No.1, 2003.

[2] M. Hosseini, M. Iravani, M. Younesian, S.J. Kass, Cognitive failures, Driving Errors and Driving Accidents, International Journal of Occupational Safety and Ergonomics (JOSE), Vol.14, No.2, pp. 149–158, 2008.

[3] M. Miyatani, Effects of number of targets on visual negative event-related potentials reflecting memory search, Bulletin of Faculty of the Education (Hiroshima University), pp. 113–120, 1991.

[4] S. Matsumoto, H. Sakuma, Analyzing the components of ERP related to competition result using a principal component analysis, Japanese Society of Baiofeedback Research, Vol.36, No.1, 2009.

[5] David Embrey, Understanding Human Behaviour and Error, Human Reliability Associates, http://www.humanreliability.com/articles/Understanding%20Human%20Behaviour%20and%20Error.pdf, (14/4/2014 access).

[6] L. Mayaud, M. Congedo, A.V. Laghenhove, M. Figère, E. Azabou, F. Cheliout-Heraut, A Comparison of Recording Modalities of P300 Event Related Potential (ERP) for Brain Computer Interface (BCI) Paradigm, Clinical Neurophysiology 43, pp. 217–227, 2013.

[7] Fujiwara Naohito, ImashioyaHayao, Comparison of event-related potentials from oddball tone sequence paradigm in the non-task condition and the task condition, Japanese Journal of Physiological Psychology and Psychophysiology Vol.12, No.1, 1994.

[8] A. Hyvärinen and E. Oja, A Fast Fixed-Point Algorithm for Independent Component Analysis, Neural Computation, 9(7) pp. 1483–1492, 1997.

[9] A. Hyvärinen, J. Karhunen, E. Oja, Independent Component Analysis, John Wiley & Sons INC, 2001.

[10] P.J. Huber. Projection pursuit, The Annals of Statistics,13-(2) pp. 435–475.

New Ergonomics Perspective – Yamamoto (Ed.)
© *2015 Taylor & Francis Group, London, ISBN 978-1-138-02751-0*

The reliability & applicability of mobile health care apps

Shi-Feng Huang & Chuan-Jun Su

Department of Industrial Engineering and Management, Yuan Ze University, Taiwan

ABSTRACT: There has been increasing awareness and emphasis on the importance of sustaining personal health, raising the demand for innovative, more convenient ways for individuals to monitor and manage their health pervasively. The advancements and proliferation of smart mobile devices equipped with camera and sensors have made readily available many types of mobile apps which enable ubiquitous acquisition of vital signs. With the prevalence and massive availability of health related mobile apps, people have been widely using apps running on smart phones or tablets for collecting such vital signs as heart rate, blood pressure, body temperature, Electrocardiogram (ECG), etc. An approximate assessment of health level can then be achieved based on the variation trend of collected data. The quality of health level assessment relies on the accuracy of vital signs acquired. Inaccurate physiological data could deduce misleading conclusion on health level and derive severe consequences. In this paper, we confront this issue by looking into the accuracy of the heart rate acquired from top-ranked smart phone based apps with respect to different exercise intensity. This paper intends to provide some insightful information about health monitoring apps by using heart rate apps as an example. The information can be used as a reference for people who are interested in adopting mobile apps as their health monitoring tools.

Keywords: Health self-management; Smartphone; App; Health monitoring

1 RESEARCH BACKGROUND AND MOTIVATION

Health Self-Management (HSM) is an important prerequisite for building endurance in health and acknowledged by virtually everyone. As stated by Lorig and Holman, "Whether one is engaging in a health promoting activity such as exercise or is living with a chronic disease such as asthma, he or she is responsible for day-to-day management" [1]. People who conduct HSM generally enjoy the benefits of enhancing their communication with their doctor, helping them to focus on the treatment plan and be adherent, growing their level of self-confidence, and better managing their well-being [2]. Despite these benefits, HSM is in general a long-term and complicated process which requires medical knowledge, convenient and friendly devices, and continuous endurance health monitoring to achieve.

Mobile devices, such as smart phones and tablet PCs, have evolved over recent years. They come with camera(s) and a variety of embedded sensors such as GPS, gyroscope, accelerometer, etc., which enable the development of novel physiological signs monitoring applications. Various vital signs can be measured conveniently and ubiquitously including heart rate, blood pressure, body temperature, Electrocardiogram (ECG), etc., as illustrated in Fig. 1.

In recent years, due to smartphone's unique features of versatility and universality, researchers have been using smartphones as tools for encouraging physical activity and healthy diets, for symptom monitoring in asthma and heart disease, for sending patient reminders about upcoming appointments, for supporting smoking cessation, and for a range of other health

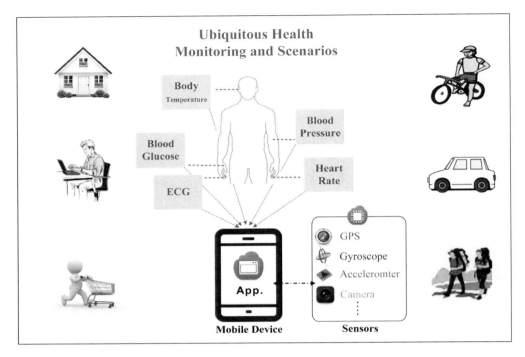

Figure 1.   Ubiquitous health monitoring and scenarios.

problems. Smartphone apps allow physicians to monitor patients with chronic heart failure [3] and to detect early signs of arrhythmia or ischemia that can indicate an imminent heart attack [4]. In addition, patients themselves can use smartphone apps and sensing and measurement devices to keep track of their physical activities [5] and to monitor physiological markers relevant to their health status [6].

With the massive availability and convenience of smart mobile devices based vital sign monitoring applications, people are generally willing to continuously collect their physiological data which can then be used to classify their health levels based on the variation trend. However, in order to have a reliable assessment on health levels, the accuracy of physiological data must be assured. Without accurate physiological data as input, the quality assessment of health level cannot be achieved regardless the analysis algorithm used. In this paper, we are mainly concerned with the issues of the accuracy and reliability of physiological data acquired from popular smart phones based apps available in the market before making substantial efforts on the analysis of health level assessment. We address this issue by analyzing physiological data generated by a set of popular apps. The results of this paper can be useful for users to understand the boundaries and limitations of smart phone based health monitoring apps and for developers to make improvements on future developments.

2   EVALUATING THE USABILITY OF MOBILE HEALTHCARE APPLICATIONS

With the proliferation of smartphones, there are new possibilities for using mobile phones as tools for health promotion. Smartphones have powerful operating systems that can run computer programs or apps, in addition to the standard features of mobile phones. There have been previous studies that set out to examine the content of the 47 iPhone apps for smoking cessation that were distributed through the online iTunes store. They come to the conclusion that iPhone apps for smoking cessation were found to have low levels of adherence to key guidelines from the US Public Health Service's 2008 Clinical Practice Guidelines [7]. A research was conducted for reviewing diabetes apps running on Android smartphones [8]. A list of 80 free and paid apps was compiled by searching the Android Market in April 2011 and 42 of them were filtered for this study. Despite the variations in functionality, usability, price, and number of downloads, only 4 of the 42 apps studies had up to standard usability score. The research revealed that few apps available in the market provide effective means for diabetes management.

This paper aims to evaluate the quality of physiological data generated by top ranked five heart rate monitor apps at the point of study, which utilize only smartphone embedded sensors without any other external devices. This work intends to provide some insight into the appropriateness of using smartphone as a device for HSM.

3 RESEARCH METHODOLOGY

3.1 *Purpose of the experiment*

Heartbeat is the most basic sign of lives, which reveals people's physiological and psychological states. Vital information regarding heart's level of exertion and overall health of a person can be obtained by trailing his/her heart rate. A research finding published in European Heart Journal [9] suggested that tracking heart rate over time can provide a profoundly simple and important marker of health issues that could become lethal but which also might be prevented with diagnosis and treatment.

The advancement and proliferation of smart mobile devices enable people to track such vital signs as heart rate, temperature, respiratory rate, blood oxygenation, blood pressure, etc., conveniently and ubiquitously in contrast to using cumbersome devices traditionally. According to [10], a smart phone application that can measure heart rate, heart rhythm, respiration rate, and blood oxygen saturation using the phone's built-in video camera was developed. It was also claimed that the app can yield vital signs as accurate as standard medical monitors now in clinical use.

In order to have a reliable assessment on heath levels, the vital signs need to be measured with high degree of accuracy. Due to the importance of heart rate serving as an important indicator of people's health, this research aims to examine and analyze the accuracy of heart rate readings generated by a set of top-ranked mobile apps.

3.2 *Experimental subject*

There are 30 volunteer adults participate in this study serving as experimental subjects. They would be asked to fill out a health survey to ensure their healthy level prior to taking the experiment. Moreover, subject was required to read subject conditions to understand the purpose of the experiment, experimental procedure and potential risks involved. Following informed written consent, subjects had their heights and weights measured and provided a brief health history of chronic diseases and medication usage.

3.3 *Experimental design*

As an initiation, the heart rate was categorized into five zones which represent various levels of exercise intensity based on Sally Edward's work [11]. Each heart rate is defined by the percentage of Max HR. Sally Edward further suggested two formulas for estimating Max HR of different genders.

$$\text{Male: } Max \ HRM = 210 - 0.5A - 0.05W + 4$$
$$\text{Female: } Max \ HRF = 210 - 0.5A - 0.05W \qquad (1)$$

where A: Age (year), W: Weight (pound)

The five heart rate zones are detailed in the following:

- Zone 1, the heart rate is between fifty percent Max HR and sixty percent Max HR during exercise. In this level, the exercise intensity is just like warm-up.
- Zone 2, the heart rate is between sixty percent Max HR and seventy percent Max HR during exercise. In this level, the exercise intensity is just like easy jogging.
- Zone 3, the heart rate is between seventy percent Max HR and eighty percent Max HR during exercise. In this level, the exercise intensity is just like brisk walking.
- Zone 4, the heart rate is between 80% Max HR and 90% Max HR during exercise. According to the American Council on Exercise and the American Heart Association, even for people in excellent physical condition usually work out at no more than 85 percent of their Max HR [12]. This zone will not be performed in our experiment.
- Zone 5, the heart rate is between 90% Max HR and 100% Max HR during exercise. This zone will not be performed either due to the same reason as Zone 4.

In the course of pre-test, we found the most of subjects were difficult to achieve the required heart rate standard in zone 3, even if a few subjects finally reached standard, but just only kept a few minutes. At first we slowly increased exercise intensity, but when subjects achieved the standard that they had to pay a lot of effort, and also just kept a few time that we could only finish one or two apps measurement. This problem is not only to be the physiological burden on subjects, but also brings the pressure of time and safety to our research, therefore we decide to adjust the experiment as described below.

- Normal, take a rest for 5 minutes.
- Zone 1, fifty meters walking.
- Zone 2, jogging one hundred meters.

As an initial phase, the subjects were asked to complete the exercise prescribed in one of the three activities. A data acquisition process that comprises three types of heart rate measurements was subsequently performed to attain the benchmark, Borg, and app heart rate data as illustrated in Fig. 2.

An FDA approved heart rate measuring device Beurer BC20 was used to acquire the heart rate readings of the subjects and record them as the benchmark data which serve as the accurate measurement of heart rate. The data would also be checked for consistency against the Max HR percentage defined in the zones.

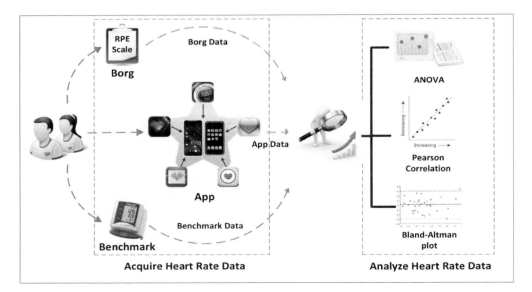

Figure 2. The overall process of acquiring and analyzing heart rate data.

An RPE scale form would be completed by the subjects to derive Borg heart rate data which serve as a subjective estimate of heart rate after exercise based on the following formula:

$$Heart\ Rate = RPE \times 10 \qquad (2)$$

where RPE: RPE scale reading

The selected top-ranked apps would then be used to acquire the subjects' heart rate readings and establish the app data.

The smart phone was held in the right hand with the index finger covering the les and LED as illustrated in Fig. 3. (a). The BC20 was placed on the subject's left wrist as illustrated in Fig. 3. (b), and the RPE scale was placed between right hand and left hand. The HRM apps selected include Heart Rate, Heart Fitness, Heart Beats, Heart Rate Calculator, and Instant Heart Rate. The selection was made on 10th of May, 2013, by searching applications featured in apple app stores and Google Play.

All data were analyzed with the SPSS Advanced Statistics software package and Med-Calc complete statistical program. In this study, one-way ANOVA was utilized to examine heart rate differences between the measuring instruments. The differences between the mean values for all the heart rates at the three measuring instruments were tested for significance by means of ANOVA for repeated measurements with post hoc Scheffe's test which is the more appropriate test to use when predicted differences are small, and the consequences of a Type II error outweigh the consequences of a Type I error. The Scheffe's test assumes equal sized experimental and control groups in the ANOVA. An alpha level of 0.05 was set prior to analysis.

The Pearson's correlation coefficient has been the most common measure of correlation in statistics for assessing reliability. The idea is that if a high (>0.8) and statistically significant correlation coefficient is obtained, the equipment is deemed to be sufficiently reliable [13]. The Pearson correlation coefficients (r) comparison were done in pair between two groups among benchmark data, Borg data, and app data.

In the limits of agreement analysis, the first step is to present and explore the test-retest data with a Bland-Altman plot, which is the individual subject differences between the tests plotted against the respective individual means [14]. As tools are expected to be highly correlated measuring the same signal agreement of the benchmark data and app data were assessed for each activity using Bland-Altman plots for repeated measures with 95% limits of agreement.

 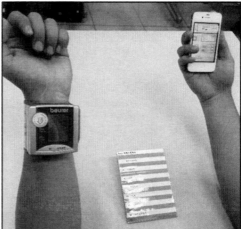

Figure 3.   (a) Placement of index finger with smart phone; (b) Placement of BC20, RPE scale and smartphone during data acquisition.

4 STATISTICAL ANALYSIS AND RESULTS INTERPRETATION

4.1 *Descriptive characteristics of the subjects*

A total of 30 healthy adults, aged from 20 to 33, volunteered to participate in this study. The participants consist of 15 males and 24 of them have been using smartphone. Among these 24 people, there are 9 people who have experience in using health related Apps in the past. About 10% of the subjects regularly measure their vital sign and half of the subjects perform regular exercise. Following informed written consent, subjects had their weights measured and revealed their information regarding age and sex. The Sally Edward's formula is then used for estimating their MHR (Max Heart Rate). The mean of MHR is 192.67 ± 2.006 (bpm).

4.2 *Result and analysis*

ANOVA Post hoc comparisons using Dunnett's T3 or Scheffe procedures were used to determine which pairs of the three group means differed. These results are given in Table 1 and indicate that heart rate data which measured by RPE (M = 110.00) significantly higher on the standardized test than heart rate data which measured by App: What's My Heart Rate in iPhone (M = 103.73) and in Android (M = 103.73) in zone 1 situation. The heart rate data which measured by BC20 (M = 125.93) significantly higher on the standardized test than heart rate data which measured by App: What's My Heart Rate in iPhone (M = 121.47) and in Android (M = 119.90) in zone 2 situation. For App: Cardiograph, the heart rate data which

Table 1. Summary of result and analysis of five HRM apps.

		ANOVA Post hoc	Pearson correlation	The range of the 95% LoA	The width of the 95% LoA
App: What's My Heart Rate					
Normal	BC20 versus iPhone	0.133	−0.821**	−0.1 ± 8.75	17.5
	BC20 versus Android	0.367	−0.731**	−0.4 ± 8.75	17.5
	RPE versus iPhone	5.000	−0.094	–	–
	RPE versus Android	4.767	−0.096	–	–
Zone 1	BC20 versus iPhone	0.500	−0.676**	−0.5 ± 7.30	14.6
	BC20 versus Android	0.500	−0.593**	−0.5 ± 8.40	16.8
	RPE versus iPhone	6.267*	−0.062	–	–
	RPE versus Android	6.267*	−0.304	–	–
Zone 2	BC20 versus iPhone	−4.467*	−0.457**	4.5 ± 11.15	22.3
	BC20 versus Android	−6.033**	−0.564**	6.0 ± 8.90	17.8
	RPE versus iPhone	−0.133	−0.006	–	–
	RPE versus Android	1.433	−0.204	–	–
App: Cardiograph					
Normal	BC20 versus iPhone	−0.967	−0.896**	1.0 ± 6.25	12.5
	BC20 versus Android	−1.333	−0.869**	1.3 ± 7.05	14.1
	RPE versus iPhone	4.767	−0.114	–	–
	RPE versus Android	5.133*	−0.106	–	–
Zone 1	BC20 versus iPhone	0.333	−0.741**	−0.3 ± 8.20	16.4
	BC20 versus Android	−2.900	−0.709**	2.9 ± 8.80	17.6
	RPE versus iPhone	5.133	−0.378*	–	–
	RPE versus Android	8.367*	−0.060	–	–
Zone 2	BC20 versus iPhone	N	−0.664**	0.3 ± 9.65	19.3
	BC20 versus Android	N	−0.445*	8.0 ± 9.60	19.2
	RPE versus iPhone	N	−0.205	–	–
	RPE versus Android	N	−0.012	–	–

(*continued*)

Table 1.   Continued.

App: Heart Beat Rate					
Normal	BC20 versus iPhone	N	−0.773**	1.9 ± 8.15	16.3
	BC20 versus Android	N	−0.760**	1.7 ± 9.10	18.2
	RPE versus iPhone	N	−0.072	–	–
	RPE versus Android	N	−0.200	–	–
Zone 1	BC20 versus iPhone	N	−0.789**	−1.1 ± 6.30	12.6
	BC20 versus Android	N	−0.780**	0.5 ± 6.80	13.6
	RPE versus iPhone	N	−0.416*	–	–
	RPE versus Android	N	−0.472*	–	–
Zone 2	BC20 versus iPhone	N	−0.681**	−0.1 ± 10.0	20.0
	BC20 versus Android	N	−0.476*	0.9 ± 12.6	25.2
	RPE versus iPhone	N	−0.146	–	–
	RPE versus Android	N	−0.183	–	–
App: Runtastic					
Normal	BC20 versus iPhone	N	−0.892**	2.0 ± 7.00	14.0
	BC20 versus Android	N	−0.803**	1.0 ± 9.25	18.5
	RPE versus iPhone	N	−0.278	–	–
	RPE versus Android	N	−0.073	–	–
Zone 1	BC20 versus iPhone	N	−0.715**	−1.7 ± 8.30	16.6
	BC20 versus Android	N	−0.872**	−1.8 ± 5.40	10.8
	RPE versus iPhone	N	−0.307	–	–
	RPE versus Android	N	−0.323	–	–
Zone 2	BC20 versus iPhone	N	−0.517**	−2.3 ± 10.20	20.4
	BC20 versus Android	N	−0.638*	1.6 ± 8.80	17.6
	RPE versus iPhone	N	−0.508**	–	–
	RPE versus Android	N	−0.411*	–	–
App: Instant Heart Rate					
Normal	BC20 versus iPhone	N	−0.861**	0.7 ± 8.55	17.1
	BC20 versus Android	N	−0.803**	1.0 ± 10.0	20.0
	RPE versus iPhone	N	−0.432*	–	–
	RPE versus Android	N	−0.675**	–	–
Zone 1	BC20 versus iPhone	N	−0.416**	−0.6 ± 7.05	14.1
	BC20 versus Android	−3.400*	−0.344	3.4 ± 7.50	15.0
	RPE versus iPhone	N	−0.098	–	–
	RPE versus Android	4.533	−0.218	–	–
Zone 2	BC20 versus iPhone	N	−0.649**	−3.1 ± 9.30	18.6
	BC20 versus Android	−7.700**	−0.099	7.7 ± 16.2	32.4
	RPE versus iPhone	N	−0.071	–	–
	RPE versus Android	10.733*	−0.180	–	–

**p < 0.01; *p < 0.05; LoA: Limits of Agreement; N: No significant differences.

measured by App: Cardiograph in Android (M = 116.10) significantly lower on the standardized test than heart rate data which measured by BC20 (M = 124.10) and RPE (M = 123.33) in zone 2 situation. For App: Instant Heart Rate, the heart rate data which measured by BC20 (M = 110.20) significantly higher on the standardized test than heart rate data which measured by App: Instant Heart Rate in Android (M = 106.80) in zone 1 situation. The heart rate data which measured by App: Instant Heart Rate in Android (M = 116.27) significantly lower on the standardized test than heart rate data which measured by BC20 (M = 123.97) and RPE (M = 127.00) in zone 2 situation.

The Pearson's correlation coefficient provides information about the magnitude of correlation as well as the direction of the relationship. The heart rates obtained from these five

HRM apps were highly correlated with those from BC20 in normal situation. However, the correlation will decrease as exercise intensity increase. The app is modestly correlated with RPE scale's data during all conditions.

The Bland-Altman method calculate the mean difference between two methods of heart rate measurement, and 95% limits of agreement as the mean difference. It is expected that the 95% limits include 95% of differences between the two heart rate measurement methods.

5 CONCLUSIONS

In this paper, we are mainly concerned with the issue of trustworthiness of popular smart phones based health monitoring apps available in the market. We address this issue by analyzing HR data generated by a set of popular apps. The results of this paper can be useful for users to understand the boundaries and limitations of smart phone based health monitoring apps.

The present findings are based on a limited-size group but results provide preliminary evidence that the 5 popular smart phone based health monitoring apps provided valid measurement of HRs across three different tasks which varied in exercise intensity. Heart rate monitoring has played an integral role in the field of health promotion and disease prevention. Medical device BC20 is deemed to be the gold standard for HR measurement in this paper.

According to the results of a single factor analysis of variance, most of the significant differences appeared in the higher exercise intensity, but also had two Apps didn't have any significant differences, which were App: Heart Beat Rate and App: Runtastic. In Pearson correlation, the lower exercise intensity had much higher correlation coefficient, but the correlation decreased as exercise intensity increased. For agreement analysis, the App: Runtastic in Android versus BC20 showed a good agreement that mean difference closed to 0 and the width of the 95% limits of agreement was the smallest.

Although each App's width of the 95% limits of agreement is not smaller than clinically acceptable error, the convenience and ubiquity are indispensable factors for measurement vital signs. When people use this kind of health care applications for the convenience, they also lose the accuracy of the data. The value of this paper, is that it provides a reference to the users who use the heart rate monitoring apps that they must take into consideration of error values.

REFERENCES

[1] Lorig, K.R., & Holman, H.R. (2003). Self-management education: History, definition, outcomes, and mechanisms. Annals of Behavioral Medicine, Vol. 26(1), pp. 1–7.

[2] Norgall, T., & Wichert, R. (2012). Towards interoperability and integration of personal health and AAL ecosystems. Studies in Health Technology and Informatics, 177, pp. 272–282.

[3] Scherr, D., Zweiker, R., Kollmann, A., Kastner, P., Schreier, G., & Fruhwald, F.M. (2006). Mobile phone-based surveillance of cardiac patients at home. Journal of Telemedicine and Telecare, 12(5), pp. 255–261.

[4] Rubel, P., Fayn, J., Nollo, G., Assanelli, D., Li, B., Restier, L., … Chevalier, P. (2005). Toward personal eHealth in cardiology. Results from the EPI-MEDICS telemedicine project. Journal of Electrocardiology, 38(4), pp. 100–106.

[5] Consolvo, S., Klasnja, P., McDonald, D., Avrahami, D., Froehlich, J., Legrand, L., … Landay, J. (2008). Flowers or a robot army?: Encouraging awareness & activity with personal, mobile displays. UbiComp 2008—Proceedings of the 10th International Conference on Ubiquitous Computing, (pp. 54–63).

[6] Mohan, P., Marin, D., Sultan, S., & Deen, A. (2008). MediNet: Personalizing the self-care process for patients with diabetes and cardiovascular disease using mobile telephony. 30th Annual International Conference of the IEEE Engineering in Medicine and Biology Society, (pp. 755–758).

[7] Abroms, L.C., Padmanabhan, N., Thaweethai, L., & Phillips, T. (2011). iPhone apps for smoking cessation: A content analysis. American Journal of Preventive Medicine, 40(3), pp. 279–285.

[8] Demidowich, A.P., Lu, K., Tamler, R., & Bloomgarden, Z. (2012). An evaluation of diabetes self-management applications for android smartphones. Journal of Telemedicine and Telecare, 18(4), pp. 235–238.

[9] Okin, P.M., Kjeldsen, S.E., Julius, S., Hille, D.A., Dahlöf, B., Edelman, J.M., & Devereux, R.B. (2010). All-cause and cardiovascular mortality in relation to changing heart rate during treatment of hypertensive patients with electrocardiographic left ventricular hypertrophy. European Heart Journal, 31(18), pp. 2271–2279.

[10] Scully, C.G., Lee, J., Meyer, J., Gorbach, A.M., Granquist-Fraser, D., Mendelson, Y., & Chon, K.H. (2012). Physiological parameter monitoring from optical recordings with a mobile phone. IEEE Transactions on Biomedical Engineering, 59(2), pp. 303–306.

[11] Edwards, S., Snell, M., & Sampson, E. (1996). Sally Edwards' Heart Zone Training: Exercise Smart, Stay Fit, and Live Longer. MA: Adams Media Corporation.

[12] Target Heart Rates. (2013, May 14). Retrieved from American Heart Assocuation: http://www.heart.org/HEARTORG/GettingHealthy/PhysicalActivity/Target-Heart-Rates_UCM_434341_Article.jsp.

[13] Coolican, H. (2009). Research Methods and Statistics in Psychology. New York: Routledge.

[14] Bland, J.M., & Altman, D.G. (1995). Comparing methods of measurement: Why plotting difference against standard method is misleading. Lancet, 346(8982), pp. 1085–1087.

New Ergonomics Perspective – Yamamoto (Ed.)
© 2015 Taylor & Francis Group, London, ISBN 978-1-138-02751-0

Estimation of useful field of view by machine learning based on parameters related to eye movement

Keisuke Morishima
Technology Center, Yamaha Motor Co., Ltd., Iwata, Shizuoka, Japan

Hiroshi Ura & Takanori Chihara
Graduate School of System Design, Tokyo Metropolitan University, Hino, Tokyo, Japan

Hiroshi Daimoto
Technology Center, Yamaha Motor Co., Ltd., Iwata, Shizuoka, Japan

Kimihiro Yamanaka
Graduate School of System Design, Tokyo Metropolitan University, Hino, Tokyo, Japan

ABSTRACT: The aim of this study was to propose a method for estimating the size of Useful Field of View (UFOV) using machine learning methods. To examine effective parameters related to eye movement for estimating UFOV in driving, we measured eye movements and blinks during simulated driving. In the experiment, head and eye movements were measured simultaneously to calculate the fixation point along a global coordinate. Three machine learning methods—support vector machine, boosting, and radial basis function networks— were applied to the data of eye movements and blinking so as to discriminate 'Wide UFOV' from 'Narrow UFOV'. A cross-validation result showed that the boosting method has the highest accurate discrimination rate (approximately 85%). The discriminate function of the boosting method may estimate UFOV based on the parameters of eye movements and blinking without affecting driver operation.

Keywords: Useful Field of View; Eye Movements; Machine Learning; Mental Workload; Vehicle

1 INTRODUCTION

Drivers of four-or two wheeled vehicles constantly use their sensory organs to perceive the driving environment so as to avoid accidents. Drivers can safely control their vehicles while they pay careful attention to the movement of other vehicles and pedestrians, and traffic signals or signs. However, there is a limit to the amount of information that a driver can perceive at a time. An increase in the amount of information that the driver must process, such as heavy traffic or conversations, may lead to human errors (i.e., oversights, misinterpretation, misjudgment, or inappropriate driving behavior). These errors also contribute to traffic accidents.

Recent years, smart phones are in widespread use and consequently the demand for high-speed mobile communications standards (e.g., Long Term Evolution) is growing. Information communication services for vehicles may also be growing; therefore, the amount of information that drivers must process will increase. Information communication service for vehicle must be designed to prevent an excessive burden for drivers; thus the level of burden should be defined for decision-making of them. This paper defines the level of burden when a driver processes various types of information as the Mental Workload (hereafter referred to as MWL), and considers techniques to quantitatively estimate MWL while driving.

The increase of information processed by drivers, or the increased MWL, may contribute to traffic accidents. 83% of traffic accidents resulting in injury or death are caused by cognition failures and delayed responses, approximately 90% of these cognition failures are related to visual information [1, 2]. Therefore, we assert a causal relationship between the Increase in MWL and the increase in the cognition of visual information. The Useful Field of View (UFOV) can be considered as a factor playing an important role in the cognition of visual information. UFOV is the range within the field of view where visual information useful in visual cognition tasks can be collected [3]. The extent of UFOV cannot be clearly defined, and no standard evaluation method has been established for UFOV. This is because its range varies depending on the type of object seen (e.g., size and shape) and environment (e.g., brightness of surroundings) [3]. Therefore, few quantitative studies deal with both MWL and UFOV, whereas previous researches focused on only MWL when driving a vehicle [4, 5]. In our previous research, a quantitative manner has confirmed that UFOV narrows with an increase in MWL [6]. Therefore, MWL can be evaluated quantitatively by estimating UFOV. Furthermore, conventional UFOV evaluation methods affected the driving task, because they required the participant's head to be stationary, and visual targets to be presented in the field of view. However, Morishima et al. [7, 8] studied a technique to evaluate UFOV without affecting the driving task. They also indicated that UFOV can be evaluated without impacting the driving task, by using various parameters obtained through the measurement of eye movements and blinks.

This study considers a technique to estimate the range of UFOV using parameters obtained through the measurement of eye movement and blinks. In particular, we propose a technique to accurately detect the narrowing of a user's UFOV, while operating a Driving Simulator (DS) by applying eye movement and blink parameters to machine learning.

2 UFOV, EYE MOVEMENT AND BLINK PARAMETERS EVALUATION VALUES

2.1 *Experimental method*

Subjects are instructed that the order of priority in their task execution is the DS operation task, visual target detection task, and MWL task. The subjects were 18 healthy males with a driving license, aged 21–24 years (average age 22.4 years, standard deviation 0.92 years). All of them were university students, and received a full explanation of the experiment's content, and participated in the experiment after a written consent was obtained.

The experimental procedure of this study was almost same as our previous research [7, 8], and we simultaneously obtained UFOV evaluation values, eye movement and blink parameters. Figure 1 illustrates the experimental layout. The subjects wore a device that measures eye movement, blinks, and head movement with the corneal reflection method. High speed cameras were placed 60 cm from the back and side of the driver's seat in the DS, respectively. The subjects executed an instructed driving task in an eye—and hands-free condition. Figure 2 shows a drive route with an image of the user's forward view. The route is a city course without any dangerous events that would lead to accidents, as is generally used by driving schools in risk prediction teaching. The subjects are instructed to keep to a maximum speed of 40 km/h and observe traffic regulations.

The subjects simultaneously performed a visual target detection task while operating the DS. The visual target detection task involves responding, using a button on the steering wheel, to detected visual targets that appear for 2 s at 3-second intervals on the DS screen, within a range of 80° horizontal and 22° vertical. The visual targets were shown in random locations on the DS screen. Targets were in the form of a white Landolt broken ring, with a size of 1° to the visual angle, on a red background so they can be detected without being affected by the color of other objects on the DS screen when they appeared (e.g., the color of a car in

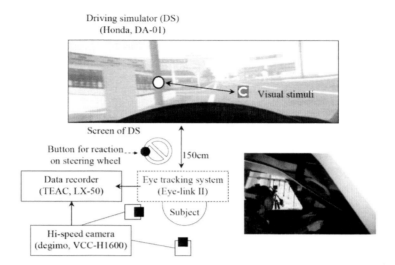

Figure 1. Experimental layout [7, 8].

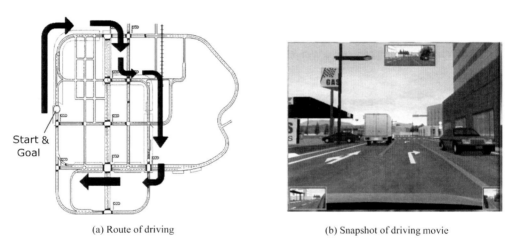

(a) Route of driving   (b) Snapshot of driving movie

Figure 2. Route of driving and snapshot of driving movie [7, 8].

front of the DS). We measured the UFOV using the score in the response task following the method used by Yamanaka et al. [9]. Figure 3 shows the distance from the fixation point to the visual target, and the distance represents the strength of the stimulus. We estimated the psychometric curve from the non-detection probability using the probit method; and a non-detection probability of 50% (Point of Subjective Equality, or PSE) [10] is defined as the edge of the UFOV. For instance, considering the time when the visual target appeared, the distance from the fixation point to the visual target is a psychophysical quantity, which expresses the strength of the stimulus (Figure 3). The higher value of the psychometric curve represents weaker stimulus. In other words, the subjects cannot detect the target when the distance between the fixation point and the target exceed a stimulus threshold distance. The threshold indicates the edge of the UFOV. The psychometric curve shown in Figure 3 expresses a functional relationship between the stimulus threshold and the detection probability. As in the present experiment, the DS visual target display area was small in the vertical direction but can secure sufficient width. Measurements are taken only for the horizontal UFOV, which allows for the accurate field of vision range measurements. Specifically, the UFOV

55

was calculated with the data from when the visual target appeared in a horizontal direction ($0° \pm 15°$ range for horizontal direction $0°$) to either the left or the right of the fixation point. To realize this, the subject's fixation point coordinates in the experimental space and a coordinate system must be obtained for the time that the visual target appears. In this study, the fixation point coordinates in the experimental space coordinate system were obtained using the same method of our previous study [8]. Figure 4 shows the center of the DS front screen; the visual target has an initial value of f_0, and the eye rotation angle and the head rotation angle are f_{et} and f_{ht}, respectively. The sum fixation point coordinate in the experimental space coordinate system is f_{eht}. In this experiment, the eye rotation angle and head rotation were obtained by the eye movement measuring device and the high speed cameras, respectively.

As a secondary task (MWL task), the subjects were required to solve numerical tasks (e.g., repeating aloud a list of numbers, simple addition). Thus, three conditions were created as follows:

- Repeat aloud: A list of numbers was relayed to the subject by voice at 3-second intervals, and the subject repeated them.
- Addition: The subject added two consecutive numbers and answered only the first digit in reply.
- No task: The subjects did nothing as the secondary task.

For these numerical tasks, the MWL quantification was already performed in previous research [11] through simultaneous measurement with the event related potential. "Addition" had the highest burden followed by "Repeat aloud" and "No task". The numerical tasks were presented through a speaker placed on the top of the display, with the volume set at 70 dB (corresponding to the level of normal conversation in a vehicle).

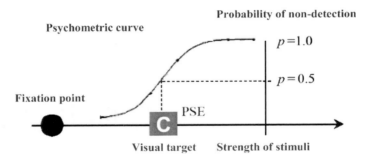

Figure 3.   Useful field of view and psychometric curve [8].

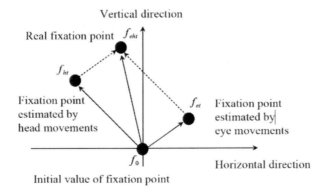

Figure 4.   Image of real fixation point [7].

The experimental schedule is shown in Figure 5. The subjects perform the experiment under one set of MWL conditions per day, and they fill out an NASA-TLX [12] evaluation sheet for a subjective evaluation index at the end of each day. MWL tasks were performed in a random order with two scenarios, i.e., one with a visual target detection task and one without. This enables studying the relation between the UFOV range and the parameters related to the eye movement measured under general driving conditions, without visual targets being shown. In addition, as the two-lap DS course (2×3 min.) is repeated under conditions with visual target detection, one UFOV is computed from the response to visual targets over a total of four laps (12 min.).

2.2 Experimental results

Figures 6 and 7 show the correct answer rate for the MWL task and AWWL scores [12] calculated using the responses to the NASA-TLX and employed as a subjective evaluation index. Both graphs show averages for all subjects under all conditions, and the error bars in the graphs indicate the standard deviation. Subsequent graphs use the same tabulation. The correct answer rate in the MWL task is almost 100% for the Repeat aloud condition, whereas it significantly drops to approximately 80% for the Addition condition ($p < 0.01$, t-test). Moreover, starting from the highest, AWWL scores were approximately 75 points for Addition, 45 points for Repeat aloud, and 30 points for No Task Significant differences were found between conditions in an analysis of variance and a multiple comparison test ($p < 0.05$, Holm method).

Figure 8 shows measurement results for the UFOV in the horizontal direction. In the MWL task, the UFOV is the narrowest for the Addition condition, and the widest for the No task condition. On performing an analysis of variance and Holm's multiple comparison test with the values in the graph as attributes, the MWL task factor ($p < 0.01$) is the affecter. Significant differences were found between No Task and Addition ($p < 0.05$), and between Repeat aloud and Addition ($p < 0.05$) in the multiple comparison test.

Practice run, 30 min. MWL task: No task Visual target detection task: No task (VTD task)	Break 10 min.	Experiment, 18 min. (3 min. × 6 laps)			NASA –TLX
		VTD task Conducted	VTD task No task	VTD task Conducted	
		2 laps	2 laps	2 laps	
		MWL task: No task or Repeat aloud or Addition			

Figure 5.   Experimental schedule [8].

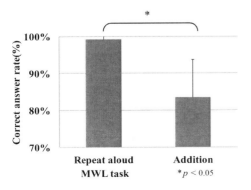

Figure 6.   MWL task and correct answer rate.

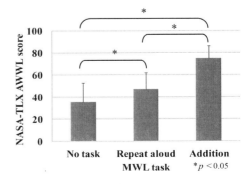

Figure 7.   MWL task and NASA-TLX AWWL scores.

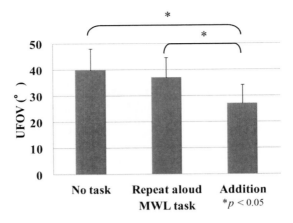

Figure 8.   MWL task and UFOV.

Previous research [6] revealed a constant relationship between MWL and the UFOV, and that the UFOV decreases as the MWL increases. The AWWL score representing subjective work load levels in this study is higher for Addition compared to No Task, and the correct answer rates are also low. Based on this we can assume that these tasks form a heavy MWL for the subjects and that these individuals markedly narrow their UFOV. However, the Repeat aloud condition did not cause a MWL that affected task performance because the correct answer rate is almost 100%. Therefore, it can be assumed that no marked narrowing of the UFOV was seen for the Repeat aloud condition.

3   VISUAL FIELD ESTIMATION METHOD USING MACHINE LEARNING METHODS

The experimental results described in section 2 showed a marked narrowing of the UFOV under the Addition condition compared to the No Task condition. We will therefore attempt to discriminate between the sizes of the UFOV: a Narrow_UFOV for the Addition condition, and a Wide_UFOV for the No Task condition. The eye movement and blink parameters used for discrimination based on machine learning methods described below.

3.1   *Calculating eye movement and blink parameters*

Previous research [8] suggested that parameters of eye movement and blinks may be used for UFOV evaluation. The present study calculated eye movement and blinks as shown in Table 1, and uses these in combination to attempt to discriminate between Wide_UFOV and Narrow_UFOV. As shown in Table 1, we used two combinations of four parameters. Saccade is a rapid eye movement to change the fixation point. Moreover, corrective saccade is defined as an eye movement to correct the distortion created when the fixation point is matched to the viewed object through saccade [13]. Under these saccades, cognition of visual information is suppressed [14]. Cognition of visual information is mainly performed under the pursuit and stationary movement. Pursuit, which is an eye movement to pursue the moving visual target, is defined as the eye movement with rotational angular velocity of less than 30°/s [15]. If the velocity exceeds 30°/s, saccade is performed to pursue the visual target. Therefore, we define saccade as the eye movement that has rotation velocity of over 30°/s and travel distance of over 3.5°, corrective saccade as eye movement with rotational angular velocity of over 30°/s and a travel distance of 1.0–3.5° occurring between 100–150 ms after the saccade, eye movement with rotational angular velocity of less than 30°/s without saccade or blinks as pursuit and stationary movement.

Table 1. Parameters of eye movements and blinks.

Parameters	Contents	Combination 1	Combination 2
Number of blinks	Number of blinks per minute	O	
Frequency of pursuit and stationary	Time of stationary and pursuit state/Experimental time	O	O
Saccade length	Average length of saccade movements	O	O
Frequency of Corrective Saccade	Number of saccades with corrective saccade/Number of all saccade	O	O
Saccade time	Average time of saccade states		O

For the respective parameters, two laps are driven without any visual target detection tasks, and one parameter is calculated per lap. As there is concern that eye movement and blinks are significantly affected by individual differences, z-scores are calculated and used for each parameter for all subjects in the application to machine learning. Z-scores are standardization values, which are calculated using average and standard deviation of each condition and laps per subjects.

3.2 *Application to machine learning*

Seventy-two datasets were applied to the machine learning methods, formed of the subjects (18) × UFOV size (two conditions) × runs (two laps), for which cross-validation was performed. Specifically, the four data sets of one subject (UFOV (two conditions) × runs (two laps)) were assigned as test data, and the other 68 datasets as training data. We evaluate the discrimination accuracy for the UFOV size when discrimination functions are obtained by the application of machine learning methods to the training datasets and applied to the test datasets. Discrimination accuracy is defined as the ratio of the number of test datasets out of the total number, for which the UFOV size was correctly distinguished. This evaluation is repeated for the 18 subjects, and the average discrimination accuracy for each subject is set as the discrimination accuracy evaluation index for each learning method. The present study uses the following three machine learning methods.

3.2.1 *Boosting*

This study uses the AdaBoost method, conceived by Y. Freund et al. [16]. In this method, strong classifiers are formed using a combination of weak classifiers, which have been formed through the combination of features (in this case, eye movement and blinks) and thresholds. In the AdaBoost method, the reliability of each weak classifier is determined using the identification error rate for weak classifiers. Using this reliability to update the weight of the samples for learning correspondingly controls the increase in computational complexity and the drop in generalization performance. The ultimate strong classifier is determined on the majority of weight for reliability given to the output of each weak classifier. The analysis performed here used four weak classifiers forming a strong classifier, and 1000 steps for the threshold of each feature (feature distribution range is divided by 1000); in other words, 4000 weak classifier candidates (number of features × threshold).

3.2.2 *Support Vector Machine (SVM)*

SVM was first proposed by Vapnik [17] and has since attracted a high degree of interest in the machine learning research community. Several recent studies have reported that SVMs generally are capable of delivering higher performance in terms of classification accuracy than other data classification algorithms [18]. SVM is set of related supervised learning methods used for classification and regression. A special property of SVM is, SVM simultaneously minimizes the empirical classification error and maximizes the geometric margin. SVM map input vector to a higher dimensional space where a maximal separating hyperplane is constructed. Two parallel hyperplanes are constructed on each side of the hyperplane that separate the data. The separating hyperplane is that which maximizes the distance between the two parallel hyperplanes. An assumption is made that the larger the margin or distance between these parallel hyperplanes the better the generalization error of the classifier will be. In this study, LIVSVN which is an integrated software for SVM [19] is used to analysis the experimental data.

3.2.3 *Radial Basis Function (RBF) network*

The RBF network [20] is a type of neural network that yields a response surface via a superposition of basis functions. The learning of the RBF network involves obtaining appropriate weights for each basis and is identical to the energy minimization of the RBF network. The main procedure of obtaining the weights is calculating the inverse matrix. Therefore, the RBF network can be evaluated quickly, and additional analysis can be easily conducted when new data sets are added. The RBF network performs well in terms of accuracy and robustness, irrespective of the degree of nonlinearity. Additionally, it is robust against experimental errors or noise [21]. In this study, the parameters proposed for the RBF network by Kitayama and Yamazaki [22] are adopted.

3.3 *Application results*

Table 2 shows the discrimination accuracy for UFOV size using the three machine learning methods. The discrimination accuracy for training data and test data indicate the average accuracy of the subjects based on the discrimination function of all subjects' data and the cross-validation, respectively.

The discrimination accuracy for training data does not vary substantially for combinations 1 and 2, and is over 80% for all methods. An extremely high accuracy is realized at over 90% for RBF in particular. In contrast, no significant difference is seen in the discrimination accuracy for test data between the combinations for RBF and SVM. However, the discrimination accuracy is high for the Boosting method for combination 1, where blinks rather than saccade time is used for the explanatory variable. The highest accuracy obtained for this combination of parameters and machine learning was approximately 85%. On the other hand, RBF can achieve the highest accuracy for training data; but it achieved the lowest accuracy for the test data irrespective of the combinations. This may be explained by the fact that the training data may exist outside the distribution area (i.e., extrapolated area). RBF creates a response surface of which objective value reaches zero in a extrapolated region, thus it has low accuracy for the test data. However, the discrimination accuracy of RBF is the highest in the interpolated domain. Therefore, the accuracy for the current test data might be improved if the interpolated domain was expanded and it contained the test data by increasing the amount of training data.

We now consider the application to the Repeat aloud discrimination function through the Boosting method using the parameters of combination 1, which has the highest discrimination accuracy for test data. Although the UFOV size for the Repeat aloud condition obtained in the experiment was slightly smaller than the Wide_UFOV (No Task condition), there was no significant difference and no marked narrowing of the field of vision. Therefore, we predict that the Repeat aloud condition will result in a Wide_UFOV if the discrimination function is sufficiently accurate. However, approximately 60% of results for the Repeat aloud condition

Table 2. Discrimination accuracy for UFOV size using each machine learning method.

	Training data		Test data	
	Combination 1	Combination 2	Combination 1	Combination 2
Boosting	85.38%	83.82%	84.72%	77.78%
SVM	83.99%	83.42%	76.39%	76.39%
RBF	96.08%	92.89%	47.22%	50.00%

was judged as Wide_UFOV and approximately 40% as Narrow_UFOV. Possible reasons for this may be that the discrimination function accuracy was in sufficient, or that narrowing the UFOV for the Repeat aloud condition was not correctly evaluated. For the present experiment the UFOV size is determined only in a horizontal direction, and no consideration is given to the UFOV in a vertical direction. Previous research [6] has shown that the vertical UFOV will narrow more than the horizontal UFOV under light MWL. It is also possible that a significant narrowing of the field of vision occurs for the Repeat aloud condition if the vertical UFOV is considered in combination with the horizontal UFOV. This is also suggested by the fact that the AWWL score, which is a subjective workload index, is significantly higher for the Repeat aloud condition than that for the No Task condition. In other words, it is also conceivable that UFOV narrowing occurred in part of the trial run under the Repeat aloud condition, and the percentage judged as Narrow_UFOV consequently increased. Verifying this is an issue for future research.

4 CONCLUSION

The present study aimed to propose a method for detecting the narrowing of the UFOV while driving a vehicle. Under experimentally controlled conditions for the UFOV for drivers operating a driving simulator, we studied methods for detecting the narrowing of the UFOV through machines learning methods using parameters relating to eye movement and blinks. As a result, discrimination functions were found through machine learning methods that allow us to distinguish whether there is strong narrowing of the UFOV with an accuracy of approximately 85%, under set experimental conditions. However, because the study of discrimination accuracy for the weak narrowing of the UFOV was insufficient, we propose this topic for future study, combined with accuracy improvement.

REFERENCES

[1] Hollnagel E. Human Reliability Analysis Context and Control, Academic Press, 1993, 145.
[2] Fujimori M, Uesako H, Kawamura M, Measurement in Consideration of Head Movement for Visual Direction of Driver on Expressway, Transactions of the Society of Instrument and Control Engineers, Vol. 35, No. 4; 1999, pp. 473–479.
[3] Miura. T, Active function of eye movement and useful field of view in a realistic setting, In R. Groner, et al. (eds.) From Eye to Mind: Information Acquisition in Perception, Search, and Reading, Elsevier Science Publishers, 1990, pp. 119–127.
[4] Ball K, Owsley C, The useful field of view test: A new technique for evaluating age-related declines in visual function, Journal of the American Optometric Association, Vol. 64, No. 1; 1993, pp. 71–79.
[5] Owsley C, Ball K, Keeton D.M, Relationship between visual sensitivity and target localization in older adults, Vision Research, Vol. 35, No. 4; 1995, pp. 579–587.
[6] Morishima K, Minochi A, Hayashida Y, Takamine K, Yamanaka K, Daimoto H, The Relation Between Mental Workload and Useful Field of View during Pursuit Eye Movement, The Japanese Journal of Ergonomics, Vol. 49, No. 5; 2013, pp. 203–210.
[7] Morishima K, Minochi A, Abe K, Furuki S, Yamanaka K, Daimoto H, Measurement of Useful Field of View in Eye and Head-Free Condition while Driving, Transactions of the Japan Society of Mechanical Engineers, Series(C), Vol. 79, No. 806; 2013, pp. 3561–3573.
[8] Morishima K, Minochi A, Ishii M, Yamanaka K, Daimoto H, The estimation of useful field of view using parameters related to eye movements, The transactions of Human Interface Society, Vol. 15, No. 2; 2013, pp. 121–130.
[9] Yamanaka K, Nakayasu H, Miyoshi T, Maeda K, A Study of Evaluating Useful Field of View at Visual Recognition Task, Transactions of the Japan Society of Mechanical Engineers, Series(C), Vol. 72, No. 719; 2006, pp. 244–252.
[10] Messe T.S., Using the standard staircase to measure the point of subjective equality. A guide based on computer simulations, Perception & Psychophysics, Vol. 57, No. 3; 1995, pp. 267–281.
[11] Daimoto H, Takahashi T, Fujimoto K: Assessment of mental workload using event-related potentials, Transaction of Human Centered Design Organization, Vol. 5, No. 1; 2009, pp. 29–37.

[12] Haga S, Mizukami N, Japanese version of NASA Task Load Index: Sensitivity of its workload score to difficulty of three different laboratory tasks, The Japanese Journal of Ergonomics, Vol. 32, No. 2; 1996, pp. 71–79.

[13] Becker W. and Fuchs A.F.: Further properties of the human saccadic system. Eye movements and correction saccades with and without fixation point, Vision research, 9, 1969, pp. 1247–1258.

[14] Zuber, B.L. Stark L. and M. Lorber: Saccadic suppression of the papillary light reflex, Experimental Neurology, 14; 1966, pp. 351–370.

[15] J.A. Sharpe and T.O. Sylvester: Effect of aging on horizontal smooth pursuit, Investigative Ophthalmology and visual science, 17, 1978, pp. 465–468.

[16] Y, Freund and R, E. Schapire, "A decision theoretic generalization of on-line learning and an application to boosting", Journal of Computer and System Sciences, Vol. 55, No. 1; 1997, pp. 119–139.

[17] V. Vapnik, The Nature of Statistical Learning Theory, NY:Springer-Verlag, 1995.

[18] S.K. Durgesh, B. Lekha, Data Classification using Support Vector Machine, Journal of Theoretical and Applied Information Technology, 12(1), pp. 1–7, 2010.

[19] R.E. Fan, P.H. Chen, and C.J. Lin, Working set selection using second order information for training SVM, Journal of Machine Learning Research, 6, pp. 1889–1918, 2005.

[20] Orr MJL, Introduction to radial basis function networks. available from [http://www.anc.ed.ac.uk/rbf/papers/intro.ps.gz] 1996; (accessed on 16 May, 2014).

[21] Jin R, Chen W, Simpson TW. Comparative studies of metamodeling techniques under multiple modeling criteria. Structural and Multidisciplinary Optimization 2001; 23 (1), pp. 1–13.

[22] Kitayama S, Yamazaki K. Simple estimate of the width in Gaussian kernel with adaptive scaling technique. Applied Soft Computing 2011; 11(8); pp. 4726–4737.

New Ergonomics Perspective – Yamamoto (Ed.)
© 2015 Taylor & Francis Group, London, ISBN 978-1-138-02751-0

A protocol for developing a quantitative index for human error prevention

Hyeon-Kyo Lim & Hyunjung Kim
Department of Safety Engineering, Chungbuk National University, Cheongju, Chungbuk, Korea

Tong-il Jang & Yong Hee Lee
Division of I&C and Human Factors, Korea Atomic Energy Research Institute, Daejeon, Korea

ABSTRACT: A quantified index denoting human error possibility, if any, is indispensable Unfortunately, however, there exist enormous kinds of factors that may influence on human behaviors, and their causal relationships are not clear so that quantification of error possibility is not easy even in modern days. Therefore, there is no way but adapt subjective judgment and experience of skillful experts who gave his service as operators in NPP's.

This study tried to introduce a protocol for developing a quantitative index not for estimating human error probability but for quantifying human error possibility. The major reason for adopting intuitive approach rather than probabilistic one was lack of probabilistic data for human behaviors.

In order to comprehend the traits of performance shaping or influencing factors that may influence on work performance, a literal survey was conducted first, and review on the possibility of modeling their causal relationships and work performance of human operators in nuclear power plants was followed with diverse modeling techniques such as Influence Diagram, System Dynamics, Analytic Hierarchical Process, Fuzzy Inference, and Neural Networks. The results implied that, among them, Fuzzy AHP technique seemed to be an appropriate technique for quantifying human error possibility since it could reflect uncertainty and evaluator's experience and point out managerial aspects for counter plans.

Thus, after that, technical research based on AHP model was conducted with influencing factors. And, at the end, all the results were reviewed in the aspect whether the human error possibility would be predicted quantitatively.

Keywords: Human Error; quantitative index; safety management; AHP; Fuzzy Theory

1 INTRODUCTION

As the computer technology is developed, most judgment and analysis have the tendency to utilize the computer skills, and eventually quantitative analysis techniques. Safety assessment of nuclear facilities is also one of them. To assess the safety level of nuclear power plants, not a few trials have been made for last several decades. Unfortunately, however, most trials neglected human factors partly because there exist enormous kinds of factors that may influence on human behaviors and, further, partly because causal relationships are not clear so that quantification of error possibility is not easy even in modern days.

Nevertheless, quantification of human error possibility is indispensable for comprehensive safety assessment of nuclear power plants. According to previous research results, human errors occupy large portion of accidents in nuclear power plants and chemical plants.[1, 2] Among them, work activities carried out in a control room in a nuclear power plant are crucial to maintain safety. It means that work performance of human operators working in main control room is a quite major factor to predict safety level of the plants. Therefore, developing a quantitative index standing for error possibility of human operators working in main control room is still an attractive research theme for understanding the safety level and estimating reliability of nuclear power plants as well as eliciting points for safety management implementation. Thus, this study was carried out to arrange the procedure for developing a quantitative index predicting error possibility of human workers working in main control room of nuclear power plants.

2 SELECTION OF STRESS FACTORS

Human performance level is a function of lots of factors such as physical, mental, psychological, and psychosocial factors. Therefore it can fluctuate without ceasing due to internal factors as well as external factors. As the first trial, however, among lots of factors which are known to influence on human performance, this study concentrated only on personal stress and fatigue factors with reference to NUREG-0711, 10CFR26, and R.G.5.73 by USNRC which ask strict management of stress and fatigue in the sense of Fitness-for-Duty of human operators.

Occupational stress can occur when people are aware of a mismatch between the occupational and organizational demands at work and their ability to overcome, circumvent or find other ways of adequately coping with those demands.[3] Various approaches are possible as a source of an occupational workload stress classification scheme.[4] Especially, however, working in demanding or threatening situations such as nuclear power plants taxes cognitive capabilities such as information processing, memory, knowledge, skills, etc.[5] It is well known that various stress factors for human workers in nuclear power plants were summarized as Performance Shaping Factors by Swain and Guttman.[6] Therefore, after literal survey on stress factors with diverse viewpoints, major stress factors were elicited as shown Table 1. To simplify the modeling process, only personal factors were taken for further consideration.

Though psycho-social factors are generally known to make effects on occupational stress levels, psycho-social factors were excluded because partly because what is required is stress factors that may direct influence on human operators in nuclear power plants,[7] and partly because human operators working in main control room in nuclear power plants were regarded as homogeneous through long-term training and education in so far as their career and experience except personal affairs.

3 SELECTION OF MODELLING TECHNIQUE

For eliciting a numerical value from the selected stress variables, a study on modelling techniques with the special view point of ergonomists were carried out. Modelling techniques surveyed include Analytic Hierarchical Process (AHP), Influence Diagram, System Dynamics, Neural Network, and Fuzzy Inference Technique. Table 2 shows comparison summary of techniques considered. The results were cited from previous researchers (Taroun et al., 2011; Chivatá et al., 2012; Mariken et al., 2013), and modified by part.

Originally, Influence Diagram and System Dynamics technique were developed for analyzing social phenomena graphically and have been widely applied to human decision making problems. However, these techniques have weak points such that identification of influence factors is subjective.

Table 1. Performance shaping factors in consideration.

Level 1	Level 2	Level 3
Human	Personal Factors	Motivation/Attitudes, Emotional State, Concentration, Confidence, Fatigue, Time Stress, Load Stress, Time Sharing, Duration of Stress
	Training, Knowledge or Experience	Suitability of Training and Education, Operator Experience, Task Knowledge, Recency of Training
Task	Procedures and Documentation	Structure, Recency in Use, Availability, Access, Location, Accuracy, Clarity, Complexity, Comprehensiveness/Completeness, Format(Electronic), Up-to-Date
	Information	Structure of Information, Availability, Clarity, Complexity, Correctness/Completeness
	Communication	Arousal, Mental Health, Physical Health, Accent, Dialect, Protocol, Tool, Clarity, Accuracy, Correctness/Completeness
System	Human Machine Interaction	Interface Elements, Interaction Elements, Screen Layout, Visual Coding, Alphanumeric Characters, Feedback, Conspicuity, Consistency of Displays
	Workplace Design	Workplace Layout, Spatial/Movement Incompatibility, Location of Control, Coding of Control, Labeling of Control, Control Resistance
Environment	Quality of Environment	Vibration, Lighting, Noise, Housekeeping, Reverberation
Teamwork	Team Factors	Team Morale (Preventing Conflict/Resolving Conflict), Team Leadership (Maintaining Standards/Monitoring Performance/Promoting Participation), Team Cooperation (assisting Others/Positive Atmosphere), Team Adaptability (Cross Checking/Prioritizing Action/Flexibility), Team Risk Management (Managing Pressure/Planning for Threats), Allocation of Function and Responsibility

Fuzzy Inference and Neural Network are useful techniques to deal with human based uncertainty. Especially fuzzy inference allows vague information to be approximately summarized with natural languages which consist of words in human communications.[8]

Analytic Hierarchy Process is now a well known technique for providing a systematic ranking of problem or system components. This technique is amenable to a wide variety of organizational and non-organizational decision-making problems but is inadequate to represent the causal relations. However, it is quite attractive to find out relative importance or weight with quantitative numerical values.

Furthermore, if fuzzy technique is added, Fuzzy AHP technique can be as an appropriate technique for human error possibility since it could reflect uncertainty and evaluator's experience and point out managerial aspects for counter plans.

Throughout a comprehensive comparative analysis, it was concluded that AHP could be a practical candidate since it can point out the factors upon which managerial efforts should concentrate, and it can give the relative weight or importance of factors in the form of numerical values. In addition, if Fuzzy Inference Technique is applied, vagueness of human native languages can be expressed as fuzzy membership functions which promote integration of lots of people into a conclusion numerically.

As a consequence, for estimation of human error possibility, an AHP model could be developed with influencing factors. Since the overall goal was 'Human error possibility', on the first level of hierarchy two factors were located—stress and fatigue, which directly contribute to the goal, and on the second level, three factors could be located which contribute to the factors on the upper level, stress and fatigue. All the influencing factors were listed under relevant categories.

Table 2. Comparison of modeling techniques.

	Influence diagram	System dynamics	AHP	Fuzzy inference	Neural networks
Format	Specific technique	Specific technique	Specific technique	Mathematic model	Mathematic model
Year	1973	1961	1970	1965	1980
Domain	Aviation, finance	Nuclear, economic, others	Nuclear, defense, others	Computer	Aviation, others
Application	Human	Human	Human	Software, human	Hardware, human
Qualitative	O	O	O	O	O
Quantitative	O	O	O	O	O
Cognitive	△	△	X	O	O
Systematic	X	△	O	X	X
Weight	X	X	O	O	△
Hierarchy	X	X	O	X	△
Subjective	O	O	O	O	O
Trainable	O	O	X	X	X
S/W Compatibility	X	X	X	O	O
Data Inference	X	X	X	O	X

4 FUZZY INFERENCE WITH AHP MODEL

It is well known that Analytic Hierarchical Process (AHP) technique is one of the most convenient techniques to determine relative weights of related factors. Taking stress and fatigue, and subsequent influencing personal factors into consideration, modeling of a hierarchical tree structure was tried in the present study.

Then, the following problem would be assigning numerical values to individual factors for their relative weights. It would be a conventional process of AHP, but it would be straightforward as far as the model concerned is psychophysiological one. The problem of integrating lots of factors into a unified value can be interpreted as the problem of determining relative weights among the factors and summing up the individual functions of factors, if exist, with the assigned weights. Assigning appropriate weights to the diverse factors can be vague and ambiguous. However, human decision making is intrinsically ambiguous. Even when persons assign a same numeric value to a hazard factor for risk assessment, confidence level may differ person to person. Thus, a single numeric value obtained through deterministic analysis such as AHP can be of question. To overcome errors from ambiguity and clearness, fuzzy inference technique was applied in the present study.

As Karwowsky indicated [9], human beings usually express their impressions and feelings with natural languages. Personal assessment of stress factors with reference to frequency, severity, work endurability, and so on would entirely depend upon personal judgments. In addition, Onisawa [8] suggested a method to formulate subjective judgement and verbal expressions with Fuzzy Inference technique, and Karwowski et al.[9] showed practical methods for modelling human decision-making. Personal language expressions could be transformed into fuzzy input variables according to the method of Karwowski et al. Judgment on a factor relative to comparative one is quite intuitive, so far as natural word of human beings are restrictive to a few words, such as "negligible", "low", "high", and "severe." Thus, adoption of natural language variables would give a way of progressive modeling that can overcome the difficulty or errors from crisp numeric values.

Despite, a set of numerical values is necessary for further computation, because input functions corresponding to the influencing factor should be expressed with weights.

A method developed by Chang (1996) can help normalization of fuzzy weights [10]. Utilizing his technique, normalized weights corresponding to natural expressions can be obtained after a few steps.

Another factor should be considered is the confidence level on the weight of factors concerned. It stands for the level how firmly a judgment is sure his/her judgment is correct or not. It can be also a fuzzy number so that it can be expressed as a fuzzy membership function. In that case, Efficient Fuzzy Weighted Average (EFWA) Technique will be helpful.[11]

Consequently, it can be said that all the factors would be formulated as fuzzy numbers, and the resultant output—implicative judgment—can be obtained by multiplying corresponding fuzzy membership functions—input characteristics, relative weights, and confidence of a person. The numerical expression using Fuzzy notation will be as follows;

$$\mu_0(x, y, z) = \mu_{I \times W \times C}(x, y, z)$$
$$= min \{\mu_I(x), \mu_W(y), \mu_C(z)\} \tag{1}$$

Fig. 1 shows an integration process of multiple judgments concurrently considering characteristics of a factor concerned—input function, relative weight, and confidence level. All the numerical functions and arithmetic computations were tried and depicted by using MAT-LAB fuzzy toolbox version 7.12.0.

In all the graphs the horizontal axis stand for dimension concerned such as weight from 0 to 1, confidence level, and influence from "negligible" to "severe," and the vertical axis stands for the degree of conviction on judgment.

Small graphs in the first column denote fuzzy membership functions which stand for weights on an individual factor which make influence on human performance level, and graphs arranged in the second column denote confidence levels on their weights, and the graphs in the third column denote characteristic input function of the individual factor. Finally, the rightmost graphs could be obtained by fuzzy Cartesian product. In Fig. 1, based on fuzzy inference rules corresponding to natural languages and related expressions, appropriate individual membership functions corresponding to each linguistic variable were modelled as triangle functions as shown. The fact that four kinds of different graphs were located in the first column in the case of Fig. 1, depict that linguistic weights could be expressed with four kinds of different expression. It means that even about the same factor, different linguistic expression would be possible with different confidence, and implies the logical base of Fuzzy rules.

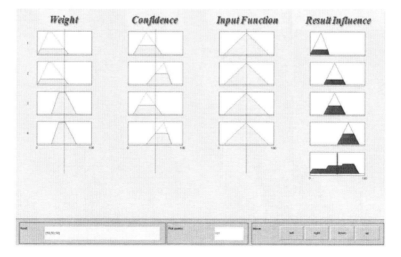

Figure 1.   Integration of individual judgment on the influence of a stress factor.

5 DEFUZZIFICATION AND SENSITIVITY ANALYSIS

Finally, after aggregating four different situation-dependent level functions by applying max-min composition technique suggested by Mamdani,[12] the final output distribution function can be obtained, as shown at the lowermost graph in the fourth column. The final graph denotes the integrated linguistic output expression on an individual factor influencing on human performance level. Eventually, The final output functions would be also a fuzzy membership function that can be reluctant to ordinary people.

The fuzzy results generated cannot be used as such to the applications, hence it is necessary to convert the fuzzy quantities into crisp quantities for further processing.[12] To promote comprehension of ordinary managers who are unfamiliar with Fuzzy expressions, defuzzification could be tried, which means induction of a crisp numeric value standing for the whole linguistic expressions and practical meaning. Among several defuzzification methods, centroid method is one of the most widely used methods. It can be defined by the algebraic expression:

$$W^* = \int \frac{w\mu_c(w)dw}{\mu_c(w)dw} \tag{2}$$

where W is a fuzzy set, and w^* is a crisp center of mass, and is the membership in class c at value w. Otherwise, weighted average method can be utilized since it is one of the more computationally efficient methods. Unfortunately, however, it is usually restricted to symmetrical output membership functions.

6 DISCUSSION

To express the safety level with a single numerical index may be meaningless, but it still is a requirement for the comprehensive analysis of system reliability and managerial comparison. For the purpose, a great numbers of researchers have studied to model and analyze the safety level of the whole system. However, few research results took great advance from mere numerical analysis because of lack of appropriate statistical data.

Traditionally human reliability studies concentrated on statistical estimation of human error rate. The major reason was that workers were considered as parts of the nuclear power plant system so that human error possibility was analyzed with a concept of mechanical failure rate. When statistical data is sufficient, it can be utilized to estimate appropriate index as a function of them. Unfortunately, however, even with this effort, it is impossible to clarify when a human error occurs and to suggest any improvement on system design or human error possibility in specific circumstance.

As an alternative, the present study adopted somewhat different approach for modelling. The final objective of human factor research related with nuclear power plants is accident prevention. Therefore the main question in safety activities is how to prevent accidents, and what should be implemented. Consequently, though there may exist somewhat awkward process in eliciting points that managerial activities should concentrate, next-to-best solution, at least, should be identified before accident happen. Those rationales are already prevalent in the area of developing Human Error Identification techniques. Likewise, the present study agree that the hierarchical structure is awkward in explaining stress and performance though, the authors tried to find out weight values with which managerial activities should be distributed.

Another point that can be embroiled in controversy is the causal relationship among stress factors and managerial factors. It can be said that stress is one of causal factors for fatigue, and vice versa. Conclusively speaking, there is no definite evidence which is the origin of the whole mechanism for human performance related with stress and/or fatigue. Many people regard stress and fatigue similar factors in the sense that both can degrade human performance of workers, but there are researchers who had somewhat different viewpoint such as

Bultmann et al.[13] According to their cohort research results, it can be said that stress is different from fatigue because their functions and concepts do not coincide each other, even though they have high correlation in their variation.

The fact that fuzzy language variable and fuzzy logic was applied may be controversial. Basically, all the decisions in industrial fields are made by human workers. Human beings usually express a great much of their opinions and feelings with verbal language. Therefore, it is quite natural there exists systematic consistency in the use of language variables.[9] Not to disturb original intention or impression of human workers, fuzzy language variables were adopted in the present study, and fuzzy logic was utilized which is known to express quite naturally the sense of human beings. Likewise, the reason why the term 'possibility' was adopted rather than 'probability' was due to the fact that fuzzy approach based upon the conceptual image of human beings, not on statistical data.

7 CONCLUSION

In order to assess the whole safety of a nuclear power plants, success probability of every component should be expressed as a finite reliability value, even if it is a human being. Conventionally, human errors were evaluated using probability, but causal relationship with job stress and fatigue factors were difficult to analyze quantitatively because the factors were numerous, and their causal relationships were not clear, and observed data were not sufficient. Furthermore, all the decision makings in the work process are made by human workers. For these reasons, an integrative model that reflect inambiguity of human beings was tried in the present study.

To maintain trait of human decision making and sensation, fuzzy linguistic variables were adopted and fuzzy inference logic was applied. A model for eliciting a numerical value standing for human error possibility resulted from various factors including stress and fatigue was constructed utilizing AHP technique which can produce a single numerical value as the final output.

After repeated experiments and field applications, the utility of the error possibility index can go higher. Until then, field study will go on and statistical data will be compared to the result of the intuitive fuzzy approach tried in the present study. However, in the mean time, the effect of managerial activities which were elicited from the present model to prevent human errors and accidents in nuclear power plants will compensate time-consuming efforts.

Several techniques were reviewed in the aspect whether the human error possibility would be predicted quantitatively. Schematic model of influencing factors is formulated for developing human error possibility index by using Fuzzy AHP. To summarize, it can be said that, if AHP technique is taken and fuzzy inference would be added a quantitative human error possibility index could be obtained in the aspect of safety management.

REFERENCES

[1] Meister, D. *Human Factors: Theory and Practice*, John Wiley & Sons, Inc., 1971.
[2] Stanton, N., *Human Factors in Nuclear Safety*, Taylor & Francis, 1996.
[3] Wallis, D., "Satisfaction, Stresses, and Performance: Issues for Occupational Psychology in the Caring Professionals," *Work & Stress*, Vol. 1, No. 2, pp. 113–128, 1987.
[4] Cooper, G., and Payne, R., *Causes, Coping, and Consequences of Stress at Work*, John Wiley & Sons Ltd., New York, 1988.
[5] Gaillard, A.W.K., "Concentration, Stress, and Performance", in *Performance under Stress*, edited by Hancock, P.A., and Szalma, J.L., Ashgate, pp. 59–75, 2008.
[6] Swain, A.D., Guttman, H.E., *Handbook of Human Reliability Analysis with Emphasis on Nuclear Power Plant Applications*, NUREG/CR-1278, U.S. Nuclear Regulatory Commission, 1983.
[7] Moray, N.P., Huey, B.M., "Human Performance", in *Human Factors Research and Nuclear Safety*, National Academy Press, 1988.
[8] Onisawa, T. "An Application of Fuzzy Concepts to Modeling of Reliability Analysis", *Fuzzy Sets and Systems*, Vol. 37, No. 3., pp. 67–86, 1990.

[9] Karwowski, W., Marek, T., Noworol, C., Ostaszewski, K., "Fuzzy Modeling of Risk Factors for Industrial Accident Prevention: Some Empirical Results", in *Applications of Fuzzy Set Methodologies in Industrial Engineering*, edited by Evans, G.W., Karwowski, W., Wilhelm, M.R., Elsevier Science Publishers B.V., 1989.

[10] Chang, D.Y., "Applications of The Extent Analysis Method on Fuzzy-AHP", *European Journal of Operational Research*, 95, pp. 649–655, 1996.

[11] Lee, D.H, Park, D. "An efficient algorithm for fuzzy weighted average," *Fuzzy Sets and Systems*, 87, pp. 39–45, 1997.

[12] Mamdani, E.H., Assilian, S., "An Experiment in Linguistic Synthesis with a Fuzzy Logic Controller", *International Journal of Man-Machine Studies*, Vol. 7, No. 1, pp. 1–13, 1975.

[13] Bultmann, U., Kant, I.J., van Amelsvoort, L.G., van den Brandt. P.A., Kasl, S.V., "*Differences in fatigue and psychological distress across occupations: results from the Maastricht Cohort Study of Fatigue at Work*", J. of Occupational Environmental Medicine, 43(11), pp. 976–983, 2001.

New Ergonomics Perspective – Yamamoto (Ed.)
© 2015 Taylor & Francis Group, London, ISBN 978-1-138-02751-0

Facial reflexology and work related musculo-skeletal disorders

Nguyen Ngoc Nga & Khuc Xuyen
Vietnam Occupational Health Association, Hanoi, Vietnam

Duong Khanh Van
National Institute of Occupational and Environmental Health, Hanoi, Vietnam

ABSTRACT: *Objective*: The objective of this study is to evaluate primarily promising application of Vietnamese method named "Dien Chan" in treatment of Musculo-Skeletal Disorders (MSDs).

Method: Descriptive study.

Subject: 25 cases suffered from musculo-skeletal disorders.

Results: Facial reflexology has been known for a long time in all far Eastern countries. "Dien Chan" is the original method of facial reflexology established by Prof. Bui Quoc Chau. It mainly developed in the 1980s as named "Facytherapy". This unique method has been taught and applied larger and larger in Vietnam and many countries.

According to "Dien Chan", the face, as a part of ourselves, symbolizes us and therefore represents us wholly. Certain areas of the human body correspond to different parts of the body elsewhere.

"Reflexology treatments apply pressure to specific reflex points or areas of the face, hands, or feet. Since areas of the face or feet correspond to different parts of the body, and applying pressure to these areas can affect the corresponding parts of the body".

Work-related MSDs are very popular in many countries, including Vietnam. Many researches illustrated high cost and DALY of MSDs.

25 cases of work-related MSDs (neck pain, shoulder pain, back pain, knee pain…) were treated with "Dien Chan". Among that there were 14 acute cases and 11 chronic ones. The result showed that "Dien Chan" is a very effective therapy with their advantages: take effect quickly (less than 1 day to 5 days), no side effects, very cheap, and especially patients can treat themselves easily.

Application of "Dien Chan" is not only to MSDs, but also in many diseases. Hundreds of therapeutic sessions have been designed by Prof. Bui Quoc Chau and his students and trainees. Some of them can be applied for incurable diseases.

Recommendation: Further researches on advantages of Dien Chan should be supported in order to expand "Dien Chan" application in treatment of work-related diseases and Occupational diseases.

Keywords: Dien Chan; Facytherapy; musculo-skeletal disorders treatment

1   INTRODUCTION

In the world, the diseases now are more and more complex. In order to overcome difficulties in fighting diseases many scientists and physicians have been trying to find out new and effective treatment method. When people came to doctors, most of them were advised to use medicine or drug. But medicine or drug may have adverse effects. The invention of modern medicine sometimes requires very high cost. In certain cases people could avoid using medicine and drug by acupuncture. But acupuncture required using needles which sometime hurt patients, and require good disinfection. Facial reflexology has long been known in all far Eastern countries. "Dien Chan" is the original method of facial reflexology established successfully by Prof. Bui

Quoc Chau. Its main developments were made in the 1980s, and named "Facytherapy". The full name of "Dien Chan" in English is "Face Diagnosis and Cybernetic Therapy" (FACY), the other names are "**Multi-reflexology**", Facial Reflexology, Vietnamese Reflexology. "Dien Chan" is a method which supports people's health and treats diseases with simple devices and tools, without medicine, drug and needles. Some of characteristics of "Dien Chan" can be listed such as: Everyone (Any people can learn and apply), Every time (Can apply at any time), Everywhere (can do any where, don't' require special place), Everything (apply for almost symptoms, diseases), Easy, Effective, Excellent (give surprising results), Economic, Equal (everyone can be treated by Dien Chan regardless of the wealth, social status…), Elastic (flexible application), Entertainment (doing like games, recreation). This unique method has been taught and applied larger and larger in Vietnam and many countries. According to "Dien Chan" approach, the Face, as a part of ourselves, symbolizes us and therefore represents us wholly. Certain areas of the human body correspond to different parts of the body elsewhere. "Reflexology treatments apply pressure to specific reflex points or areas of the face, hands, or feet. Since areas of the face or feet correspond to different parts of the body, and applying pressure to these areas can affect the corresponding parts of the body". Work-related MSDs are very popular in many countries, including Vietnam. Many researchs illustrated high cost and DALY of MSDs. Up to now hundreds of regimens of treatment with Dien Chan have been developed by Prof. Bui Quoc Chau and his students. Those regimens have shown their effectiveness in practice. The regimens for MSDs treatment were ones of those.

2 OBJECTIVES

This study aims at primarily evaluation of "Dien Chan" application in treatment of MSDs and its promising application in a broader field of occupational health.

3 RESULTS

3.1 *Characteristics of the subjects*

Characteristics	Males	Females
Number	9	16
Age (years)	30–70	30–65
Disease:		
Acute	7	7
Chronic	2	9
MSDs classification		
Neck-shoulder pain	4	7
Back and low back pain	2	6
Shoulder joint, elbow pain	1	1
Injury:		
– Ankle	2	
– Knee		2
– Shoulder joint	1	1

25 patients with different types of MSDs, including neck and shoulder pain, back pain, sacrum-coccyx pain, both acute and chronic ones. Besides, some cases of injury were taken into account. These MSDs are common in workplace.

3.2 *Treatment tool*

There are many kinds of "Dien Chan" devices. To treat these MSDs cases only some simple tools were used. Those were: "point searching stick"—and moxa.

Figure 1. Point searching stick.

Figure 2. Moxa.

Figure 3. To heat up with moxa.

3.3 *Conditions and treatment*

3.3.1 *Neck and shoulder pain*

Code of patient	Age	Duration of the pain	Frequency	Number of treatment	Results
Acute					
01	30	3 days	Continue	5	Cured
02	41	1 day	Continue	2	Cured
03	55	1 day	Continue	2	Cured
04	65	1 day	Continue	1	Cured
05	31	3 days	Continue	4	Cured
06	34	1 day	Continue	3	Cured
07	32	2 days	Continue	4	Cured
Chronic					
08	61	1 year	Often	2	Good, sometime repeat
09	65	1 year	Sometime	3	Last year: 2 times
10	66	3 years	Often	1	Better, not follow
11	54	2 years	Often	5	Cured

11 of 25 cases were neck-shoulder pain. There were 7 acute cases. The patients get continuous pain for 1–3 days. Most of them were treated by massage or attachment of Salonpas pain patch. But their condition improved not remarkably. With "Dien Chan" after 1–4 times of treatment, all of those 7 cases were cured. One treatment lasted for less than 5 minutes. In common 3 treatments were done in 1 day. That means they need only about 1–1.5 day with total of less than 15 minutes.

Case study 1

A 31-years-old man suddenly got neck and shoulder pain. In five days he used pain patches and heat balm for treatment. It seemed that there was no remission of pain.

Therapy: To press on points: 65, 106, 16. Heat apply above back of wrist, back of ankle (on the left of the body) by moxa. The duration of each treatment was about 5–10 minutes. Applied the treatment 3 times per day.

After being treated by "Dien Chan" with 4 times of treatment in 1.5 day, his pain disappeared.

Code of patient	Age	Duration of the occurrence	Frequency	Number of treatment	Results
Acute					
12	54	2 hours	Continued	2	Cured
13	49	1 day	Continued	3	Cured
Chronic					
14	62	7 years	Frequent	6	Better, continue
15	70	5 years	Often; Recent 2 months continue pain	3 × 7 days	Better, continue
16	41	4 years	Often	6	Better, continue
17	50	3 years	Often Recent 4 months continue pain	6	Better, continue
18	34	1 year	Often	2	Better, not follow
19	30	2 months	Frequent	5	Cured

There were 8 cases of back pain, low back pain, sacrum-coccyx pain. Among that there were 2 acute cases, 6 chronic ones. These 2 acute cases were due to carrying heavy objects. After one day of treatment, 2 acute cases were cured. The real spent time for treatment was about 10–15 minutes.

Up to when the data was collected 6 chronic cases have undergone 1–7 days of treatment. Their conditions were improved remarkably, 5 of them are continuing the therapy.

Case study 2
A 30-year-old woman worked in an office. She had to work with computer every day. She got back pain and shoulder pain sometimes. After a rest the pain was disappear. In last year, she got sacrum-coccyx pain. The pain occurred every day after working for about 30–40 minutes.

Therapy: To press on points: 53, 63, 19
– The 1st day: 1 time. The patient felt better
– The 2nd day: 1 time. The patient felt better
– The 3rd day: 2 times.

And 4 months have been passed since the last treatment and the patient has no pain any more.

3.4   *Injury*

Code of patient	Age	Duration of the pain	Frequency	Number of treatment	Results
Shoulder pain					
20	30	2 days		2	Cured
21	56	1 month		1	Cured
Ankle pain					
22	39	1 hour		2	Cured
23	49	1 day		6	Cured
Knee pain					
24	41	1 week		2	Better—not follow
25	40	5 years ago	Sometimes	Treatment when pain	Better

Above table list 6 cases called "Injury". Among that, there were 2 cases of shoulder pain, 2 cases of ankle pain, and 2 cases of knee pain. One of shoulder pain was resulted from lifting bucket of water; the other resulted from tennis playing. Ankle pain and knee pain were resulted from sliding After treatment 2 cases of knee pain were better, but we could not follow because the patient had to go to other province. Other cases were cured after 1–6 treatments that means that 20 minutes were spent for treatment in 2 days.

Some typical case study as bellows:

Case study 3

A 56-year-old woman suffered from severe pain from shoulder to elbow that she could not move her right hand. She had to do everything by left hand. Her right arm always had to fold. She had got treatment with medicine and traditional acupuncture for 1 month. It seemed that the situation was improved not significantly.

Therapy:
- The 1st time: To find pain points on her face. Massage pain points for 2–3 seconds per one point. She felt better (the pain reduce by 20–30%).
- The 2nd time: To massage pain points as above again. She could outstretch her right arm.
- The 3rd time: To massage pain points as above again. She could raise her hand and waved.

6 months have been passed. No pain occurred.

Case study 4

A 30-year-old man suffered from severe shoulder and shoulder joint pain on the right due to playing tennis. 2 days later he came to get treatment.

Therapy:
- The 1st day:

In the morning: Press points: 477, 97, 99, 98, 106, 34. Knocked the point 65. He felt better remarkably.
   At noon: Do the same (second time). He could rotate his arm with a little pain.
In the afternoon: Do that again (third time). He didn't pain any more.

- The 2nd day: He felt a little uncomfortable on in his shoulder. He was treated as above, then continued for 3 days more.

2 months have been passed. There was no any more pain.

Case study 5

A 38-year-old man suffered from pain in the ankle of left foot due to slipping.
Treatment:
- To heat up the ankle of left arm by moxa for about 2 minutes. The pain was greatly reduced.
- 3 hours later: One more the same treatment. The pain was over

---

4 DISCUSSION

MSDs are very common in the life and in occupational activity. MSDs may be acute, chronic and resulted from different reasons. The result of treating 25 cases as above description had showed amazing effectiveness of "Dien Chan".

It takes very short time for treatment but the results come very quickly, for example Case 3, Case 5.

In general acute cases would be cured quicker than chronic ones. The sooner patients come the quicker they are cured.

As above mentioned "Dien Chan" is a therapy without drug, medicine and don't need complex instrument, so that it is very economic.

The tools were quite simple, and the operation is quite easy. The patient can learn and his/herself became his/her own physician.

According to "Dien Chan" theory, our face is a wonderful medicine garden. Prof. Bui Quoc Chau and his students found out sets of points which has been using as medicine in treatment and preventing of diseases, improving people health. Those sets of points could be applied in occupational health.

RECOMMENDATION

Further study should be done in order to expand application of Dien Chan. We hope that this application should not be in MSDs but in other conditions, such as other work-related diseases, even Occupational diseases.

APPENDIX. REFLEX POINTS

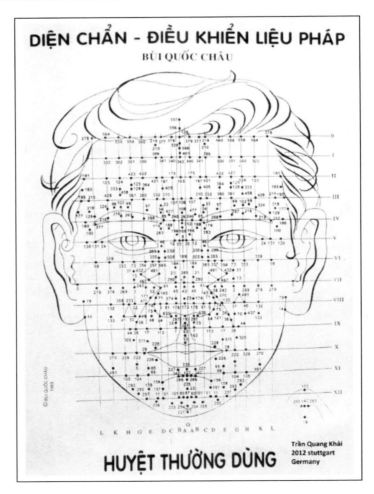

76

REFERENCES

1. Bui Quoc Chau Dien chan dieu khien lieu phap (Face Diagnosis and Cybernetic Therapy, Da Nang Press, 2011.
2. Bui Quoc Chau Chua benh bang do hinh phan chieu va dong ung (Treatment of Disease by reflex diagram) Viet Y dao Centre, 2011.
3. Bui Quoc Chau Học thay khong tay hoc ban, Trung tam DC-DKLP, 1992.
4. Marie-France Muller Facial Reflexology—A self-care manual, Healing Arts Press—Roschester, Vermont.

New Ergonomics Perspective – Yamamoto (Ed.)
© 2015 Taylor & Francis Group, London, ISBN 978-1-138-02751-0

Self-screening test for upper extremity musculoskeletal disorders in computer users

Pichayada Suwandee, Keerin Mekhora, Chutima Jalayondeja &
Petcharatana Bhuanantanondh
Faculty of Physical Therapy, Mahidol University, NakornPrathom, Thailand

ABSTRACT: ***Background***: Work-related Musculoskeletal Disorders (WMSDs) are common among computer users. Prevention is necessary in the process of minimizing the occurrence of these disorders. Apart from ergonomic workstation modification, exercise to maintain a fitness level sufficient to cope with a specific workload is the responsibility of each individual. Before performing an exercise, one must know whether one is capable of that exercise or needs other types of management.

Objective: This study aimed to develop and validate an instrument for screening upper extremity WMSDs in computer users to ensure they use exercise as a prevention method. The agreement will be measured between the results of a self-screening questionnaire and a clinical examination.

Materials and Methods: Several methods were used to develop the self-screening questionnaire. First, available screening algorithms were reviewed, and ideas were shared between groups of experienced physical therapists to plan the scope of the algorithm. The algorithm was designed and pretested. After that, the algorithm was sent to 107 experienced office computer users, aged 23 to 60 years. Additionally, the subjects were all physically examined by experienced physical therapists. The results from the algorithm were compared with the collaborative suggestions of the group of physical therapists.

Results: A descriptive analysis shows that the agreement between the opinions of physical therapists and the self-screening test scores was 89.72%. Analysis of agreement using the Kappa coefficient showed that the scores from the screening tool agree with the clinical examination at a level of 0.768. The total set of scores was in substantial agreement.

Conclusion: These measures indicated that this self-screening algorithm is an instrument exhibiting substantial agreement in terms of scoring between therapists and individual computer users with a questionnaire across time. The algorithm can be used to screen cases who can perform exercise by themselves or need further investigation or treatment.

Keywords: WMSDs; WRULDs; Ergonomics; Computer users; Self-screening test

1 INTRODUCTION

Computers have become common tools for most workplaces. After using computers for a period of time, computer users are reported to have Work-related Musculoskeletal Disorders (WMSDs). More than 90% of computer users have those disorders as a result of their work.[6,15] The involved structures include the muscles, tendons, ligaments, nerves, joints, and bones in the spine and extremities. The disorders could be acute or chronic conditions,[9,14] affecting the physiological, anatomical, and psychological systems.[6]

Many studies have shown the high prevalence of WMSDs in computer users.[2,8,12] For instance, Ranasinghe (2011) reported prevalences of 56.9% of upper arm pain, 42.6% of lower arm pain, 36.7% of neck pain, and 32% of shoulder pain.[16] This group of users needed treatment for their conditions and had to suspend work as a result of the symptoms disturbing their daily lives. The common risk factors of musculoskeletal pain are increasing hours of computer use, sustained awkward postures, improper working habits, stress, and lack of ergonomic information.[3,4,10]

Prevention was advocated to be the best choice for management of WMSDs.[17] However, for those with symptoms, treatment may be necessary. Exercise was also commonly added to the intervention program,[5,13] for the purposes of improving physical performance and the modality of treatment. However, those who have highly severe symptoms may not benefit from exercise without any combination of other treatments.

We believe that early detection of symptoms can help manage WMSDs because the symptoms can be managed by the users.

We also agree that clinical experts in physical therapy are essential to assess upper extremity symptoms but may be inconvenient for office workers. Therefore, it is important to develop a self-assessment test to help users evaluate themselves to check whether the symptoms they are having can be resolved by only exercise and prevention or need further investigation.

2 METHODS

2.1 Research design

A cross-sectional study was performed that aimed to develop and validate an instrument for screening upper extremity WMSDs in computer users. The agreement was used between results of a self-screening questionnaire and a clinical examination, including a subjective and objective examination conducted by experienced physical therapists.

2.2 Participants

The participants consisted of 107 office computer users, including 45 males (42.0%) and 62 females (58.0%), who had experience using a computer. They were selected and contacted via email by their company to ascertain their willingness to participate in this study. The average age of the participants was 35.74 years old (SD = 8.74, range = 23–60).

2.3 Instrumentation

2.3.1 The self-screening test
The self-screening test consisted of two sections: musculoskeletal symptom data and a test for severity screening. Details of the test included answers about the red flag questions, the characteristics of symptoms, the stage of severity, and the signs and symptoms of each area of disorder such as functional active range of motion, self nerve tension test, self muscle length test, and other special tests. The objective of this test was to categorize the participant into two groups: a group of participants who needed complete evaluations or appropriate treatment by a physician or physical therapist and a group of participants who could manage their symptoms by themselves.

2.3.2 Physical examination
The physical examinations were performed in this study by seven experts or experienced physical therapists. The examiners had more than five years of experience in musculoskeletal physical therapy or acquired not only undergraduate training but also postgraduate knowledge and skills (in formal education, participation in certification courses, apprenticeship in specialized training institutions, physical therapy research, or education). All of the physical therapists had a meeting for discussion on the measurement criteria. Everyone received the same physical examination form and was shown the method of examination.

2.4 Data analysis

Data were analyzed using the Statistical Package for Social Sciences (SPSS) for Windows, release 17.0. The level of significance was set at a probability level (p-value) less than 0.05

($p < 0.05$). Kappa analysis was used to analyze the agreement between the results from the physical therapist assessments and the results from the screening test.

3 RESULTS

3.1 *The results from the self-screening test and suggestions from physical therapists*

3.1.1 *Suggestions from physical therapists*

After a physical examination, including recording patient history, observation, palpation, physical functional tests, and special tests, the therapists divided the participants into two groups. The first group consisted of participants who were able to take care of themselves using methods such as exercise or adjusting their workstations to relieve their symptoms. The second group consisted of participants who needed to meet with a physician or physical therapist as soon as possible for further investigation or treatment. The physical therapists suggested that 78 participants (72.89%) were able to take care of themselves, and 29 participants (27.11%) needed treatment from a physician or physical therapist (Table 1).

3.1.2 *The results from the self-screening test*

After the participants finished the questionnaire, the results were used to divide them into a group of participants who were able to take care of themselves and a group who needed to meet with a physician or physical therapist. The results from the self-screening test showed that there were 67 participants (62.62%) in the first group and 40 participants (37.38%) in the second group (Table 1).

The results from the physical examinations by physical therapists and the self-reported questionnaires are shown in Table 1.

For the statistical analysis, the percentages of agreement and disagreement were 89.72% and 10.28%, respectively. The data were also analyzed by using the Kappa measurement of agreement via SPSS version 17.0, yielding a value of 0.768 ($p < 0.01$).

3.2 *Agreement and the signal detection matrix*

The agreement between the physical therapists and the self-screening test can categorized into two major groups and four subgroups. The major groups are the group of agreement and the group of disagreement, and the subgroups are as follows. Group A (62.62%) is defined as the "correct rejection" group (physical therapists and the questionnaire indicated participants were able to take care of themselves). Group B (29.10%) is also in the agreement group and is defined as the "hit" group, but the participants need to see a doctor or physical therapist for proper treatment. Group C (10.08%) is in the disagreement group and is defined as the "false alarm" group (the physical therapists suggested the participants can exercise or adjust their workstations to relieve their symptoms, but the results from the questionnaire indicated the opposite). The last subgroup, Group D, is in the disagreement group and is defined as the "miss" group, i.e., the physical therapists decided the participants needed to see a physician or receive physical therapy treatment as soon as possible, but the results from the questionnaire showed that the participants are not at risk and that they can take care of themselves. The results of each group are shown in Table 2.

Table 1. The results from the physical examinations by physical therapists and the self-screening test.

The results	Suggestions from physical therapists	Results from the test
Take care of themselves	78 (72.89%)	67 (62.62%)
Call a physician or PT	29 (27.11%)	40 (37.38%)

Table 2. The results of agreement and disagreement between the physical therapists and the self-reported questionnaire.

| | | Physical therapist opinion | |
		Call a physician or PT	Take care of themselves
The results from the test	Call a physician or PT count (Percent)	*B: Hit* 29 (27.10%)	*C: False alarm* 11 (10.28%)
	Take care by themselves count (Percent)	*D: Miss* 0 (0%)	*A: Correct rejection* 67 (62.62%)

4 DISCUSSION

This cross-sectional study determined the development of self screening for upper extremity WMSDs in computer users. The results from this screening test were compared with physical examinations by experienced physical therapists

Then, we categorized the participants. The agreement group between the results from the screening test and the suggestions from the physical therapists are shown in two categories. In the first subgroup in this group, the outcomes of the screening test and the examination by a physical therapist agreed that the computer users did not need to seek treatment. In other words, they could manage themselves by exercise and/or adjust their workstations. In another subgroup, there was agreement in the suggestion that participants needed any type of treatment to relieve their symptoms. In contrast, this study also found that there was not agreement on a set of outcomes. The data show that the results from the self-screening test for 11 participants suggested that they receive treatment from a physician or physical therapist, but these results were not in agreement with suggestions from physical therapists.

The results were in accordance with the theory of signal detection.[1,7,11,18] The field of signal detection is devoted to the detection of signals and the control of criteria that are used for this purpose. This theory is often used in various domains, including psychology, medical diagnosis, statistical decision theory, and human–machine interaction. In this study, the musculoskeletal disorder or symptoms related to work are the signal, and the self-screening test uses responses from computer users to detect this signal and discover their health status. This signal can be used as an alarm when computer users work beyond their limits. One objective of early detection is to prevent risk factors from WMSDs.

There are four terms used in signal detection theory to classify outcomes: hit, false alarm, correct rejection, and miss. For this study, a "hit" means the agreement between the self-test and the suggestion from the physical therapist. The test showed that the subject needed proper treatment, which was in line with the suggestion from the physical therapist. A "false alarm" means that the test recommended that a subject call a therapist, but the suggestions from the physical therapist are not in agreement. A "correct rejection" means that the self-test indicates that the subject does not need immediate treatment, and the physical therapist is in agreement. A "miss" means that the subject was diagnosed with a problem that requires professional attention by a physical therapist, but the self-test did not identify this problem.

For the diagnostic tool, it is unacceptable to have many subjects in the "miss" group. To reduce the number of subjects in the "miss" group, the criteria to judge must be strong to separate the signal from the noise. The tool should never miss a patient when there are signs or symptoms present, and there should be a very high hit rate. On the other hand, reducing the number of misses will increase the number of false alarms. However, we find this acceptable because early detection is more important in terms of safety to users. There were only 11 participants with outcomes that were not in agreement. That group consisted entirely of users who were identified by the self-screening test as needing treatment, but the physical therapists suggested that they did not need to seek treatment and were able to manage their problems themselves. It is fortunate that no one in this study who needed the appropriate treatment by a physician or physical therapist was identified by the self-screening test as someone who did not need to seek treatment.

For the Kappa analysis of measurement, this study found that agreement between suggestions from physical therapists and the self-screening outcomes was 0.768. Overall, the scores were in substantial agreement.

5 CONCLUSION

This study found substantial agreement between a self-screening test and a physical clinical examination. This demonstrated that the test is useful for application to computer users because it can screen the severity of symptoms to alert them to prevent the occurrence of WMSDs.

REFERENCES

1. Adbelhamid T, Narang P, Schafer D. Quantifying Workers' Hazard Identification Using Fuzzy Signal Detection Theory. *The Open Occupational Health & Safety Journal,* 2011, *3,* (Suppl 1-M3) 18–30.
2. Battevi, N., Menoni, O., & Vimercati, C. The occurrence of musculoskeletal alterations in worker populations not exposed to repetitive tasks of the upper limbs. Ergonomics; 1998: 41, 1340–1346.
3. Britt L, Karen S, Lars R. Work-related neck-shoulder pain: a review on magnitude, risk factors, biochemical characteristics, clinical picture and prevention interventions. Best practice & Research Clinical Rheumatology 2007, 21(3):447–463.
4. Cagnie B, Danneels L, Tiggelen D, Loose V, Cambier D. Individual and work related risk factors for neck pain among office workers: a cross sectional study. Eur Spine J (2007), 16:679–686.
5. Department of Labor and Employment Division of workers' compensation. Cumulative Trauma Disorder (CTD) Medical Treatment Guidelines. State of Colorado; 2006.
6. European Agency for Safety and Health at Work. Work-related musculoskeletal disorders: Back to work report 2007; ISBN 978-92-9191-160-8.
7. Green D, Signal Detection Theory and Psychophysics. New York Wiley, 1996.
8. Hadler, N.M. Work-related disorders of the upper extremity: I. Cumulative trauma disorders a critical review. Occupational Problems in Medical Practice; 1989: *4,* 1–8.
9. Hales TR, Bernard BP. Epidermiology of work-related musculoskeletal disorders. Orthop Clin North Am 1996; 27(4): 679–709.
10. Hau A, Feuerstein M, Grant D. Job Stress, Upper Extremity Pain and Functional Limitations in Symptomatic Computer Users. AMERICAN JOURNAL OF INDUSTRIAL MEDICINE;2000, 38:507–15.
11. Heeger D. Signal Detection Theory. 1997.
12. Ireland, D.C.R. The Australian experience with cumulative trauma disorders. In L.H. Millender, D.H. Louis, & B.P. Simmons (Eds.), Occupational disorders of the upper extremity; 1992. London: Churchill Livingstone.
13. Marjorie L. Baldwin. Reducing the costs of work-related musculoskeletal disorders: Targeting strategies to chronic disability cases. Journal of Electromyography and Kinesiology 14 (2004): 33–41.
14. Mary F, Barbea b, Barra A. Inflammation and the pathophysiology of work-related musculoskeletal disorders. *Brain Behav Immun.* 2006 September; 20(5): 423–429.
15. Ministry Of Information and Communication Technology National Statistical Office. The 2012 Information and Communication Technology Survey. 2012: 1905.
16. Priyanga R, Yashasv S, Dilusha A, Supun K, Naveen J, Senaka S, Prasad K. Work related complaints of neck, shoulder and arm among computer office workers: a cross-sectional evaluation of prevalence and risk factors in a developing country. Environ Health. 2011; 10: 70.
17. The Australian Safety and Compensation council, Research on the prevention of work-related musculoskeletal disorder stage 1–literature review; 2006.
18. Wickens, T.D. Elementary signal detection theory. Oxford: Oxford University Press, 2002.

New Ergonomics Perspective – Yamamoto (Ed.)

The factors of nurses that influence Accurate Nursing Records' Inputting on Electronic Medical Record System

Ayako Kajimura

Graduate School of Applied Informatics, University of Hyogo, Hyogo, Japan

ABSTRACT: The Ministry of Health, Labour and Welfare Japan in 1999 announced "Recommendation of 3 years preservation for Electric Medical Treatment and Consultation Records" as the matter of 3 principles "authenticable, readable and preservatives.

Although there isn't any stipulation about Nursing Records on present Act on Public Health Nurses, Midwives and Nurses, Accurate Nursing Records is essential for medical safety and evidence for quality on nursing care to guarantee. And Nursing Support System which includes supporting systems of Nursing Care Needs evaluation and recording" to accurate and preserve was developed.

For the previous study of mine, I had some nurses used this system to evaluate "the Nursing Care Needs". And there were some mistakes of the Nursing Care Needs and Nursing Record Leakage.

So in this study, it aims to consider the factors of nurses that influence Accurate Nursing Records' Inputting on Electronic Medical Record System.

The concordance rate of the questionnaire's model answer on this system was 86.89 (±8.65). "Nursing Care Needs evaluation time" was 7 m 14.95 s (±4 m 29.37 s). And "the number of years using computers for nurses" was 7.56 years (±5.09 y). The results of multi-regression analysis, there neither were significant factors "Nursing Care Needs evaluation time ($\beta = -0.371$ p < 0.05)" nor "the number of years using computers for nurses ($\beta = -0.341$ p < 0.05)". It is assumed that the both factors of "Nursing Care Needs evaluation time" and "the number of years using computers for nurses" negatively influenced the concordance rate. That reveled that there are no relation between its evaluation time and Accurate Nursing Records' Inputting. In addition, the number of years using computers for nurses also influenced the concordance rate negatively. And it didn't confirmed that there were relations between the number of years using computers for nurses and Accurate Nursing Records' Inputting.

Accurate Nursing Records' Inputting on this system is for Accurate Inputting of Nursing Care Needs evaluation. The number of years using computers is one of the scales to know about the knowledge of computer use. So it should be possible that nurses input wrong evaluation of the Nursing Care Needs. It is necessary to consider about Influences between Knowledge of computer use and Accurate Nursing Records' Inputting for further study.

Keywords: Accurate Nursing Records; multi-regression analysis; Nursing Care Needs evaluation time; Input; Nursing Support System; Electronic Medical Record System

1 BACKGROUND

In 1999, the Japanese Ministry of Health, Labour and Welfare announced the "Recommendation of 3-year Preservation for Electronic Medical—Treatment and Consultation Records" [1] to demonstrate that securing "the three principles of electronic preservation"—Authenticity, Readability, and Preservability—is the standard for the preservation of medical treatment and consultation records via electronic media.

Later, in 2001, the e-Japan Strategy [2] was proposed, facilitating the introduction of Electronic Medical Records in medical institutions. This caused the electronic conversion of medical treatment and consultation records to develop further. However, the current Act on Public Health Nurses, Midwives and Nurses has no stipulation regarding Nursing Records, nor the responsibility regarding preservation. Nonetheless, accurate Nursing Records are essential as evidence of the provision of medical safety and high-quality nursing care.

Therefore, I developed a nursing support system (hereafter referred to as "the system") installed with functions designed to support the evaluation and recording of Nursing Care Needs (one type of Nursing Record) to accurately input and preserve Nursing Records. The Nursing Care Needs were newly stipulated as an institutional reference for the 7(patients)-to-1(Nurse) basic hospitalization rate as per the Revision of Remuneration for Medical Care in 2008 [3]. Moreover, in the introducing of Nursing Care Needs, it is necessary to staff some nurses who were trained to be evaluator of Nursing Care Needs. So it is essential for nurses to be trained of Nursing Care Needs [4].

For that reason, nurses are required to accurately evaluate and record Nursing Care Needs.

In this study, it aims to clarify the nurse factors that influence the accuracy of Nursing Records within computer-based Electronic Medical Records. These factors will be used as part of future system development.

2 PURPOSE

This study aims to clarify the nurse factors that influence Accurate Nursing Records' Inputting through the evaluation and input of Nursing Care Needs by nurses who are using the system.

3 METHOD

Nurses who provided consent to participate in the study evaluated the Nursing Care Needs of five model patients and input the evaluation results using the system. The results were analyzed according to the procedure below by using SPSS Statistics Version 21.

1. The Nursing Care Needs data that were input into the system were extracted to calculate the concordance rate with the model answer (hereafter referred to as "the concordance rate") in each of the model patients.
2. The concordance rate calculated in "1" is used as an explanatory variable in multi-regression analysis to examine the nurse factors that influence the concordance rate. The objective variables were evaluation time (operation time to evaluate the Nursing Care Needs of the model patient and finish the input; hereinafter referred to as "evaluation time") and years of computer use by the nurses who consented to participate in the study. A significance level of 5% was adopted.

4 RESULTS

4.1 *Study participants*

Nine nurses provided consent to participate in the study. They had no experience of evaluating Nursing Care Needs. The nurses' years of computer use were 7.56 years (± 5.09 years).

4.2 *The concordance rate and evaluation time of Nursing Care Needs*

The concordance rate of Nursing Care Needs that were input into the system was 86.89 (± 8.65). The time required for finishing input post-evaluation of the Nursing Care Needs

Table 1. Results of the multi-regression analysis of the concordance rate and nurse factors.

	Standardizing coefficient, β	t-value	p-value
Evaluation time	−0.371	−2.592	0.013
Years of computer use	−0.341	−2.378	0.022
R^2	0.153		

of the model patients (hereafter referred to as "evaluation time") was 7 min and 14.95 sec (±4 min and 29.37 sec). The multi-regression analysis (Table 1) revealed that both the evaluation time of the Nursing Care Needs and years of computer use by nurses negatively influenced the concordance rate.

5 DISCUSSION

According to the results, there were possibly two causes of mistake in this system. One of them is missing input itself of the evaluation of Nursing Care Needs. And the other is wrong input of the evaluation of Nursing Care Needs. I consider the mistake in this system was caused by a human error. Temporal factors—background factors of human errors—include operation time [5]. Therefore, the influence of "evaluation time"—estimated as operation time in the temporal factors—on the concordance rate was investigated. Evaluation time ($\beta = -0.371$, $p < 0.05$) was revealed to negatively influence the concordance rate. No margin in operation time causes frequent paraphasia and/or forgetfulness because of one's attempts to reach the final goal early by skipping the required checks due to impatience and hurrying (Sato et al., 2007) [5]. However, nurses who took their time could not accurately input the Nursing Records. An internal factor relating to the background factor of human error might have arisen because this experiment used evaluation and input of Nursing Care Needs for model patients; the nurses may have unconsciously had lower motivation to input the Nursing Records as securely and accurately as they would in their daily recording operation.

In addition, it was understood that years of computer use as an index showing habituation of nurses to computer operation. The years of computer use ($\beta = -0.341$, $p < 0.05$) were also revealed to negatively influence the concordance rate. History of computer use is reported as significantly involved in the degree of burden on nurses after the introduction of Electronic Medical Records [6]. It was considered that an increase in feelings of work-related burden would lead to an increase in the internal factor (the background factor of human error) resulting in a higher likelihood of causing errors. However, in the present study, years of computer use showed habituation to computer operation negatively influencing the concordance rate. This is likely due to a "mistake" in the classification of human errors caused by "habituation" rather than the influence of stress from the work-related feelings of burden in nurses. Erroneous operation and so forth caused by habituation to computer operation were likely to have influenced the concordance rate. Because the slow input of Nursing Records and the presence or absence of habituation to computer operation cannot be considered factors that necessarily support accurate input of Nursing Records, the main causes of error that nurses make must be further investigated.

6 CONCLUSION

From the previous study of mine, I had some nurses used this system to evaluate "the Nursing Care Needs". And there were some mistakes of the Nursing Care Needs and Nursing Record Leakage.

So this study aims to clarify the nurse factors that influence Accurate Nursing Records' Inputting through the evaluation and input of Nursing Care Needs by nurses who are using the system.

The concordance rate of the questionnaire's model answer on this system was 86.89 (±8.65). "Nursing Care Needs evaluation time" was 7 min and 14.95 sec (±4 min and 29.37 sec). And "the number of years using computers for nurses" was 7.56 years (±5.09 years). The results of multi-regression analysis, there neither were significant factors "Nursing Care Needs evaluation time ($\beta = -0.371$ p < 0.05)" nor "the number of years using computers for nurses ($\beta = -0.341$ p < 0.05)". It is assumed that the both factors of "Nursing Care Needs evaluation time" and "the number of years using computers for nurses" negatively influenced the concordance rate. Evaluation time and years of computer use negatively influenced the concordance rate, revealing that a longer time for inputting Nursing Records and habituation to computer operation are not factors that support accurate input of Nursing Records. Because years of computer use are a scale to determine habituation to computer operation, it is also likely that nurses input correctly but evaluated the Nursing Care Needs incorrectly. The influence of habituation to computers on accurate record input in this system needs to be investigated further.

REFERENCES

[1] Recommendation of 3-year Preservation for Electronic Medical Treatment and Consultation Records http://www1.mhlw.go.jp/houdou/1104/h0423-1_10.html.
[2] e-Japan Strategy. http://www.kantei.go.jp/jp/singi/it2/dai1/1siryou05_2.html.
[3] Notice Pertaining to Revision of Remuneration for Medical Care in FY 2008. http://www.mhlw.go.jp/topics/2008/03/tp0305-1.html.
[4] Iwasawa K, Tsutsui T. Nursing Care Needs second edition—New evaluation standards of Nursing Services Japanese Nursing Association Publishing Company; 2007.
[5] Sato S, Sato K. Medical human engineering utilized for medical safety. Iryo Kagaku-Sya; 2007.
[6] Negishi A, Yoshioka H, et al. Consideration on the burden after introduction of Electronic Medical Records—From the comparison of consciousness survey results between outpatient nurses and ward nurses—Proceedings of 44th Japanese Nursing Association Nursing Academy—Nursing Management—Annual Meeting 2013; 221.

2 *Workplace ergonomics*

New Ergonomics Perspective – Yamamoto (Ed.)
© *2015 Taylor & Francis Group, London, ISBN 978-1-138-02751-0*

Required forces in manually inserting a small object

Kwan Suk Lee & Kwan Mo Gu
Department of Industrial Engineering, Hongik University, Seoul, Korea

ABSTRACT: ***Objective***: It is important for assembly workers to be able to exert appropriate forces for assembling parts together. The objective of this study is to find acceptable forces and maximum forces in manually inserting small objects in the main body of a part.

Background: In assembling parts, there are many tasks of inserting a small subpart in a hole or to the main body of a part. But there has not been any study of the force when the small object needs to be gripped and assembled together.

Method: A laboratory experiment was conducted to measure the acceptable and maximum forces which can be exerted by subjects using their fingers. 30 young college students participated and 18 different postures and four different sizes of objects were used for the experiment.

Results: The acceptable forces were significantly low with a small object and in some postures. As the size of the object gets bigger, the acceptable forces and maximum forces get bigger.

Conclusion: It needs to be noted that the size of small objects should be considered in work design for the assembly tasks.

Application: For the future design of work for the assembly including insertion tasks, forces found in this study should be considered to make workers free from overexertion which can be led to potential musculoskeletal disorders. The appropriate forces to workers could minimize the assembly error and malfunction of a system.

Keywords: Assembly; Insertion; Acceptable force

1   INTRODUCTION

There have been many tasks of inserting a part in a hole or to another part in industry. Many workers have complained of fatigue due to the excessive force required to perform such tasks. The mismatch of forces required and forces which can be exerted by workers and the overexertion by workers were main complaining issues for the assembly workers. Connector mating is one of inserting tasks of which workers complained quite often. Lee and Ha (2013) reported the problem by automobile assembly workers although there has been a guideline which suggested the maximum insertion force required for the connector mating to be less than 7.5 Kgf. since many connectors exceed this limit. Even Mazda Motor company was trying to change their connector design workers in the Ford Motor company complained of high strength required in mating connectors manufactured by Mazda Motor company. Sander and McCormick (1993) focused in finding the optimum sizes which can allow the maximum strength to be exerted. Other study reported the finger strength (Kim et al., 2009; Choi et al., 2006). However, there has not been any study which shows the acceptable and maximum force depending upon the size and shape of the object to be held by fingers. Thus, the objective of this study is to find acceptable forces and maximum forces in insertion tasks.

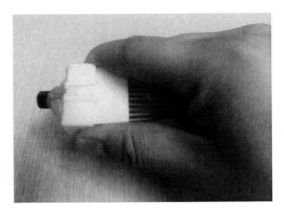

Figure 1.   The holding method of an object.

Table 1.   Size and shape of objects held by subjects: **unit (mm)**.

	Width	Length	Height
Round bar type	Ø12	25	–
Small square bar type	28	38	14
Medium square bar type	39	34	16
Big square bar type	40	27	27

2   METHOD

An experiment was conducted in the laboratory where subjects inserted objects whose sizes are shown in Table 1. Thirty young healthy college students (age 25 ± 1.65 years) who have not had an history of musculoskeletal disorder participated in this experiment. Their height was 177 ± 3.77 cm and weight was 75 ± 7.78 kg. Connectors were used as square bar type objects to be held and a round tip for the force gauge was used as a round bar type object.

R A subject held an object with his/her finger and pushed against the force gauge which was held by the other hand. The force gauge used in this experiment was the model FGP-50 made by SHIMPO Co., and maximum force limit for the measurement was 50 Kgf.

18 different body postures were used in this experiment and all measurements were repeated. Three trunk angles (0, 45, 90 degree), three upper arm angles (45, 90, 135 degrees) and two lower arm angles (180, 30 degree) were used for different postures.

A training session was given for subjects for understanding the concept of the acceptable force and the maximum force and to teach how to hold the object as shown in Figure 1. Postures were randomized to avoid the potential fatigue effect although three minutes rest period was given between each trial.

3   RESULTS

The measured data were analyzed using Microsoft Excel 2010. Table 2 and 3 show means and standard deviations of acceptable forces and maximum forces at each posture. As expected, as the size of the object gets bigger, acceptable and maximum forces get bigger. The round type bar shows significantly lower acceptable forces and maximum forces than square type bar. This may not be due to the shape of the bar since the diameter of the round bar was significantly smaller than sizes of square bar type objects. It was also found that means of acceptable forces and the maximum forces for the square type bars were higher than 7.5 Kgf limit suggested by the connector design guideline. However, in many postures, only 1/2 of

Table 2. Acceptable forces (unit: Kgf.).

Trunk angle	Upper arm	Lower arm	Means ± Standard Deviation			
			Round	Small square	Medium square	Big square
0	45	180	3.66 ± 1.30	5.3 ± 1.44	5.52 ± 1.65	5.74 ± 1.6
		30	4.96 ± 1.94	9.11 ± 2.38	9.65 ± 2.24	10.82 ± 3.83
	90	180	4.11 ± 1.73	6.07 ± 1.77	6.15 ± 1.48	6.82 ± 1.77
		30	5.58 ± 2.81	10.33 ± 2.84	11.03 ± 3.13	11.88 ± 4.26
	135	180	4.08 ± 1.77	5.84 ± 1.5	5.86 ± 1.46	6.03 ± 1.61
		30	5.73 ± 3.62	10.19 ± 2.6	10.26 ± 3.13	11.7 ± 3.7
45	45	180	4.17 ± 2.03	6.06 ± 1.83	5.77 ± 1.35	6.35 ± 2.22
		30	5.42 ± 2.87	9.89 ± 2.98	9.69 ± 2.35	10.66 ± 3.5
	90	180	4.22 ± 2.14	5.83 ± 1.81	6.02 ± 1.48	6.2 ± 1.84
		30	5.51 ± 2.89	10.64 ± 3	10.87 ± 3.33	12.4 ± 4.65
	135	180	4.47 ± 2.38	6.13 ± 1.67	5.97 ± 1.48	6.67 ± 1.88
		30	6.06 ± 4.12	10.87 ± 3.29	11.36 ± 3.38	12.65 ± 5.22
90	45	180	4.20 ± 1.85	5.64 ± 1.85	6.1 ± 1.75	5.97 ± 1.67
		30	6.14 ± 3.58	9.95 ± 3.22	10.13 ± 2.89	10.31 ± 2.8
	90	180	4.54 ± 2.45	5.99 ± 1.79	5.89 ± 1.47	6.26 ± 1.39
		30	6.12 ± 3.94	10.8 ± 3.43	10.95 ± 3.71	12.56 ± 4.85
	135	180	4.56 ± 2.77	5.78 ± 1.51	6.58 ± 2.1	6.45 ± 1.59
		30	5.81 ± 3.85	11.92 ± 3.86	12.02 ± 3.96	12.98 ± 4.64
Total mean			4.96 ± 2.67	8.13 ± 2.38	8.32 ± 2.35	9.03 ± 2.95

Table 3. Maximum forces (unit: Kgf.).

Trunk angle	Upper arm	Lower arm	Means ± Standard Deviation			
			Round	Small square	Medium square	Big square
0	45	180	7.49 ± 1.25	9.74 ± 2.59	9.43 ± 2.47	9.97 ± 2.33
		30	9.87 ± 1.75	16.97 ± 4.29	17.08 ± 3.20	19.37 ± 5.04
	90	180	7.17 ± 1.27	9.58 ± 2.37	9.95 ± 2.49	10.72 ± 2.52
		30	10.22 ± 1.87	18.30 ± 4.06	19.36 ± 4.27	20.68 ± 4.73
	135	180	7.06 ± 1.24	9.41 ± 2.18	9.51 ± 2.40	9.99 ± 3.17
		30	9.26 ± 1.90	18.34 ± 4.38	18.18 ± 4.05	21.48 ± 5.28
45	45	180	7.32 ± 1.98	9.77 ± 2.40	9.25 ± 2.12	9.98 ± 2.92
		30	9.38 ± 1.73	17.66 ± 4.23	17.15 ± 3.81	19.32 ± 5.08
	90	180	6.54 ± 1.76	10.09 ± 2.80	10.01 ± 2.58	10.28 ± 3.11
		30	9.24 ± 2.47	19.09 ± 5.60	19.17 ± 4.85	20.88 ± 5.65
	135	180	6.99 ± 2.30	10.35 ± 2.71	9.90 ± 2.23	10.59 ± 2.83
		30	8.81 ± 2.60	19.63 ± 6.46	20.16 ± 5.84	21.80 ± 6.03
90	45	180	6.86 ± 1.92	9.64 ± 2.62	9.55 ± 2.47	9.67 ± 2.62
		30	9.52 ± 1.91	18.01 ± 4.77	17.95 ± 4.39	19.20 ± 5.32
	90	180	6.71 ± 1.41	9.88 ± 2.53	9.79 ± 2.40	10.55 ± 2.90
		30	9.50 ± 1.97	19.35 ± 6.86	20.05 ± 6.23	21.96 ± 6.10
	135	180	6.77 ± 2.07	9.89 ± 2.83	10.68 ± 3.39	10.73 ± 3.04
		30	10.03 ± 3.20	20.92 ± 6.03	21.83 ± 6.32	23.50 ± 5.27
Total mean			8.26 ± 1.92	14.26 ± 3.87	14.39 ± 3.64	15.59 ± 4.11

subjects' acceptable forces were higher than 7.5 Kgf limit. It also needs to be noted that there are many postures in which means of subjects' acceptable forces were lower than 7.5 Kgf limit. It was found that in a posture in which the trunk angle 90 degree, the upper arm angle 135 degree and the lower arm angle 30 degree, subjects could exert the highest acceptable force and the maximum force followed by the posture of the trunk angle 45 degree, the upper arm angle 135 degree, and the lower arm angle 30 degree. A posture with the trunk angle 0 degree, the upper arm angle 45 degree and the lower arm angle 180 degree, subjects could exert the smallest acceptable force and the maximum force followed by the posture of the trunk angle 90 degree, the upper arm angle 45 degree, and the lower arm angle 180 degree. Further, it was found that the acceptable force was about 60% of the maximum force in the same posture for all object sizes.

4   CONCLUSION

It is concluded that the size of an object and a posture in an insertion task affect a person's acceptable force and maximum force. In some postures, acceptable forces were significantly lower than the design guidelines for connectors. Thus, if the object to be held is small, the force required to perform any insertion task should be considered to be very low.

CONTACT

More information is available at: kslee@hongik.ac.kr

REFERENCES

[1] Lee K.S. and Ha, C.H., The evaluation of the connectors to prevent the half-lock. Report, Hyundai Motor Co., 2013.
[2] Sanders, M. and McCormick, Human Factors in Engineering and Design, Wiley, 1993.
[3] Kim D.M., Han J.G., Jung M.C., Lee K.S., and Son S.T.: A Study on the relationship between Individual finger forces and Subjective ratings by using an "Multi Finger Force Measurement (MFFM) System", Journal of the Ergonomics Society of Korea, 470–473, 2009.
[4] Choi C.M., Shin M.H., Kwon S.C., and Kim J.: EMG-based Real-time Finger Force Estimation for Human-Machine Interaction, Journal of the Korean Society for Precision Engineering, 26(8):132–141, 2006.

New Ergonomics Perspective – Yamamoto (Ed.)
© 2015 Taylor & Francis Group, London, ISBN 978-1-138-02751-0

Weight bearing on foot under sitting postures in foot scanning

Hsin-Hung Tu
Department of Industrial Engineering and Management, Hsiuping University of Science and Technology, Dali District, Taichung City, Taiwan, R.O.C.

Chi-Yuang Yu
Department of Industrial Engineering and Engineering Management, National Tsing Hua University, Hsinchu, Taiwan, R.O.C.

ABSTRACT: The purpose of this study is to measuring the weight bearing on foot under different sitting postures in foot scanning. Weight bearing on feet of 34 subjects (17 males and 17 females) were measured under sitting postures in foot scanning. The experiment was a three factor design with three replications, in which three factors are gender (male, female), foot side (left foot, right foot), and posture of upper trunk (erective, forward). A hydraulically adjustable chair were used to adjust the subject's sitting posture as 90 degrees in knee joint and in ankle joint. Half of subject's thigh is sitting on the chair to simulate the sitting posture usually adopted in foot scanning. Two balance boards of Wii fit were used to measure the weight bearing on foot of subjects in sitting posture. The result of weight bearing on foot is normalized by subject's Body Weight as a ratio (%BW) and analyzed further. The results of ANOVA showed that weight bearing on foot were significantly different in gender factor and trunk posture factor, while foot factor and interactions of factors were not. In total population of subjects with erective sitting posture (erective posture of upper trunk), the weight bearing of left foot is 9.2% BW, and right foot 8.9% BW. The results of this study provided important reference for foot researches.

Keywords: foot scan; weight bearing; posture

1   INTRODUCTION

Weight bearing on foot is an important reference in foot measurement researches. Many studies had reported that Weight Bearing of Foot (WBF) would affect the foot dimensions, therefore conditions of weight bearing on foot were usually given or setup in studies (Xiong et al., 2009; Houston et al., 2006).

Studies related to foot measurement usually had the subject allocate his/her weight bearing on the foot measured in standing postures by monitoring the weight bearing on the other foot with a weight scale. Cornwall and McPoil (2011) had subjects put WBF as 50% of Whole Body Weight (WBW) by standing equally on both feet. Cobb et al., (2011) had subjects put WBF as 10% WBW and 90% WBW respectively. Mall et al., (2007) had subjects allocate WBF as 90% WBW. William and McClay (2000) had subjects allocate WBF as 10% WBW and 90% WBW respectively. The same way could not be applied when subjects was asked to be measured in sitting posture in study, therefore researchers usually assumed or gave the WBF of subjects directly without scaling. Richards et al., (2003) had subjects allocate their WBF as 50% WBW while WBF in sitting posture seemed to be assumed as 10% WBW without report. Zifchock et al., (2006) had subjects put WBF as 50% WBW and assumed subjects' WBF in sitting posture as 10% WBW in estimating equation. Pohl and Farr (2010) had subjects allocate their WBF in conditions of standing equally as 50% WBW and 90% WBW respectively, while WBF of subject in sitting posture was taken as 10% WBW without weight scale.

Therefore, the purpose of this study is to measuring the weight bearing on foot under different sitting postures in foot scanning.

2 METHOD

2.1 *Subjects*

Thirty four subjects were recruited in this study, including seventeen males and seventeen females. Basic anthropometric data of subjects are shown in Table 1. Mean of their ages is 22.47 years with Standard Deviation (S.D.) as 5.19 years. Mean of their height is 165.94 cm with S.D. as 7.69 cm, and mean of their body weight is 69.47 kg with S.D. as 16.20 kg.

2.2 *Apparatus*

A hydraulically adjustable chair and two Wii fit balance board were used in this study. Two Wii fit balance board were set in parallel and the distance between them is about 32 cm. Sampling rate of Wii fit balance board is about 120 times per second. Measurement data is calculated by an average during certain time interval. The measurement accuracy in different measuring intervals of Wii fit is as follows: in measuring interval between 0~68 kg, the measurement accuracy is 0.8 kg; in measuring interval between 68~100 kg, the measurement accuracy is 1.2 kg; in measuring interval between 100~150 kg, the measurement accuracy is 2 kg.

2.3 *Experimental design and procedures*

A factorial experiment with three factors and three repetitions was conducted in this study. Three factors included gender as male and female, foot as right and left, trunk posture as erective and bending.

Each subject was first asked to stand as each foot on one Wii fit balance board respectively. Subject maintained his/her standing posture with equal body weight on each foot for 20 seconds. Weight bearing on each foot was calculated as the average of the measurement data in 10 seconds, which were in the middle of the measuring time interval. The summation of the weigh bearing of both feet was used as subject's whole body weight.

After measuring whole body weight of subject, subject was asked to sit on the hydraulically adjustable chair to take the sitting posture in which half of the thigh was on the chair, and the chair height was adjusted to maintain subject's knee and ankle joint both 90 degrees. In the first measuring cycle, subject was asked to maintain his/her trunk erect for 20 seconds, and then to bend his/her trunk forward to his/her maximum extent for 20 seconds, as shown in Figure 1. Measurement data of 10 seconds in the middle of the measuring time interval, 20 seconds, were used in calculation. The 10-second-time-averaged weight measured by each balance board was used as weight bearing on each foot respectively. The measuring cycle was then repeated for another two times, therefore each subject was measured for three repetitions.

2.4 *Data analysis*

Measured weigh bearing on each foot would be normalized by subject's own whole body weight as Body Weight Percentage (BW%). ANOVA analysis would then be conducted

Table 1. Basic anthropometric data of subjects.

	Number	Age (yr)	Stature (cm)	Weight (kg)
Male	17	22.41 (3.71)	172.00 (4.36)	76.48 (14.18)
Female	17	22.53 (6.47)	159.88 (4.99)	61.97 (14.89)
Total	34	22.47 (5.19)	165.94 (7.69)	69.47 (16.20)

in terms of BW% to see whether there were significant difference among the factors or not.

3 RESULTS

3.1 *Weight bearing on foot*

Results of original weight bearing on each foot of subjects were shown in Table 2, and the normalized Weight Bearing (BW%) were also shown in Table 3. In total population, it could be seen in Table 3 that BW% on left foot in erective trunk posture was 6.33% as average with Standard Deviation (S.D.) as 1.65%, while on right foot was 6.09% as average with S.D. as 1.63%. In bending trunk posture, BW% on left foot was 14.76% as average with S.D. as 3.71%, while on right foot 15.30 as average with S.D. 3.89.

3.2 *ANOVA analysis*

Result of ANOVA analysis was shown in Table 4. No significant difference of weight bearing on foot could been observed in foot factor, but significant difference of weight bearing on foot could be seen in gender factor and trunk posture factor. There were no significant difference among all interactions of three factors.

Figure 1.   Subject bended her trunk forward.

Table 2.   Weight bearing on each foot of subjects in male, female and total population.

Population	Trunk posture	Left foot	Right foot
Male	Erective	7.04 (1.33)	6.77 (1.4)
	Bending	16.32 (3.59)	17.08 (3.81)
Female	Erective	5.61 (1.64)	5.41 (1.57)
	Bending	13.20 (3.15)	13.53 (3.11)
Total	Erective	6.33 (1.65)	6.09 (1.63)
	Bending	14.76 (3.71)	15.30 (3.89)

Table 3. Weight bearings as BW% on each foot of subjects in male, female and total population.

Population	Trunk posture	Left foot	Right foot
Male	Erective	9.4 (2.0)	9.0 (1.7)
	Bending	21.7 (5.1)	22.6 (5.1)
Female	Erective	9.1 (1.7)	8.8 (1.5)
	Bending	21.6 (4.1)	22.1 (3.6)
Total	Erective	9.2 (1.8)	8.9 (1.6)
	Bending	21.6 (4.6)	22.4 (4.4)

Table 4. ANOVA analysis by SPSS.

Source of variance	SS	df	MS	F	Sig. level
Gender	0.021	1	0.021	5.901	0.017*
Foot	0.000	1	0.000	0.016	0.899
TrunkPosture	1.971	1	1.971	556.913	0.000*
Gender × Foot	0.010	1	0.000	0.001	0.971
Gender × TrunkPosture	0.000	1	0.010	2.775	0.098
Foot × TrunkPosture	0.002	1	0.002	0.524	0.470
Gender × Foot × TrunkPosture	0.000	1	0.000	0.000	0.998
Residual	0.453	128	0.004	☐	☐

*Statistically significant difference.

4 CONCLUSION AND DISCUSSION

The purpose of this study is to measuring the weight bearing on foot under different sitting postures in foot scanning. 34 subjects (17 males and 17 females) were measured under sitting postures in foot scanning. A three-factor factorial experiment with three replications were conducted, in which three factors are gender (male, female), foot side (left foot, right foot), and posture of upper trunk (erective, forward).Weight bearing on foot was normalized by subject's Body Weight as a ratio (%BW) and analyzed further. The results of ANOVA showed that weight bearing on foot were significantly different in gender factor and trunk posture factor, while foot factor and interactions of factors were not. In total population, BW% of subjects from erective trunk posture to bending trunk posture varied from 8.9% to 21.7% in right foot and from 9.2% to 22.6% in left foot. The results of this study provided important reference for researches of foot measurements.

Weight bearings on foot of subject were 8.9% in right foot and 9.2% in left foot in erective trunk posture in total population. These data were lower than 12.9%~20.9%, the range of leg weight in proportion of whole body weight collected and shown by Chaffin et al., (2006), and were thought to be reasonable for half of subject's thigh sat on the chair under measuring in erective trunk posture.

There was significant difference between BW% in erective trunk posture and BW% in bending trunk posture. The possible reason is the angle of pelvic differs from erective trunk posture to bending trunk posture, which resulted in the rotation of iliac tuberosity and the center of gravity of the whole body in sitting posture transferring forward to foot. The further study could focus on the relationship between bending angle of trunk posture and the weight bearing on foot.

ACKNOWLEDGEMENTS

This study was partially supported by NSC project in Taiwan (Project No: NSC 102-2221-E-007-107-MY2).

REFERENCES

[1] Xiong S, Goonetilleke RS, Witana CP, Weerasinghe TW, Au EY. Foot arch characterization: a review, a new metric, and a comparison. J Am Podiatr Med Assoc 2010;100(1):14–24.

[2] Houston VL, Luo G, Mason CP, Mussman M, Garbarini M, Beattie AC. Changes in Male Foot Shape and Size with Weightbearing. J Am Podiatr Med Assoc 2006;96(4): 330–343.

[3] Cornwall MW, McPoil TG. Relationship between static foot posture and foot mobility. J Foot Ankle Res 2011;4(4).

[4] Cobb SC, James CR, Hjertstedt M, Kruk J. A Digital Photographic Measurement Method for Quantifying foot posture: validity, reliability, and descriptive data. J Athl Train 2011;46(1):20–30.

[5] Mall NA, Hardaker WM, Nunley JA, Queen RM. The reliability and reproducibility of foot type measurements using a mirrored foot photo box and digital photography compared to caliper measurements. J Biomech 2007;40:1171–1176.

[6] Williams DS, McClay IS. Measurements used to characterize the foot and the medial longitudinal arch: reliability and validity. Phys Ther 2000;80(9):864–871.

[7] Richards CJ, Card K, Song J, et al. A Novel Arch Height Index Measurement System (AHIMS): Intra- and Interrater Reliability. In Book of Abstracts, Toledo, Ohio; The Annual Meeting of the American Society of Biomechanics; September 2003.

[8] Zifchock RA, Davis I, Hillstrom H, Song J. The Effect of Gender, age, and lateral dominance on Arch Height and Arch stiffness. Foot Ankle Int 2006;27(5):367–372.

[9] Pohl MB, Farr L. A comparison of foot arch measurement reliability using both digital photography and caliper methods. J Foot Ankle Res 2010;3:14.

New Ergonomics Perspective – Yamamoto (Ed.)
© 2015 Taylor & Francis Group, London, ISBN 978-1-138-02751-0

Development of a Production Management Self-diagnosis System for small and medium-sized enterprises and case study using this system

Kenichi Iida
Hokkaido Industrial Institute, Sapporo, Japan

Koki Mikami
Hokkaido University of Science, Sapporo, Japan

Masahiro Shibuya
Tokyo Metropolitan University, Tokyo, Japan

Toshifumi Sakai
Hokkaido University of Science, Sapporo, Japan

ABSTRACT: In Hokkaido, aging is remarkable in Japan. Particularly, a manufacturing company of Hokkaido requires productivity increase by improvement of the production technique in QCD (Quality, Cost, & Delivery) in order to survive in an aging society. Therefore, the author, et al., developed a Production Management Self-diagnosis System which makes it possible for a company to understand its strong and weak points and work on KAIZEN activities by itself. This system is composed of the two of the Checklist for understanding of the company's strong and weak points and the Manual for explanation of the evaluation items and criteria. It is also usable for the leading staff training to foster the ability of the leading staff (persons in charge of the evaluation) who can conduct the evaluation and KAIZEN in the company. We would like to promote the KAIZEN of the manufacturing industry by use of this system and strengthen the competitive power.

Keywords: Production Management; QCD; Checklist; Self-diagnosis; KAIZEN

1 INTRODUCTION

For manufacturing enterprises in Hokkaido to develop market roots to the other main islands of Japan or foreign countries, it is necessary to step up their tackling for strengthening QCD abilities. Therefore, in this study the author, et al., developed a production management self-diagnosis system which made it possible for small and medium-sized manufacturing enterprises to understand their strong and weak points synthetically by themselves and promote their independent improvement activities. We would like to, by establishing an effective system for small and medium-sized manufacturing enterprises, advance field improvement activities and strengthen their competitive power.

2 RESEARCH BACKGROUND

2.1 *The existing conditions and problems*

Automobile-related enterprises such as TOYOTA Motor Hokkaido Inc., and Denso Electronics Inc., have opened new factories in Hokkaido, and their accumulation has been accelerated. Automobiles are composed of 20,000 to 30,000 parts such as molding, metal-press parts and rosin. This industry is a wide-bottom one where many kinds of makers take a multistory-type labor-division system.

Further, their influences on the local communities, such as employment, creation of related enterprises' demand and the ripple effects of production or production management techniques, are great. However, the local provision rate of the new comers is approximately 12%, which is much lower than the rate 50% of Kyushu, the southern main island of Japan and the rate 30% of the Tohoku district on the north of the Tokyo district. The main cause is shortage of QCD-related production management techniques, which demands quick measures.

2.2 *Our previous tackling*

To improve the production management techniques of the local enterprises, in cooperation with the government, we have held KAIZEN seminars which focus on "field practice" concerning the TOYOTA production method. These seminars have been practiced with the help of TOYOTA Motors Hokkaido Inc., and attended by a total of 44 enterprises in 6 years from 2006. The contents of the seminars consist of understanding of the two chief ideas of the TOYOTA production method, that is, "Just-in-time" and "Automation", and practicing of improvement activities by all the participants at a manufacturing field chosen from the participating enterprises. At first, the seminars were conducted in the central area of Hokkaido such as Sapporo City and Tomakomai City and produced many good results such as higher productivity and reduction of goods in process. From 2009 we have held such seminars in other areas in Hokkaido (Asahikawa, Obihiro and Hakodate).

64% of the enterprises which participated in the seminars are tackling with KAIZEN activities by use of the seminar results, but the other enterprises (more than 30%) are not. Some of them say, "we do not know how to put it in practice."

3 METHOD

The procedure of this study is as follows:

4.1 Concept planning of a production management self-diagnosis system
4.2 Development of a check list composed of two kinds of lists
4.3 Making of a manual for each item
4.4 Making of the curriculum for core staff training
4.5 Practice of the above self-diagnosis system

4 RESULTS AND DISCUSSION

4.1 *Concept planning of the production management self-diagnosis system*

4.1.1 *The characteristics of the system*
To promote improvement activities in the manufacturing field, examination from the viewpoints of the two of "field management" and "production management "are required. The former is composed of "equipment management" for increasing the working ratio of equipment and "work management" for increasing workers' performance. The latter is composed of "production planning" and "stock management" to realize the former. Therefore, we developed this system so that it would make it possible for small and medium-sized manufacturing enterprises to diagnose totally the levels of their "field management" and "production management" and conduct improvement.

Concretely, the diagnosis is conducted by use of the production management self-diagnosis check list. And then, for items with low evaluation values examination of their factors and measures is conducted. In this way it is possible to find good improvement measures and advance their effective improvement.

The characteristics:

- The evaluation results are shown on the radar chart.
- The radar chart makes it possible to examine their strong and weak points classified by divisions.
- It is easy to find their next improvement aim, because the one-up evaluation criterion becomes their step-up aim.

4.1.2 *The objective production forms*

Taking into consideration the characteristics of many labor intensive-type manufacturing enterprises with the production of many models in small quantities, evaluation items and criteria were set for the production form (thick letters in Table 1) of "Made-to-order production".

4.1.3 *The structure and operation of the production management self-diagnosis system*

This system is made up of the two of a "check list" for grasping the enterprise's strong and weak points and a "manual for each item" for explaining the evaluation items and criteria of the check list. Core staff training was also practiced so as to foster core human resources (responsible for evaluation) who would practice evaluation and improvement inside their enterprises for the operation of the system. As shown in the following and Figure 1, the operation is conducted in line with the PDCA cycle.

1) Conduct diagnosis/evaluation by use of the check list analyze, their strong or weak points, and clarify their problems.(A)
2) Make an improvement plan and lay out an improvement schedule.(P)
3) Conduct improvement, and then conduct diagnosis/evaluation by use of the check list again.(Do)
4) Repeat this cycle regularly.(C)

4.2 *Development of the check list*

The evaluation items of the check list were decided in light of the opinions of outside knowledgeable people and executives of the local enterprises. As shown in Figure 2, we turned our attention to the two divisions of "the field" and "the management" supporting the field, and decided the evaluation items. The evaluation items for "the management division" are ① Safety/Hygiene, ② Quality, ③ Cost, ④ Process management, ⑤ Subcontract, ⑥ Materials, ⑦ Equipment integrity and ⑧ Staff training. The ones for "the field division" are ① 2S's, ② More multi-skilled workers, ③ Production technique, ④ Standardization, ⑤ KAIZEN, ⑥ Arrangement, and ⑦ Layout. They are 15 items in total. And also each item has 2–3 specified items (check items). The management division has 23 items, and the field division has 16. There are 39 specified items in total.

The evaluation criteria are shown in Figure 3. The tackling conditions of each evaluation item is evaluated by giving "Good, 5 points", "Ordinary, 3 points" or "Bad, 1 point", and then each item's average is calculated.

Table 1. The production form.

Item	Type
Production time	Production-to-stock, Production-to-order
Production kind, Production quantity	High-variety low-volume, Middle-variety middle-volume, Low-variety High-volume
Production directions	Push system, Pull system
Flow of processed goods	Flow type, Jobshop type
Production type	Engineer-to-order, Lot (Batch), Continuous

Figure 1.   The image of the system.

Figure 2.   The evaluation items.

Evaluation Items	Evaluation Criteria	Grade	Score	Ave.
A. Safety / Hygiene				
1) Management System	· Manager and foreman as patrol for safety health in work site, regularly permonth.	5		
	· The safety educations are carried out and the machine and work manual are maintained.	3	1	
	· There are not the safety manual and safety management are left by workers	1		
2) Safety Activity	· The activity of the morning safety assembly and the meeting of safety, etc are carried out	5		
	· The notice for safety signage and safety health are carried out	3	1	10
	· There are not activity for safety	1		
3) Work Environment	· The work environment improvement for dust, gas, noise, vibration, high temperature, etc are carried out	5		
	· The work environment measurement for dust, gas, noise, vibration, high temperature, only are carried out	3	5	
	· There are not activity for work environment	1		

Figure 3.   The evaluation criteria.

The results are shown on the radar chart (Figure 4), by which the enterprise's strong and weak points are analyzed, leading to clarification of their problems and points to be improved. Further, judging the diagnosis results and the existing conditions of the field synthetically, the analysis results are recorded on the side column of the radar chart.

4.3   Making of the manual

To make the evaluation items and criteria of the check list clearer, we made a manual (Figure 5) for each evaluation specified item (39 items in total).

In the manuals, each evaluation item and its evaluation criteria (3 grades) are illustrated concretely, easy enough for everyone.

4.4   Making of the curriculum for core staff training

To make the best use of this system, it is important to train core staff (responsible for evaluation) who can make a plan and practice KAIZEN concerning weak points obtained from the check list.

104

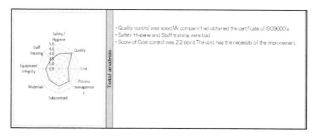

Figure 4. The results of evaluation.

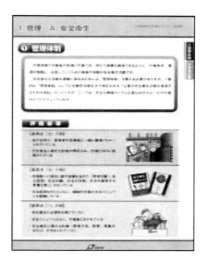

Figure 5. The manual.

Therefore, core staff training was decided for the purpose of deepening the understanding of the evaluation item contents and mastering of how to forward KAIZEN, etc. The training contents for each item (15 items) are composed of "Basic knowledge & necessity", "KAIZEN methods", "Supplementary explanation of the evaluation criteria", etc. The curriculum is about 30 minutes long for one evaluation item, totally 450 minutes long, that is, 2 days of the morning or the afternoon.

4.5 Practice of the production management self-diagnosis system

We held a core staff training course by use of the developed check list and manual for each item, which was intended for managers responsible for the field or management divisions of manufacturing enterprises. And also, we had the trainees diagnose their own enterprises and present their diagnosis results and improvement plans.

4.5.1 Practice of the core staff training
The course was attended by 13 persons from 5 manufacturing enterprises (steel-frame, boards, casting, press and canning) (Figure 6). The period was 2 half days. The lecturers were 4 outside knowledgeable persons whose specialty was production management.

The 15 items were divided among them. They explained each manual deeper, and fostered the trainees' better understanding.

4.5.2 Practice of the self-diagnosis of their own enterprises
Using the check lists for the field and management divisions, the five participating enterprises conducted self-diagnosis, and then the results and their future improvement plans were

105

Figure 6.   A scene of the core staff training course.

presented. All their average points of the 8 management items were over 3 points. The participating enterprises had obtained the certificate of ISO9000's, and especially their points concerning "Quality" and "Safety/Hygiene" were high. As for "the field", the average points (7 items) of 2 enterprises were from 2.0 to 2.9, and those of the other enterprises were from 3.0 to 3.5. Especially, the items concerning "Standardization" and "KAIZEN" had lower points.

The presentation of problems and improvement plans based on the self-diagnosis results had "Improvement of inventory management", "improvement of staff training", etc., for the management division, and "Persuasion of 2S's", "Making of a manual for work standard", etc. for the field division.

5   CONCLUSION

We developed a check-list system composed of evaluation items and indices in consideration of the characteristics of small and medium-sized manufacturing enterprises such as production of many models in small quantities. The system made it possible for those enterprises to diagnose their field management (efficiency increase, removal of waste, etc.,) and production management (achievement of QCD) totally and conduct KAIZEN. And also, we made manuals for their core staff to make good use of this system. Further, a training curriculum for fostering core staff (responsible for evaluation) was made for those enterprises to conduct KAIZEN by themselves by use of the developed system and manuals.

New Ergonomics Perspective – Yamamoto (Ed.)
© *2015 Taylor & Francis Group, London, ISBN 978-1-138-02751-0*

Development of a KAIZEN checklist tool for the Productive Aging to play an active part in smaller manufacturers

Koki Mikami
Hokkaido University of Science, Sapporo, Japan

Kenichi Iida
Hokkaido Industrial Research Institute, Sapporo, Japan

Masahiro Shibuya
Tokyo Metropolitan University, Tokyo, Japan

Tetsuya Hasegawa
Kinki University, Iizuka, Japan

Toshifumi Sakai
Hokkaido University of Science, Sapporo, Japan

Takatoshi Murakami
Sakai Factory, Daikin Industries Ltd., Sakai, Sakai, Japan

Masaharu Kumashiro
The Association for Preventive Medicine of Japan, Tokyo, Japan

ABSTRACT: In process of Japan's financial straits and "fewer-children" society, the Productive Aging in good health who can contribute to business and society are needed to play an active part. The Productive Aging have various advantages such as earning of income in terms of individuals, reduction of training time/cost by use of their experience in terms of business, and more tax revenues in terms of the nation. In this study was developed a KAIZEN checklist tool for the Productive Aging with the above advantages in smaller manufacturers which support Japan's Monozukuri technologies. The items in the checklist are grasped from the viewpoints of both industrial safety and health and production management, consisting of 10 large items and 66 small ones. In the small ones are shown cases with pictures, and the tool can be used easily as a hint for the improvement of manufacturing sites.

Keywords: Checklist; KAIZEN; Productive Aging; Industrial Safety and Health; Production Management

1   INTRODUCTION

In process of Japan's financial straits and "fewer-children" society, the Productive Aging in good health who can contribute to business and society are needed to play an active part. According to a nationwide survey conducted in Japan in May, 2014, 50.4% of the elderly hope to keep working after the age, 65 because their pension is not enough. The Productive Aging have various advantages such as earning of income in terms of individuals, reduction of training time/cost by use of their experience in terms of business, and more tax revenues in terms of the nation. In this study was developed a KAIZEN checklist tool for the Production Aging with the above advantages in smaller manufacturers which support Japan's Mono-zukuri technologies. Functional lowering caused by aging is undeniable. To make effective use of the Productive Aging, what is important is the working environment where they can work safely in good health and the viewpoint of increase in productivity for the company to continue to exist. Therefore, the items in the checklist were grasped from the viewpoints of both industrial safety and health and production management, and an action-type checklist consisting of 10 large items and 66 small ones was made. And also, each small item has a case titled "Visit to a worksite,"which is usable as a KAIZEN tool.

2 METHOD

The procedure of this study is as follows:

1. Establishment of the concept of a KAIZEN checklist tool.
2. Verification of the effectiveness of the existing checklist items for the Productive Aging to play an active part.
3. Making of new checklist items for the Productive Aging to play an active part.
4. Development of a KAIZEN checklist tool.

3 RESULTS AND DISCUSSION

3.1 *Establishment of the concept of the KAIZEN checklist tool*

Checklists are classified into two types. One is for the purpose of finding out problems about the existing work conditions and the worksite. The check items are minute and comprehensive, by which the subject task, environment and so on are judged good or bad. For example, the item, "Is the height of the work table appropriate?" or "Is the height of your chair appropriate?" is judged. This checklist needs a manual to tell how to deal with the problems found there. The other is a problem-solving-type checklist called an "action-type checklist" in which countermeasures are selectable. One famous checklist is the human engineering checklist by ILO [1], which is based on the "should-be" theory that the circumstance or condition should be like this, quoting some past successful cases.

In this study we made an action-type checklist to maintain labor safety and improve productivity for the existence of the company, and also gave it a function which can provide hints for improvement when necessary.

3.2 *Verification of the effectiveness of the existing checklist items for the Productive Aging to play an active part*

The author, et al., made an action-type checklist consisting of (1) Conveyance/handling of heavy objects (1–10), (2) Work postures (11–17), (3) Work method (18–25), (4) Hand tools/jig-meters (26–32), (5) Work using the eyes and (6) Other environments (40–46), and used it to improve worksites. In this study whether the above items were effective or not to make good use of the Productive Aging was examined. The subjects were smaller-companies' work leaders. One company is in Kita-Kyushu City, 12 are in Hokkaido and 15 are Tokushima Prefecture. The investigation was mainly conducted by use of questionnaire, and also we in person made observation of some of the worksites and conducted interviewing.

3.3 *Making of new checklist items for the Productive Aging to play an active part*

The results of the investigation revealed that some of the existing items should be deleted and that some new items should be added. Based on the results we made a new checklist consisting 10 large items and 66 small items for the Productive Aging to play an active part. Table 1 shows "Checklist for improving your worksite."

Table 1.   Checklist for improving your site.

Check item	Yes	No	Priority
1. For a worksite friendly to workers. (Are 3'S all right?: arrangement, tidiness and cleaning) (11 items)			
(1) The passage and working area are clearly distinguished with partitions or mark lines.			
(2) There are no obstacles such as products or materials on the passage.			
(3) There are no unnecessary materials, parts, equipment or machines in the working area.			
(4) The floor is cleaned up with no dust or oil.			
(5) Necessities are properly arranged, and there are plain signs for items and stocks.			
(6) Spot lights are fitted in such a way that they do not cause blurs in the illumination of their surroundings and the working area.			
(7) The building is well-lighted.			
(8) Unpleasant glare is presented by avoiding direct rays of lights or devising other means.			
(9) Parts and tools are arranged in such a way that workers can find them at first sight.			
(10) Some means are taken so that workers wearing earplugs can make themselves understood easily.			
(11) Regular cleaning is conducted.			
2. For efficient work with no wasteful actions. (8 items)			
(12) Work is devised for less going-up-and-down work (actions or movements)			
(13) Worksite (or between worksites) is laid out for less horizontal movements.			
(14) Every necessary object is placed within the circle drawn with an elbow as its center.			
(15) There is enough working space to do necessary actions. (Some devices are taken for no zigzag or quick turn actions)			
(16) Some means are taken for no unevenness in processing time.			
(17) Some means such as storage by colors are taken for no looking-around or choosing actions.			
(18) Objects are placed in such a way that workers do not need to move their head in looking at their objects (The criterion is within a 60° view)			
(19) There are work manuals.			
3. For reduction of mistakes. (3 items)			
(20) Simple signs or colors are used to avoid mistakes			
(21) An additional stock spot is set for each worker to avoid continuous work and wind up his work by himself.			
(22) The sizes and colours of letters/numbers are devised to avoid misreading.			
4. For making tools and jigs easy to use. (6 items)			
(23) Tools necessary for work are systematically placed in a safe storage space with simple signs.			
(24) Tools and materials are properly placed for easy use.			
(25) Holding equipment such as hanging tools is installed.			
(26) Tools are properly repaired, and tools worn by long use are put away.			
(27) Gripping parts are kept unslippery, and the thickness and shape are proper for easy use.			

(continued)

Table 1. Continued.

Check item	Yes	No	Priority

5. For prevention of backache. (8 items)

(29) Heavy objects/containers requiring manual handling have handles on for easy use.

(30) Containers are well sized for lifting or carrying heavy objects.

(31) Support apparatuses such as a push car, a crane, an intelligent balancer and a conveyor are used when heavy objects are moved.

(32) Some means are taken to prevent a stooping posture in standing-posture work.

(33) Worktables are devised for workers in standing posture to keep their knees straightened.

(34) Some means are taken for no twisting or bending postures.

(35) Some means are taken to keep the position of workers' hands between the positions of their shoulders and navel.

(36) Exercises for prevention of backache are practiced before work.

6. For prevention of stiff shoulders. (1 item)

(37) Worktables and so on are devised for workers in a standing posture not to need to raise their arms above their heart.

7. For prevention of fatigue-accumulation. (3 items)

(38) In repetitive or burdensome work, workers can take a spontaneous reset or a little break besides regular breaks during work.

(39) Necessary jigs are well used. And manual work is changed to work with jigs.

(40) Jigs or tools are introduced in consideration of workers' workload reduction.

8. For reduction of injuries and accidents. (11 items)

(41) The power plants and dangerous places which workers might touch during work or moving are enclosed.

(42) Proper wiring devices are used for safe connection of equipment.

(43) The floors of work areas are kept unslippery.

(44) Some means are taken to prevent workers' unsafe actions.

(45) Emergency stop buttons are noticeable.

(46) Machines are regularly inspected, and the record is kept.

(47) Inflammable or harmful objects are kept in their depository.

(48) Some measures are taken against earthquakes.

(50) Scaffolds are secured.

(51) Exercises for prevention of injuries are practiced.

9. In case of an injury or accident by any chance. (7 items)

(52) Every floor or large work room has 2 emergency exits and over.

(53) Fire escape routes are secured.

(54) Whom to contact in an emergency is in a noticeable place.

(55) How to deal with emergency is in a noticeable place.

(56) Every fire extinguisher is placed in a noticeable place.

(57) There is a first-aid kid in a noticeable place.

(58) There is an AED in a noticeable place.

(*continued*)

Table 1. Continued.

Check item	Yes	No	Priority
10. For working in good health. (8 items)			
(59) The occurrence sources of harmful objects or gas are completely isolated, or workers' exposure to them is made less by use of an extractor fan or a movable ventilator.			
(60) Radiant heat and harmful light are shut off by use of a screen, an enclosure and others.			
(61) The day's conditions of workers are grasped.			
(62) There is a comfortable and relaxing resting room suitable for refreshment.			
(63) Sound isolation measures are taken for machines causing loud noise. Earplugs are used when workers need to work in a very noisy place.			
(64) There are facilities for hand-washing and gargling.			
(65) There are separate areas for smokers and non-smokers, with a smoking room.			
(66) Necessary protective clothes or equipment are properly used, and there is a notice for that in the worksite.			

Table 2. An example of "Visit to a worksite."

(1) The passage and working area are clearly distinguished with partitions or mark lines.

[Before KAIZEN]

[After KAIZEN]

Drawing a line between the working area and the passage is the starting point to keep obstacles away from the passage, and helps make the work flow better and secure swift carriage. Shown in the left picture, the passage paint was worn out with conspicuous floor unevenness and repair marks in this factory. So, we made the passage clear by having the passage and floor painted and making them flat. This led to easy carriage by push car and easy vacuuming of dust. It also encouraged workers to clean their workpsite, and increased safety, with no stumbling.

3.4   Development of a KAIZEN checklist tool

In this study we wanted this new checklist usable as a KAIZEN checklist tool, and made it possible to use it in KAIZEN by putting a case titled "Visit to a worksite" in each item. In "Visit to a worksite" are pictures or illustration, and we added what meaning the item had and what effect it would produce. Table 2 is a concrete example of the KAIZEN checklist as a tool.

In Table 2 is shown Large item 1 "For a worksite friendly to workers." This item consists of 11 small item, and 3S' (arrangement, tidiness and cleaning) of what is called 5S' (arrangement, tidiness, cleaning, neatness and discipline) are emphasized here. Small item1

Table 3. "Visit to a worksite" of small item 20.

(20) Simple signs or colors are used to avoid mistakes

[Before KAIZEN]

[After KAIZEN]

Mistakes can bring about a trouble and a great loss. Labels and signs are a tool to provide much important information. They should be put on spots where workers can see, read and understand easily. And also it is possible to devise a hard-type contrivance for the prevention of mistakes as well as signs. In this factory when they used solvent for printing, they needed to pour from the cans using a pump. The cans were on the floor and took rather much room, which caused unnatural work postures. It was an unsafe condition with difficulty for finding the right solvent. So, we made clear cases with a label and a cock, which enabled workers to easily understand the color and residual quantity of solvent. And also, the solvent cases took only a little space, and prevented work errors from occurring.

is "The passage and working area are clearly distinguished by partitions or mark lines." The purpose of this item is "The worksite should be kept in this state." When you observe the actual worksite and find this kept, you do not need KAIZEN and check "Yes." Otherwise, you judge that the state needs KAIZEN and check "No." And then, referring to successful cases of the distinction, a KAIZEN team can make a suggestion list and promote KAIZEN activities.

Table 3 shows an example of "Visit to a worksite" of Small item 20 for reduction of mistakes.

4 CONCLUSION

In this study was developed a KAIZEN checklist tool usable as a KAIZEN tool from the viewpoints of both industrial safety and health and production management to make good use of the Productive Aging, which can produce advantages for the individual, the company and the nation in the midst of Japan's "fewer-children society." The new checklist was made, in an action type, by use of both the opinions or suggestions of workers in charge of the worksite and experts and field research, and has 10 large items and 66 small ones. Each small item has a case with pictures or illustration explaining simply the meaning and effects, and can give you hints before you take action for KAIZEN.

REFERENCE

[1] ILO: Ergonomic Checkpoints: Practical and Easy-To-Implement Solutions for Improving Safety, Health and Working Conditions, 1996.

New Ergonomics Perspective – Yamamoto (Ed.)
© *2015 Taylor & Francis Group, London, ISBN 978-1-138-02751-0*

Correlation of exercise level and electromyography of trunk stabilizer muscles during manual lifting among experienced back belt users

Nopporn Kurustien
Faculty of Physical Therapy, Huachiew Chalermprakiet University, Samut Prakan, Thailand
Faculty of Physical Therapy, Mahidol University, NakornPrathom, Thailand

Keerin Mekhora & Wattana Jalayondeja
Faculty of Physical Therapy, Mahidol University, NakornPrathom, Thailand

Suebsak Nanthavanij
Engineering Management Program, Sirindhorn International Institute of Technology, Thammasat University, Pathum Thani, Thailand

ABSTRACT: **Background**: The Exercise Level of the Trunk Stabilizer muscles (ELLS) is known to be associated with the ability to control back stability with the Transversus Abdominis (TrA) muscle. An important role of the TrA is feedforward activity necessary for preventing Lower Back Injury (LBI). Therefore, it can be postulated that lifting workers who always wear a back belt for certain periods of time are at risk for decreasing activity in the TrA or other trunk muscles during lifting with or without a back belt.

Objective: This study evaluates the correlation between the ELLS and Electromyography (EMG) of selected trunk muscles during manual lifting with and without a back belt in experienced back belt users.

Materials and Methods: Sixteen participants from a warehouse in Thailand, aged 22 to 44 years, were assessed for ELLS, which was indicated by values ranging from level 1 (weakest) to level 6 (strongest), and the EMG of selected trunk stabilizer muscles, including the Rectus Abdominis (RA), External abdominal Oblique (EO), TrA/Internal Oblique (IO), Erector Spinae (ES), and Multifidus (MF). The EMG data were recorded during manual lifting in a dynamic semisquat posture for conditions of lifting with and without a back belt.

Results: The results of the Pearson correlation coefficient between the ELLS and Normalized EMG (NEMG) of the selected trunk muscles showed a positive significant correlation between the ELLS and TrA/IO activity only during lifting without a back belt ($r_p = 0.537$, $p = 0.032$). However, there was no correlation between the ELLS and other selected muscles during lifting with or without back belt.

Conclusion: This study demonstrates that ELLS indicates the importance of workers using the TrA during lifting without a belt. Therefore, a specific exercise program to improve the strength of the TrA is necessary for back belt users.

Keywords: Transversus abdominis; abdominal support; material handling; ergonomics; lumbarstabilization; MVC

1 INTRODUCTION

The Exercise Level of Lumbar Stabilizers muscles (ELLS) is measured in a series of six exercises ranging from easy (level 1) to most difficult (level 6) on the basis of the ability to maintain the spine in a static position while increasing lower extremity loading [1]. This ELLS was modified from the series of seven exercises created by Hagins [2] and was improved to be more reliable and more appropriate for Thai people [3]. The ELLS is known to be associated with the ability to control back stability with the Transversus Abdominis (TrA) muscle [2]. Therefore, the highest level of ELLS indicated the strongest TrA, which can consequently help prevent back injury.

A previous study found an increasing back injury rate among lifting workers when they were not wearing a belt following a period of wearing a belt [4]. It was believed that the belt decreases back muscle activity by increasing intra-abdominal pressure. This effect helps a lifting worker feel safe in lifting significantly more weight. However, it was not recommended for healthy workers to wear a back belt as a protective device [5]. In spite of this, a belt is still commonly used by manual laborers. Thus, manual lifting workers who still need to wear a belt should know more about their ability to control back stability, especially with the TrA, to prevent back injuries. However, there have been no studies concerned with the level of ELLS and the relationship between the level of ELLS and EMG of back stabilizers muscles among experienced back belt users during manual lifting. The results of this study will help produce recommendations for back belt users to further exercise the TrA as an important trunk stabilizer muscle.

The purpose of this study was to clarify the correlation between the ELLS and EMG of selected trunk muscles during manual lifting with and without a back belt among experienced back belt users.

2 METHODS

2.1 Study design

The study had a quasi-experimental design. The subjects performed a lifting task three times with two belt conditions, with and without a belt, and performed an ELLS test.

2.2 Subjects

Sixteen male subjects with no previous history of back or abdominal surgery that may have affected the experiment participated in this study. All of the subjects were workers from one section of a warehouse and distribution center in Thailand, and their work involved only repetitive manual lifting tasks. They had been wearing back belts for at least six months. All subjects provided written informed consent before participating in the study. The protocol for the study was approved by the Mahidol University Institutional Review Board.

2.3 Instruments

- Back belt: The back belt used in this study was a typical industrial elastic belt with four semirigid bars aligned on the back with anterior fastening with Velcro. The belt had a posterior height of 20 cm and an anterior height of 12 cm.
- Work simulator and EMG: A Primus RS system (BTE Technologies, Inc., USA) was used to simulate lifting work tasks. A lifting box was created from wood with dimensions of $25 \times 32 \times 29$ cm^3 (W × L × H) and was placed on a wooden stand 24 cm high. The base of the box was attached to the cable system of the Primus RS. The two parameters of the lifting task, Torque (T) and Velocity (V), were synchronized with an EMG unit (Telemyo 2400 G2, Noraxon, USA, Inc.,) using the Noraxon Program on an EMG monitor.
- Pressure Biofeedback Unit (PBU): This unit, called the Stabilizer (Chattanooga, manufactured in Australia), which was used to test for the level of lumbar stabilization, consisted of a three-connected-chamber air-filled bag. The air-filled bag was inflated to fill the space between the target body area and a firm surface. The pressure gauge was marked in increments of 2 mmHg from 0 to 200 mmHg to indicate the pressure in the bag for feedback on position.

2.4 Experimental procedure

All of the experimental procedures related to EMG electrode placement, the MVC test, and the lifting task test have already been described in a previous study by the authors [6]. The EMG electrode was placed on the right trunk muscles, including the TrA/Internal Abdominal Oblique (IO), Rectus Abdominis (RA), External abdominal Oblique (EO), Erector Spinae (ES), and Multifidus (MF) [7]. The lifting task test (see Fig. 1) involved dynamic semisquat lifting from the middle of the shank to knuckle height with and without a back belt. Subjects were instructed to lift three times with 20 s of rest between each lift. The lifting condition order was randomized with 5 min of rest between conditions. Lifting speed was controlled by a metronome preset at a frequency of 48 beats/min.

The EMG signals were recorded at a sampling rate of 2 kHz and processed using MyoResearch XP EMG Application Protocols v. 1.06.54 (Noraxon Inc., USA) to reduce the Electrocardiogram (ECG) signal and smooth data using the Root Mean Square (RMS) while moving the processing window every 20 ms. Only the position of the start of the lift of the box was calculated to determine the percentage of the MVC (%MVC) in order to normalize all EMG values (NEMG) [6].

The ELLS test in this study followed the guidelines from Thongjunjua 2005 [8]. To start testing, the examiner aligned the PBU bag under the L1-S2 back level of the subject and inflated the bag to a pressure of 40 mmHg. Then, all subjects had to complete the tests in the order stated below. In each test, the subject was instructed to hollow the abdominals and maintain a pressure of 40 mmHg (±4 mmHg) for three breathing cycles. The degrees of difficulty of the test exercise were as follows [8]:

- **Level 1: Abdominal hollowing.** Participant assumed bent lying position with the hips at 70° and feet flat on the floor (Fig. 1) and performed static abdominal hollowing for three breathing cycles.
- **Level 2: Unilateral abduction.** Participant assumed bent lying position with the hips at 70° and feet flat on the floor. Then, the right leg was abducted to approximately 45° from the floor with the left knee held steady.

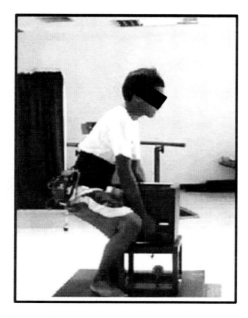

Figure 1.   Position of EMG recording in the position of start lifting.

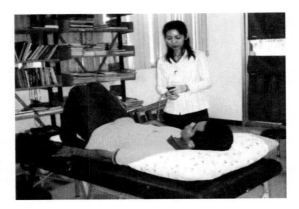

Figure 2.   Starting position of ELLS test.

- **Level 3: Unilateral knee extension**. Participant assumed bent lying position with the hips at 70° and feet flat on the floor. The right knee joint was extended to 0° while controlling the hip to remain at a constant angle.
- **Level 4: Unilateral knee raise**. Participant assumed bent lying position with the hips at 70° and feet flat on the floor. The right knee was raised toward the chest until it just passed a hip flexion of approximately 90° and then was allowed to flex naturally.
- **Level 5: Bilateral knee raise**. Participant assumed bent lying position with the hips at 70° and feet flat on the floor. The right knee was raised toward the chest until it just passed a hip flexion of approximately 90° with the knee in a flexed position and held. This was followed by raising the left knee in the same manner and holding both knees together and finished by placing the right leg in the starting position. The procedure was then repeated with the left leg.
- **Level 6: Bilateral knee raise**. Participant assumed bent lying position with the hips at 70° and feet flat on the floor. Both knees were raised toward the chest until they just passed a hip flexion of approximately 90° with natural knee flexion. Both legs were placed in the starting position to finish the movement.

The outcome of ELLS test for each subject was the last level of ELLS that the subject could completely perform three times.

2.5   *Statistical analysis*

The Pearson correlation coefficient was used to determine the association between the NEMG of trunk muscles and ELLS.

3   RESULTS

There were a total of 16 experienced manual lifting workers in the study. They had no musculoskeletal disorders or other disorders that might have affected participation in the study. Their mean age, height, weight, and Body Mass Index (BMI) were 30.17 ± 6.15 yrs, 170.5 ± 6.05 cm, 62.08 ± 8.4 kg, and 21.29 ± 2.02 kg/m^2, respectively. Most of them had experience in lifting tasks in the range of 1–7 years and had regularly worn a back belt for more than 8 h per day. They took the belts off only during resting periods.

Table 1 presents the ELLS test results among all 16 subjects in this study. A higher level indicates a greater ability of the TrA to stabilize the lumbar spine (Table 1). That is, level 1 represents the least control using the TrA muscle, and level 6, which is highest level in this

Table 1. The exercise level of the lumbar stabilization of subjects.

Level of ELLS	Number of subjects (n = 16)	Percentage
Level 1	1	6.2%
Level 2	3	18.8%
Level 3	2	12.5%
Level 4	8	50.0%
Level 5	2	12.5%

Table 2. Correlations between level of trunk stabilization and %MVC of RA, TrA/IO, EO, ES, and MF during lifting without and with back belt (n = 16).

Independent variables	TrA	RA	EO	ES	MF
Correlation coefficient with %MVC of trunk muscle activities during lifting without a back belt					
Exercise level of lumbar stabilization (1–6)	$r_p = 0.537^*$ $p = 0.032$	$r_p = 0.039$ $p = 0.887$	$r_p = -0.343$ $p = 0.194$	$r_p = -0.185$ $p = 0.493$	$r_p = 0.026$ $p = 0.924$
Correlation coefficient with %MVC of trunk muscle activities during lifting with a back belt					
Exercise level of lumbar stabilization (1–6)	$r_p = 0.303$ $p = 0.254$	$r_p = 0.106$ $p = 0.697$	$r_p = -0.386$ $p = 0.139$	$r_p = -0.070$ $p = 0.796$	$r_p = 0.177$ $p = 0.511$

Note: * = Significant difference at $p < 0.05$.
r_p = Pearson correlation coefficient.

test, represents the greatest ability of the TrA muscle to stabilize the lumbar spine. The results showed that half of the subjects in this study had an ELLS at level 4 (50%).

Table 2 presents the statistical analysis results of the Pearson correlation coefficient between the level of ELLS and NEMG (%MVC) of five trunk muscles during manual lifting with and without a belt. The results showed a significant positive correlation between ELLS and TrA/IO activity during lifting without a back belt ($r_p = 0.537$, $p = 0.032$). A higher level of ELLS corresponds to more EMG activity of the TrA muscle during lifting without a belt among belt users.

4 DISCUSSION

The hypothesis of this study was that wearing a back belt may reduce the activity of the TrA because the effect of the belt may help stabilize the spine. Thus, long-term use may decrease the role of this stabilizer. However, the results of the current study showed that more than half of the subjects had adequate levels of trunk stabilization control, and half of them (50%) had an ELLS of level 4, which meant that they had good control of the TrA muscle.

In the statistical study between the ELLS test and NEMG, it was found that NEMG of the TrA during lifting without a belt was significantly correlated with exercise level (Table 2). This means that during lifting without a back belt, the activity of the TrA/IO increased in subjects who had a high level of ELLS and decreased in subjects who had a low level of ELLS. Thus, without a belt, workers performing lifting tasks should be encouraged to exercise the TrA muscle, as it can improve ELLS, which can help prevent back injury from lifting tasks. On the other hand, inadequate support indicated that the belt decreased ELLS among belt users, as most of them had high levels of ELLS.

5 CONCLUSION

This study reports that more than half of experienced lifting workers who wore a back belt for more than 1 yr had adequate levels of trunk stabilization control (level 4). A positive significant correlation between ELLS with TrA/IO activity during lifting without a back belt was found; thus, a specific exercise program to improve the strength of the TrA is necessary for back belt users.

ACKNOWLEDGEMENTS

The authors are thankful for support from the Faculty of Physical Therapy, Mahidol University. We would also like to give special thanks to Dr. Sirikarn Somprasong and Mr. Veeranat Jeenawath for their assistance in data collection. We are thankful to all of the subjects in this study and their workplace administrators and gratefully acknowledge the research grant from Huachiew Chalermprakiet University Affairs.

REFERENCES

[1] Thongjunjuea S, Wongprasertgan M. Reference values of exercise level attained for lumbar stabilization exercises in young adults. *Thai J Phy Ther*. 2011;**34**(1):37–44.
[2] Hagins M, Adler K, Cash M, Daugherty J, Mitrani G. Effects of practice on the ability to perform lumbar stabilization exercises. *J Orthop Sports Phys Ther* 1999;**29**(9):546–55.
[3] Thongjunjua S. Effects of lumbar stabilization exercises on exercise level attained in healthy subjects. *Thai J Phy Ther*. 2006;**29**(1):1–13.
[4] Reddell CR, Congleton JJ, Dale Huchingson R, Montgomery JF. An evaluation of a weight-lifting belt and back injury prevention training class for airline baggage handlers. *Appl Ergon* 1992;**23**(5):319–29.
[5] McGill SM. Should industrial workers wear abdominal belts? Prescription based on the recent literature. *Int J Ind Ergonom* 1999;**23**:633–6.
[6] Kurustien N, Mekhora K, Jalayondeja W, Nanthavanij S. Trunk stabilizer muscle activity during manual lifting with and without back belt use in experienced workers. *J Med Assoc Thai* 2014;Suppl(**97**):s75–9.
[7] Marshall P, Murphy B. The validity and reliability of surface EMG to assess the neuromuscular response of the abdominal muscles to rapid limb movement. *J Electromyogr Kines:* 2003;**13**(5):477–89.
[8] Thongjunjua S. Effects of lumbar stabilization exercises on exercise level attained in healthy subjects. [Thesis (M.Sc.(Physiotherapy)]. Bangkok: Mahidol University; 2005.

New Ergonomics Perspective – Yamamoto (Ed.)
© *2015 Taylor & Francis Group, London, ISBN 978-1-138-02751-0*

Comparison of work characteristics related with Musculoskeletal Disorders based on company size and work types

Jin-Youb Jung
Department of Safety Engineering, Chungbuk National University, Cheongju, Chungbuk, Korea

Meiling Luo
Division of I&C and Human Factors, Korea Atomic Energy Research Institute, Daejeon, Korea

Jong-Hun Yun
Department of Administration, Korea Institute of Construction Technology, Goyang, Gyeonggi, Korea

Sang-Hun-Byun & Hyeon-Kyo Lim
Department of Safety Engineering, Chungbuk National University, Cheongju, Chungbuk, Korea

ABSTRACT: In order to prevent work-related musculoskeletal disorders, hazard assessments came to mandatory in every 3 years for all the manufacturing companies in Korea. Meanwhile, in most cases, the same assessment tools as REBA, RULA, NLE, and so on are applied without consideration of different characteristics of the work. Manufacturing work operations in workplace can be classified into two groups—cyclic routine works and acyclic works. Cyclic routine works such as assembly work operations in automobile manufacturing companies, are repetitive while acyclic works automobile repair works depend on the ask of customers which results awkward and/or prolonged work postures. This study aimed to seek differences in work characteristics and work postures with reference to company size and work types. Over a few years, around 300 subjects were selected from automobile assembly lines and automobile repair shops and around the same numbers of workers were selected from medium- and small-sized companies with consideration of work types for comparison. With their work characteristics and work types, typical work postures were elicited and compared using multivariate statistical techniques. Regression analysis was also conducted for statistical significance. The results were somewhat different for company size and work characteristics which implied the necessity of different intervention for musculoskeletal disorders from medium- and small-sized manufacturing companies.

Keywords: Musculoskeletal Disorders; Cyclic routine works; Acyclic works; Medium- and Small-sized Companies

1   INTRODUCTION

The momentum that Work-related Musculoskeletal Disorders (WMSD's) got attention of most industrial workers and officers in Korea was the event that Musculoskeletal Disorders (MSD's) of telephone operators working at a telecommunication industry obtained approval of Occupational Safety and Health Insurance Fund so that was admitted as industrial accidents by provoke a sensation in 1996. After that, WMSD became a hot issue over industrial labor area as well as the whole society.

According to a report on industrial accidents published by the Korean Department of Employment and Labor in 2012, out of 6,742 workers who suffering occupational illness, the number of workers complaining WMSD was 5,230 which occupied 77.6% of the whole number of them [1]. Especially, among them, the number of industrial workers from medium- and small-sized companies was 4,388 which occupied as high as 83.9%. This report implied the fact that WMSD in medium- and small-sized companies is much serious than that in large-sized companies.

The major reason why the phenomenon like this turns out is the poverty of companies. Due to limitation of capital, medium- and small-sized companies have paid little attention to WMSD, and less money for work improvement.

As well known, diverse factors can influence on the occurrence of WMSD's. According to OSHA(1999), major causes can be attributed to 1) work factors such as repetitive motion and awkward postures, 2) workplace factors such as workstation, devices tools as facilities used during work periods, 3) personal factors such as age, body condition, work habit, and patient's clinical history, 4) environmental factors such as vibration and cold stress.

Especially, repetitive motion is one of the most important causes that may result in WMSD's. With the criterion of repetitive motions, most works can be classified into two groups—cyclic work and acyclic work. Usually, the former is highly repetitive so that frequently used body parts can be exhausted and minute damages on them can be accumulated, which may result in WMSD's. On the contrary, the latter highly depends upon the orders of customers, but intermittently requires excessive muscle contractions due to force exertion and/or awkward postures.

A reason why this study was undertaken was that the number of studies on WMSD's in medium- and small-sized companies has been few which is known as more serious than those in large-sized companies. This study aimed to find out characteristics of WMSD's in medium- and small-sized companies, and to compare them with those in large-sized companies to ask a basic question whether the same prevention strategy can be applied.

2 METHOD

2.1 *Selection of works and workers*

For a few years, WMSD-related data were obtained from several large companies including automobile manufacturing companies and ship building companies. Data for cyclic works were selected from automobile-part manufacturing works whereas data for acyclic works were collected out of mechanics from more than 20 repair shops located over the country. On the other hand, data were also collected out of 273 work steps from medium- and small-sized companies in which less than 300 people worked. The number of companies concerned was about thirty from diverse industries, including automobile-part manufacturing companies.

The criterion to classify cyclic works and acyclic works were continuity of the same work for more than 4 hours. That is, if the same work is repeated for more than 4 hours, the work was classified as cyclic work reflecting the Korean regulation. Yet, in the case of acyclic work, a mechanic in a repair-shop can do different work from time to time so that assigning a work to a worker can be meaningless. Therefore, eventually, more works than the number of workers were selected for both cases.

As a consequence, for large-sized companies, 111 kinds of cyclic works and 137 acyclic works were included, while 169 kinds of cyclic works and 104 acyclic works were selected as research objects.

2.2 *Analysis method*

Basically, all the work sites were visited by the authors, and video recording on work procedures and measurements on characteristic dimension were committed. Recorded work postures were analyzed in a laboratory by the authors, and the major analysis technique was REBA (Rapid Entire Body Assessment) [2].

After work procedure and posture recording, interview was conducted at the work site directly with the worker recorded. He/She assessed his/her work difficulty related with WMSD with 5 point (Likert) scale in the aspect of severity and frequency. Comprehensive risk index was by multiplying both.

In parallel, a questionnaire survey was carried out for investigating complaint rates. The questions were adopted from KOSHA Code H-28-2002, and the judgment criteria were also adopted from NIOSH.

3   RESULTS

3.1   *Influencing factors on WMSD's*

To comprehend influencing factors on WMSD complaint rate, Multiple Regression Analysis was carried out. Among independent variables, 1) personal factors such as age, gender, and work career, 2) work condition factors such as handle, weight, and repetition, 3) work posture factors such as REBA codes for body parts were included [3]–[5].

The result showed that somewhat difference by company size and work type. As for cyclic works in large companies, major significant factors were body parts involved in work such as hands and wrist, but their power of explanation R^2 was negligible. On the contrary, in the case of acyclic works in large companies, work career as well as body parts were significant factors.

However, in the case of cyclic works in medium- and small-sized companies, repetition was one of the major significant factors while work condition factors such as weight and repetition were major influencing factors in the case of acyclic works in medium- and small-sized companies.

3.2   *Quantified comparison*

To compare the characteristics of WMSD related variables, The 3rd Quantification Technique [6] was applied and pictorially compared by using Multi Dimensional Scaling

Table 1.   Major influencing factors on WMSD complaint rates.

	Large-sized company		Medium- and small-sized company	
	Cyclic works*	Acyclic works	Cyclic works	Acyclic works
Neck	Wrist*	Career* Shoulder*	Neck* Leg*	Neck* Repetition*
Shoulder	Elbow*	Career*	Neck*	Shoulder* Weight/Force*
Back	–	Handle*	Weight/Force*	–
Arm/Elbow	Age*	Age* Elbow*	Back* Neck* Work hours*	Back* Handle* Work hours*
Wrist/Finger	Elbow*	Career*	Work hours* Neck* Leg*	Weight/Force*
Foot/Ankle	Wrist*	Wrist* Career* Shoulder*	Neck* Leg* Work hours*	–

* p < 0.05.

121

Technique. A part of the results are shown in Fig. 1, where the horizontal axis and the vertical axis corresponds that in Principal Component Analysis, respectively.

As for cyclic works in large-sized companies, most characteristic variables clustered around the origin so that the existence of weight or force exertion was only an important factor determining the characteristics of works. By the same token, repetition and weight/force were deterministic factors in the case of acyclic works in large companies.

However, the existence of handles was added as an important factor for works in medium- and small-sized companies, regardless of periodicity—cyclic works and acyclic works.

3.3   *Comparison of work postures*

Cluster Analysis is a statistical technique to unify multivariate identities into a group with similar characteristics. According to the result of Cluster Analysis, work postures in medium- and small-sized companies were more complicated and diverse than those in large-sized companies as expected [7].

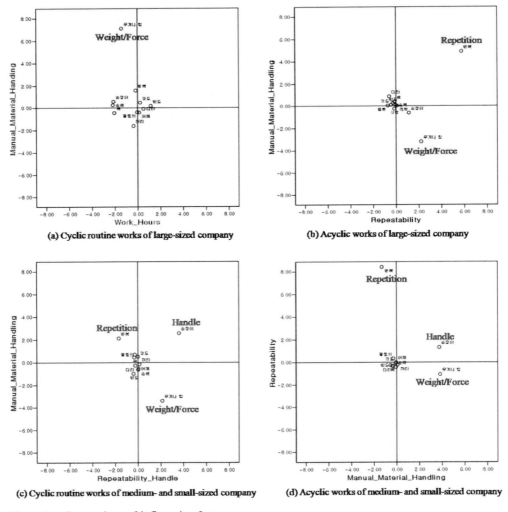

Figure 1.   Comparison of influencing factors.

122

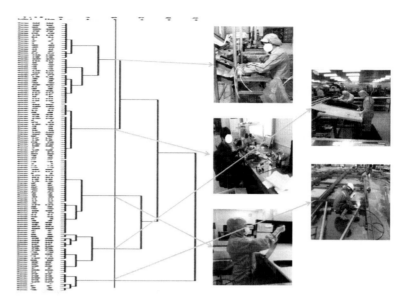

Figure 2.   Cluster analysis of work postures (Cyclic works in medium- and small-sized companies).

In general, work postures were different with respect to company size. Work postures in large companies were mainly classified by postures of lower back and neck, while those in medium- and small-sized companies were classified by shoulders, arms, and wrist.

4   DISCUSSION AND CONCLUSION

This study tried to find out whether there exist any differences between WMSD's from large-sized companies and those from medium- and small-sized companies.

The research results show that significant factors influencing on WMSD's were different, and that major principal factors describing distributions of complaining body parts and work characteristics were somewhat different. As for work characteristics related with WMSD's, most factors influencing significant effect on WMSD's were almost similar, regardless of company size work type. The common factors ranked high were mainly weight of work objects, exerted force required, repetition, and handle, in sequence. However, work postures were different. Low back and neck were major body parts that characterize work postures in large companies. In contrast, shoulders, arms, and wrist were major body parts in case of medium- and small-sized companies.

To summarize all the research results, it was not easy to find out differences between WMSD-related data from large-sized companies and those from medium- and small-sized companies. Therefore, it could not be rational to regard all the works as homogeneous and to apply the same techniques to assess hazards of musculoskeletal system. In order to assess hazards of a work related with WMSD, more careful selection and application would be required, especially for arm and wrist postures in the case of medium- and small-sized companies.

REFERENCES

[1] Ministry of Employment and Labor, *2012 Industrial Accident Analysis*, 2013.
[2] Hignett, S., McAtamney, L., "Rapid Entire Body Assessment (REBA)," *Applied Ergonomics*, 31, pp. 201–205, 2000.

[3] Kilbom, A., "Repetitive work of the upper extremity: PartI-Guidelines for the practitioner", *International Journal of Industrial Ergonomics*, Vol. 14, pp. 51–59, 1994.

[4] Jones, T., Kumar, S., "Comparison of ergonomic risk assessments in a repetitive high-risk sawmill occupation: Saw-filer", *International Journal of Industrial Ergonomics*, Vol. 37, No. 9–10, pp. 744–753, 2007.

[5] Bernard, B.P., "Musculoskeletal disorders and workplace factors: A critical review of epidemiologic evidence for work-related musculoskeletal disorders of the neck, upper extremity, and low back", Cincinnati, OH: US Department of health and human services. Vol. 7, No. 1, pp. 10–12.

[6] Huh, Myung. Hoe, *Quantification Methods I, II, III, IV*, Freedom Academy, 1992.

[7] Yun, Jong. Hun, Lim Hyeon. Kyo, "Comparison of Work Characteristics for Evaluating Musculoskeletal Hazards of A typical Works", *Proceedings of Ergonomics Trends from The East*, pp. 193–198, 2008.

New Ergonomics Perspective – Yamamoto (Ed.)
© 2015 Taylor & Francis Group, London, ISBN 978-1-138-02751-0

Characteristics of small Japanese trawlers as workplace environments

Hideyuki Takahashi

Fisheries Research Agency, Hasaki, Kamisu, Ibaraki, Japan

ABSTRACT: The number of Japanese fishermen is decreasing rapidly and those remaining are aging. According to the Food and Agriculture Organization, the number of fishermen in Japan is rapidly declining in contrast to those of other major fishing countries increasing or leveling off. One reason is thought to be the poor work environment. Although many Japanese fishing boats were mechanized in the last quarter of the twentieth century, most of the effort was directed at maximizing their catch efficiency, not improving their work environments. Therefore, fishing is still one of the hardest occupations in Japan. This study compared three case studies conducted in small trawl fisheries during actual work. The times required for the main tasks were measured, revealing that fish sorting was the most time-consuming task in all cases. The physical burdens of the main tasks were estimated using the Ovako Working-posture Analyzing System (OWAS), the simplest, best-known method for judging the demands of tasks based on work postures. The results of the OWAS analyses varied with the tasks and cases. Regardless, fish sorting imposed a large physical burden when performed on deck. Use of a table appears to be a simple solution to improve work posture.

Keywords: fishery; work population; aging; work time; work posture

1   INTRODUCTION

The number of Japanese fishermen is decreasing rapidly and those remaining are aging. There were approximately 500,000 fishermen in the 1980s, and that number has declined to approximately 200,000 today [1]. By contrast, the number of fishermen is increasing or leveling off in major countries other than Japan [2]. One of the main causes of the decrease in Japan is thought to be the poor work environment. Although many Japanese fishing boats were mechanized in the last quarter of the twentieth century, most of the effort was directed at maximizing the fish catch efficiency, not at improving the work environment. Therefore, fishing is still one of the hardest occupations in Japan. Although fishermen engage in hard work, they often cannot catch enough fish or sell them at acceptable prices; consequently, the occupation has become less appealing in Japan. To revive the fishery in Japan, the work environment must be improved.

This study focused on issues in the work environment of the small trawl fishery, one of the main coastal fisheries in Japan. Three small trawlers were studied and analyzed quantitatively to understand their features as workplaces. Based on the results of the analyses, possible measures for improving the working environment are suggested.

2   MATERIAL AND METHOD

The time required for, and the physical burden involved in, each main task were determined on three small trawlers.

2.1 Surveyed cases

Case A (Fig. 1, left) operates in Aichi Prefecture. The boat has a crew of three and a gross tonnage of 13.9 tons. Mantis and other shrimps are caught using a net with a pair of otter-boards that open the net mouth via wing surface resistance. The main fishing equipment on the boat is a net winch, a seawater spray device to keep the catch fresh, and fish holds. A fishing trip lasts approximately 12 hours. For details see Takahashi *et al.* (2012) [3].

Case B (Fig. 1, center) operates in Mie Prefecture with a crew of one and a gross tonnage of 4.9 tons. Clams are caught with two dredge nets, using a stiff frame to keep the net-mouth open. The main fishing equipment on the boat is a warp winch, a derrick crane to haul up the dredge nets, and an automatic clam sorting machine. Part of the clam sorting task is performed on the wharf by the crewmember and three additional workers with a table and a water tank. A fishing trip lasts approximately 4 hours. For details see Takahashi *et al.* (2010) [4].

Case C (Fig. 1, right) operates in Osaka Prefecture with a crew of three and a gross tonnage of 9.7 tons. Shrimps and small fish are caught using four dredge nets. The main equipment on the boat consists of two warp winches, a table and chairs for fish sorting, and fish stock containers. A fishing trip lasts approximately 10 hours. For details see Takahashi (2013) [5].

2.2 Videotaping tasks

To obtain material for analysis, tasks were videotaped continuously during a fishing trip using cameras installed to cover the places where the main tasks were performed.

2.3 Time study

The time series of the tasks performed in each case during daily operation were constructed, and then the percentage of time spent on each task was calculated to detect the most time-consuming tasks.

2.4 Estimating physical burden

The physical burdens of the main tasks in each case were estimated based on the crew members' work postures, as recorded in the video images, using the Ovako Working-posture Analyzing System (OWAS), which is one of the simplest and best-known methods [6]. In this method, work postures are categorized based on three body regions (the upper body, upper limbs, and lower limbs) and the weight of the objects handled. When a work posture has been identified precisely, the physical burden is ranked in one of four Action Categories (ACs); these serve as an index of the degree of demand for improvement. The number associated with an AC reflects the harmfulness of the work posture, with AC1 being the best posture and AC4 being the most harmful. When a harmful posture occurs with high frequency, preference should be given to improving the work procedure.

Figure 1. Appearances of surveyed fishing boats, case A (left), case B (center), and case C (right), respectively.

3 RESULTS

3.1 *Outlines of the observed tasks*

Case A (Fig. 2, left): During the fishing trip, the net was cast, towed, and hauled 12 times, and fish were sorted after each haul. During net casting and hauling, two of the crew were engaged mainly in manual tasks (*e.g.*, cast cod-end, attach/detach otter-boards, arrange hauling positions of warps on net winch, and unload fish catch), while the remaining crew member operated the net winch. The same two crew members also engaged mainly in fish sorting on the port side of the stern deck during the next tow. In the fish sorting task, target fish species were separated and thrown into tubs and baskets. The sorted fish were stored in the holds in the bow.

Case B (Fig. 2, center): The net was cast, towed, and hauled seven times during the fishing trip and the catch was sorted each time. Before casting the net, the fisherman reeled out the anchor and wire for net towing. Then, he manually pushed a dredge out on each side of the boat. The nets were towed by reeling in the wire. In the net hauling task, the crewmember lifted the dredges using a capstan, placed one each side of the boat, and then manually recovered the cod ends and unloaded the catch on the deck. To sort the catch on the deck during

Figure 2.   Main tasks performed in case A (left), case B (center), and case C (right), respectively.

the next tow, the catch was thrown in an automatic sorting machine and roughly sorted into three types: large clams, small Japanese littleneck clams, and smaller non-target species and other non-living objects. The crew member then removed any non-target clams from the machine-sorted clams. The sorted clams were packed into net bags. When the boat returned to port, the clams in the net bags were landed manually. Three workers joined in the sorting task on the wharf. The clams were washed, sorted precisely, weighed, and packed in net bags for auction.

Case C (Fig. 2, right): The net was cast, towed, and hauled 24 times during the fishing trip, and the catch was sorted. During the casting and hauling tasks, two crew members operated wire reels that cast four dredge nets on the captain's sign. In the net-hauling task, the dredges were hung up using a roughly reversed procedure from the net casting task, the cod ends were recovered manually, and the catch was unloaded into baskets on the deck. During the next tow, all three crew members sorted the catch manually using a table and chairs on the stern deck. The sorted catch was stored in an aerated water tank or cooler boxes placed on the stern deck.

3.2 Time study

To measure the time required for each task, the start and end of each task were defined (Table 1). The times required for the main tasks and the percentage of the total operating time that they comprised are listed in Table 2. Of the main tasks, sorting the catch took the greatest time, while net casting and hauling occupied 3~4% and 9~12% of the operating time, respectively.

Table 1. Definitions of start and end of tasks in each surveyed case to determine times required for each task.

Task		Case A	Case B	Case C
Net casting	Start	Throw cod-end	Push 1st dredge to cast	Operate wire reel
	End	Stop net winch	Push 2nd dredge to cast	Operate wire reel
Net hauling	Start	Start to reel up the net winch	Reel up dredge	Operate wire reel
	End	Unload catch from cod-end	Unload catch from cod-end	Unload catch from cod-end
Fish sorting (deck)	Start	Sort catch on stern deck	Use automatic sorting machine	Gather catch on table
	End	Store fish into fish holds	Pack clams into net bags	Store fish into cases
Fish sorting (wharf)	Start	–	Take clams out from net bags	–
	End	–	Pack clams into net bags	–

Table 2. Times required for main tasks in each surveyed fishing boat.

Case		Net casting	Net hauling	Fish sorting (deck)	Fish sorting (wharf)	Other	Rest	Total
A	(min.)	27.0	62.2	352.1	–	54.7	222.2	718.2
	(%)	3.8	8.7	49.0	–	7.6	30.9	100.0
B	(min.)	12.0	33.4	41.2	52.5	96.8	46.1	282.1
	(%)	4.3	11.8	14.6	18.6	34.3	16.4	100.0
C	(min.)	18.2	68.8	345.7	–	74.7	80.6	587.8
	(%)	3.1	11.7	58.8	–	12.7	13.7	100.0

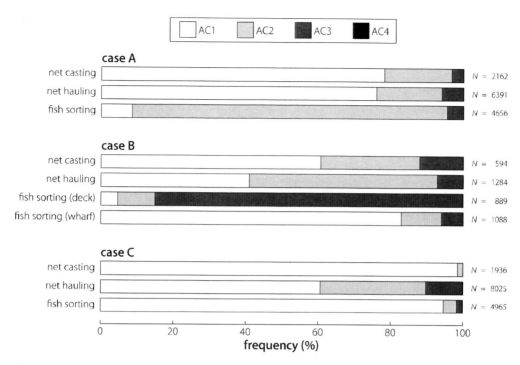

Figure 3.   Frequencies of Action Categories (AC) in the main tasks of each case study.

3.3   *Estimate of physical burden*

Fig. 3 summarizes the estimated physical burdens of the main tasks in each case. The results varied with the tasks and cases. In case A, the percentage of time that involved poor work postures, which increases the physical burden ($AC2), was high in the fish sorting task (91%), while the percentages were low in the net casting and hauling tasks (22% and 24%, respectively). In case B, the percentage of time that involved poor work postures was high during the fish sorting task on deck (95%), while the percentages were low in the net casting and hauling tasks and the fish sorting task on the wharf (39%, 59%, and 17%, respectively). For the fish sorting task on deck, the percentage of work that was AC3 or more was 85%. In case C, the percentage of time using poor work postures in the net hauling task was high (39%), while the percentages in the net casting and fish sorting tasks were less than 10%.

4   DISCUSSION

This research project conducted three case studies in the small trawl fishery and quantified the times required for, and the physical burdens involved in, the main tasks. In comDCEEmon, the longest times were needed for sorting the catch in all cases, while the estimated physical burdens differed across the cases, even for the same task. In case A, the net casting and hauling tasks were mostly performed in desirable work postures, while the physical burden of the fish sorting task was judged to be large. The physical burdens of all of the tasks on board in case B were greater than in case A, and the fish sorting task on board was performed with the poorest work postures among all of the tasks in all cases, while the fish sorting task on the wharf was performed in desirable work postures. Although some poor work postures were included in the net hauling task, most of the tasks were performed in desirable work postures in case C.

If a task performed with poor work postures is time-consuming, its effect on the worker's musculoskeletal system will be marked. The fish sorting tasks were the most time-consuming in all of the cases investigated in this study. This tendency has also been seen in other cases of small trawlers investigated by the author [7] [8]. Therefore, the fish sorting task should be improved preferentially if the physical burden is large. Observing the fish sorting task in detail, the percentage of time using poor work postures was high in cases A and B (on deck), while it was low in cases B (on wharf) and C. In the former cases (A and B (on deck)), the catch was placed on the deck and the crew sorted them while kneeling or squatting with their upper bodies bent forward. For the crew member in case B, the physical burden involved in the fish sorting task was judged to be the greatest of all tasks, because he often raised his hand above his shoulder to throw the clams into the automatic sorting machine. In comparison, in the latter cases (B (on wharf) and C), the catch was placed on tables and the crew sorted them while sitting. In summary, fish sorting tasks performed on deck cause poor work postures, while the postures are improved by using a table.

In Japan, the size of fishing boats is regulated using the gross tonnage, which roughly corresponds to the volume of the boat. The regulations are based on the idea that the fishing capability of a boat is generally proportional to its size. As a result, fishermen mostly use the limited volumes of their boats to maximize their fishing capability, leaving little space for fishermen to work safely and comfortably. In the small trawl fishery, it is obvious that the physical burden placed on the crew is improved if they use a table for fish sorting, although tables are used rarely, perhaps because there is insufficient space to install a table because the volume of the fishing boat is regulated. Therefore, from an ergonomic perspective it might be undesirable to regulate the size of fishing boats using gross tonnage. A more desirable way to regulate the size of fishing boats should be developed to match the modern circumstances of Japanese fishermen.

ACKNOWLEDGEMENTS

I am grateful for all the fishermen who kindly cooperated with my researches, and for the people who helped the implementations of the investigations. I am also thankful with my colleagues who joined some of the investigations, and Professor Nobuo Kimura for his kind advice.

REFERENCES

[1] Ministry of Agriculture, Forestry and Fisheries. *The 5–12th census of fisheries*; 1975–2010: http://www.maff.go.jp/j/tokei/census/fc/index.html. (in Japanese).

[2] Food and Agricultural Organization of the United Nations. *The state of world fisheries and aquaculture* 2012. Rome; 2012.

[3] Takahashi H, Saeki K, Watanabe K. Work analysis of a small trawl fishery at Minami-chita district, Aichi prefecture. *Fish Eng* 2012; **49(2)**: 133–140. (in Japanese).

[4] Takahashi H, Saeki K, Watanabe K. Work status of fish sorting tasks of a small trawl fishery at Kuwana city, Mie prefecture. *Proceedings of the annual meeting of the Japanese society of fisheries engineering* 2010; 177–180. (in Japanese).

[5] Takahashi H. Work analysis of a small trawl fishery (dredge net fishery) at Kishiwada city, Osaka prefecture. *J Fish Tech* 2013: **6(1)**: 89–98. (in Japanese).

[6] Karhu O, Kansi P, Kuorinka I. Correcting work posture in industry: a practical method or analysis. *Appl Ergon* 1977; **8(4)**: 199–201.

[7] Takahashi H. Work analysis on a small trawl fishery at Choshi city, Chiba prefecture. *Fish Eng* 2009; **46(1)**: 1–8. (in Japanese).

[8] Takahashi H. Work condition of the small trawl fishery in Hatsukaichi city, Hiroshima prefecture. *J Ergon Occup Saf Health* 2013; **15s**: 120–123. (in Japanese).

3   *Occupational ergonomics*

New Ergonomics Perspective – Yamamoto (Ed.)
© *2015 Taylor & Francis Group, London, ISBN 978-1-138-02751-0*

Difference between low back load caused by changing a diaper at various bed heights in female care workers

Kaoru Kyota & Keiko Tsukasaki

Department of Nursing, School of Health Science, College of Medical, Pharmaceutical and Health Science, Kanazawa University, Kodatsuno, Kanazawa, Ishikawa, Japan

ABSTRACT: *Aim*: Muscle activities and subject evaluations were compared for diaper changing across various bed heights to investigate working techniques and the reduction of load on the dorsolumbar region due to bed height.

Subject: The subjects were 21 female skilled caregivers.

Method: The bed height was adjusted to three different heights: the level at which the subject usually changes diapers, a level she prefers, and 45% of the subject's height. The activities of eight muscles were measured using surface electromyogram (TeleMyo2400) and subject evaluations were compared across the differing bed heights.

Results: The bed height at which the 21 subjects usually change diapers was 51.7 ± 7.2 cm, the bed height that they preferred was 63.4 ± 8.9 cm, and the bed height corresponding to 45% of their height was 71.0 ± 2.5 cm. Time required for changing a diaper were significantly different across the three bed heights ($p = 0.006$). The right erector spinae muscle activity level while changing a diaper was significantly different across the three heights ($p = 0.034$), and so were the subjective evaluations ($p < 0.001$). The fulcrum of diaper change movement differed across the three heights: the fulcrum was set at the thighs and knees at the routine height, thighs, knees, and lower abdominal region at the preferred height, and thighs at a height corresponding to 45% of the subject's height.

Discussion: The bed height at which care workers usually changed diapers (about 52 cm) may have been determined in order to assist with transfer of wheelchair patients, but the preferred bed height was higher than that of the routine height by 11 cm and required less time, suggesting that caregivers could more efficiently change a diaper at their preferred bed height. At 45% of the subject's height, high activity of the erector spinae muscle was observed, but subjectively, caregivers felt a low back load suggesting inconsistency between muscle activity and subjective evaluations. It was assumed that skilled caregivers made contact with the bed mattress with their bodies while changing a diaper, adjusting the fulcrum depending on bed height to reduce the low back load.

Conclusions: Skilled caregivers mastered and practiced adjusting the fulcrum of diaper change movement corresponding to the bed height. It was suggested that the bed height at which they can easily temporarily kneel on one knee is about 52 cm (height ratio: about 32%), and the 45% bed height involves a risk of low back pain.

Keywords: surface electromygram; bed height; low-back load; care worker; diaper change

1 INTRODUCTION

The incidence of low back pain among hospital nurses, care workers of nursing homes, and caregivers in families is high [1,2]. Low back pain is a major health problem for female nurses because their physical strength is generally weaker than that of males, and the strength of the upper half of their body is insufficient [3,4]. Diapers have to be changed frequently regardless of the time of day or night, and caregivers do it mostly in a forward-bent posture. More than 20 degrees of anteflexion of the trunk causes deviation from a stable posture. The resultant load on the erector spinae muscle is greater than of that in the standing position [5].

To reduce the load on the dorsolumbar muscles while nursing on a bed, the use of optimum work techniques and adjustment of the bed height to an appropriate level are recommended as ergonomic countermeasures [6]. Shogenji et al.[7] recommended kneeling on one knee on a 50 cm-high bed for caregivers. For nurses, Pheasant [8] recommended that, anthropometrically, kneeling on one knee is difficult when the bed height is 70 cm or higher. In hospitals and nursing homes, the bed height varies depending on the activity of the workers and the preference of patients requiring long-term care, and several workers care for each patient. Therefore, it is necessary to identify the bed height at which caregivers can kneel on one knee and adjust their position during each nursing session. However, the optimum bed height for kneeling on one knee has not been identified, and work techniques corresponding to the bed height have not been established.

De Looze et al.[9] observed that time-integrated compression and peak shear force significantly decreased when the nurses adjusted the bed height to one that they thought was appropriate. Caboor et al.[10] and De Looze et al.[9] compared the standard height and the height chosen by subjects to identify which height produced the smaller low back load, but the optimum bed height has not yet been identified. The bed height producing minimal low back load while nursing has not been established in Japan or in other countries. In Japan, a bed height at 45% of the subject's height is easy for administering care. However, there are no available studies with objective data that confirm that a bed height at 45% of the subject's height is the optimal height.

Skilled caregivers frequently change diapers of many patients requiring long-term care but do not develop low back pain. Skilled caregivers may conduct care movements producing minimal lumbar stress [11]. We assume that their diaper change movement is close to the optimum work technique, and they practice it while experiencing only a small load on the dorsolumbar muscles. Elucidation of the diaper change movements of skilled caregivers may clarify the work technique corresponding to the bed height and which bed height produces minimum low back load on the dorsolumbar region.

2 AIM

In this study, skilled caregivers changed a diaper at the bed height set in their routine practice, the height preferred by them, and the height corresponding to 45% of their height. Imaging analysis of the diaper change movement was performed, and dorsolumbar agonistic muscle activity and subjective fatigue of the low back were assessed to identify the bed height resulting in minimum low back load. In addition, the nursing movement producing minimum low back load corresponding to the bed height was investigated.

3 METHOD

3.1 *Participants*

The subjects meeting the following criteria were selected:

1. Female certified care workers working at health care facilities or special nursing homes for the elderly for three years or longer.
2. Frequently changing diapers (almost every day).
3. Healthy, without low back pain.

Participation in the study was publicly advertised at seven facilities near the study facility, and 21 subjects aged 24–59 years gave consent for participation (Table 1).

3.2 *Data collection*

Subjects performed their usual diaper change movements three times in a laboratory, either at Kanazawa University or a nursing home, during October to December 2011, to clarify the

Table 1. Subject attributes.

	Skilled caregiver n = 21	Study patient n = 7
Age	38.7 ± 11.6	37.1 ± 13.5
Height (cm)	159.0 ± 5.4	160.5 ± 5.2
Body weight (kg)	53.8 ± 7.0	55.0 ± 6.7
Professional experience (years)	7.5 ± 3.4	–
Right-handed	18(85.7%)	–
Left-handed	3(14.3%)	–
Frequency of diaper change in the day	5.7 ± 3.5	–
(number of times) in the night	16.2 ± 12.5	–

Mean and std. dev.

state of low back load while changing a diaper. The height of the bed was set at each of the following three levels:

1. a height corresponding to 45% of the subject's height;
2. the bed height used in their routine diaper change work;
3. the bed height the subject preferred.

The nursing movement was repeated three times at five minute or longer intervals. The bed width was 80 cm, the height included the mattress thickness, and manually removable bed fences were attached to both sides.

Seven healthy, certified care workers or nurses aged 21–54 years (see Table 1) were collected by public advertisement and gave their consent to act out the role of patients requiring long-term care. We assumed that patients' handicaps and environment would influence the nurses' lumbar load during the diaper change, and thus established a handicap (right hemiplegia) for all patients and performed measurements on the same bed. Each caregiver changed a diaper three times for the same simulated patient requiring long-term care.

The following two items were surveyed in the subjects:

1. Agonistic muscle activities of the trunk and lower limbs during diaper changing:
   A surface electromyograph, Telemyo2400 (NORAXON, USA) was attached to the subjects while diaper changing three times. The sampling frequency and frequency band were set at 1,500 and 10–500 Hz, respectively. As the agonists of the trunk and lower half of the body, the bilateral lumbar erector spinae, rectus abdominis, vastus lateral, and semitendinosus muscles (eight muscles) were selected, and muscle activity recorded. Maximum Voluntary Contraction (MVC) of each muscle was measured and normalized (%MVC). The integral of %MVC was calculated per second of diaper changing. The measurement data were analysed using Myoresearch XP software (NORAXON, USA). Video was simultaneously recorded, and the movement and muscle activity were analyzed in real-time.
2. Subjective fatigue of the low back after changing diapers:
   After changing a diaper three times, the strength of subjective fatigue of the low back was measured using the visual analogue scale (0–100 mm).

3.3 *Ethical considerations*

This study was approved by the Medical Ethics Committee of Kanazawa University (No. 324, June 7, 2011).

3.4 *Data analysis*

The time required for changing a diaper, bed height, and muscle activity were analyzed using repeated measures analysis of variance (repeated measures ANOVA) and the Bonferroni test. For comparison of subjective fatigue of the low back, Friedman's test and multiple comparison tests were employed. Statistical analysis was performed using IBM SPSS

Statistics 20. The significance level of the Bonferroni test was set employing Bonferroni correction: 0.05/3 = 0.0167. The significance level of the other tests was set at 5%.

4 RESULTS

4.1 Bed height

The bed height at which the 21 subjects usually changed diapers was 51.7 ± 7.2 cm (range: 46–65 cm), and it corresponded to 32.5 ± 4.7% (range: 27–42%) of their height (height ratio). The height preferred by the subjects was 63.4 ± 8.9 cm (range: 44–79 cm), which was 40.0 ± 5.5% (range: 28–51%) of subjects' height. The bed height corresponding to 45% of subjects' height (hereafter, the 45% bed height) was 71.0 ± 2.5 cm (range: 67–77 cm). Significant differences were noted among these three heights (repeated measures ANOVA; $F(2, 40) = 53.51$, $p < 0.001$).

4.2 Time required for changing a diaper at the three bed heights

The time required for changing a diaper was 3.2 ± 0.7 minutes at the routine height, 3.1 ± 0.1 minutes at the preferred height, and 3.4 ± 1.0 minutes at the 45% bed height, showing significant differences across the three heights (repeated measures ANOVA, $F = 5.79$, $p = 0.006$). In Bonferroni multiple comparison, the time required at the preferred height was shorter than that at the 45% bed height.

4.3 Muscle activity at the three bed heights

The right erector spinae activity level significantly differed across the three bed heights (repeated measures ANOVA, $F = 3.67$, $p = 0.034$) (see Figure 1). The muscle activity level increased in the order of the routine height, preferred height, and the 45% bed height, but no significant difference was noted following Bonferroni multiple comparison. No significant differences were noted regarding the activity level of the left erector spinae muscle, bilateral rectus abdominis, vastus lateral, or semitendinosus muscle at any of the three heights.

4.4 Diaper change movement at the three bed heights

At the 45% bed height, all 21 subjects made contact with the bed mattress with their thighs to form a fulcrum throughout diaper changing. At the routine height, the largest number of the subjects (13, 62%) contacted the mattress or bed frame with their thighs or knees throughout diaper changing, and the bed height and its height ratio were 54.5 ± 7.9 cm (range: 47–65 cm)

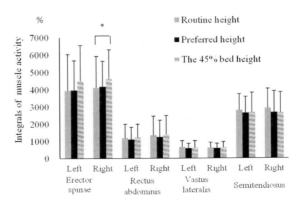

Figure 1. Comparison of activity levels of eight muscles across the three bed heights n = 21, Mean ± std. dev. Repeated measures analysis of variance. * $p < 0.05$.

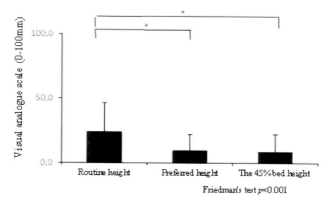

Figure 2. Comparison of subjective fatigue of the low back across the three bed heights n = 21, Mean ± std. dev. Multiple comparison. * $p < 0.05$.

and 34.4 ± 5.0% (range: 29–42%), respectively. At the preferred height, the largest number of the subjects (17, 81%) contacted the mattress or bed frame with their thighs, knees, or lower abdominal region throughout diaper changing, and the bed height and height ratio were 65.8 ± 7.4 cm (range: 54–79 cm) and 41.3 ± 4.7% (range: 34–51%), respectively. Three subjects (14%) knelt on one knee on the bed.

4.5 *Objective and subjective evaluations of low back load*

The levels of subjective fatigue of the low back after diaper changing were 24.0 ± 22.7, 9.5 ± 12.5, and 8.9 ± 13.4 out of 100 mm at the routine and preferred heights, and the 45% bed height respectively showing significant differences across the three heights (Friedman's test, $\chi^2 = 27.22$, $p < 0.001$).

5 DISCUSSION

5.1 *Bed height while changing a diaper by skilled caregivers*

The bed height in routine diaper changes was lower than that preferred by the skilled caregivers. The bed height may have been adjusted to assist with transfer to a wheelchair for those with high dependency on long-term care. However, the time required for changing a diaper was shorter when the height was adjusted to the preferred level, suggesting that caregivers can more efficiently change a diaper at their preferred bed height.

5.2 *Bed height producing minimal low back load*

The right erector spinae muscle activity level of the 21 skilled caregivers while changing a diaper was significantly different across the three bed heights (about 52, 63, and 71 cm, respectively), but no significant difference was detected in multiple comparison tests. This may have been due to the fact that the subjects adjusted their movement to the bed height to reduce the low back load. There have been few studies on the load on the spinal muscles associated with bed height during patient care, and the clarification is insufficient [9, 10]. The present study focused on skilled caregivers and investigated their diaper change movements and the lumbar erector spinae muscle activity during movements at various bed heights.

5.3 *Fulcrum while changing a diaper corresponding to the bed height*

Studies have also observed muscle activity during dishwashing [12] and other activities besides diaper changing [13] which have reported lower lumbar load when position forming a fulcrum is used. The skilled caregivers made contact with the bed mattress or frame with their thighs, knees, and lower abdominal region to form a fulcrum, through which the force loaded on the dorsolumbar region may have been dispersed, reducing the low back load.

5.4 Bed height when one knee was placed on the bed

At the routine height, seven of the 21 subjects temporarily knelt on one knee on the bed while diaper changing; the height at which they could easily kneel on one knee while changing a diaper was about 52 cm (about 32% height ratio). This study concluded that kneeling on one knee while changing a diaper is useful at the standard bed height used in nursing homes (50 cm) [7]. A 52 cm bed height corresponded to about 32% of the height of the caregivers, who may have retained their posture to improve access to the patient and move more easily.

5.5 Difference between objective and subjective evaluations of low back load

The right erector spinae muscle activity level was higher at the 45% bed height than at the routine and preferred heights, but subjective fatigue of the low back were smaller, suggesting that the low back load accumulates when caregivers change diapers at the 45% bed height because subjective symptoms were mild. It is necessary to take a sufficient rest after diaper changes in consideration of the load on the dorsolumbar muscles.

6 LIMITATIONS

As limitations of this study, it is possible that the erector spinae muscle activity level was underestimated in electromyography, because a flexion-relaxation phenomenon may occur due to the absence of muscle activity in a forward-bent posture, for which measurement of the low back load based on the trunk inclination angle [5] is necessary, in addition to electromyography.

ACKNOWLEDGEMENT

We are grateful to representatives of all the cooperating nursing homes and all the subjects who participated in this study. This study was supported by the France Bed Medical Home Care Research Subsidy Public Interest Incorporated Foundation of Japan, 2011.

REFERENCES

[1] Smith DR, Mihashi M, Adachi Y, Koga H, Ishitake T. A Detailed analysis of musculoskeletal disorder risk factors among Japanese nurses. *J Safety Res* 2006; **37**: 195–200.
[2] Yalcinkaya EY, Ones K, Ayna AB, Turkyilmaz AK, Erden N. Low back pain prevalence and characteristics in caregiver of stroke patient: a pilot study. *Top Stroke Rehabi* 2010; **17**: 389–93.
[3] Smedley J, Egger P, Cooper C, Coggon D: Prospective cohort study of predictors of incident low back pain in nurses. *Br Med J* 1997; **314**: 1225–228.
[4] Karahan A, Kav S, Abbasoglu A, Dogan N: Low back pain: prevalence and associated risk factors among hospital staff. *J Adv Nurs* 2009; **65**: 516–24.
[5] Punnett L, Fine LJ, Keyserling WM, Herrin GD, Chaffin DB: Back disorders and nonneutral trunk postures of automobile assembly workers. *Scand J Work Environ Health* 1991; **17**: 337–46.
[6] Nelson A, Baptiste AS: Evidence-based practices for safe patient handling and movement. *Orthop Nurs* 2006; **25**: 366–79.
[7] Shogenji M, Izumi K, Seo A, Inoue K: Biomechanical analysis of the low back load on healthcare workers due to diaper changing. *J Tsuruma Health Sci Soc, Kanazawa University* 2007; **31**: 57–69.
[8] Pheasant S: Some anthropometric aspects of workstation desigin. *Int J Nurs Stud* 1987; **24**: 291–97.

[9] Delooze MP, Zinzen E, Caboor D, Heyblom P, Van Bree E, Van Roy P et al.: Effect of individually chosen bed-height adjustments on the low-back stress of nurses. *Scand J Work Environ Health* 1994; **20**: 427–34.

[10] Caboor DE, Verlinden MO, Zinzen E, Van Roy P, Van Riel MP, Clarys JP: Implications of an adjustable bed height during standard nursing tasks on spinal motion, perceived exertion and muscular activity. *Ergonomics* 2000; **43**: 1771–780.

[11] Koshiono Y, Ohno Y, Hashimoto M, Yoshida M: Evaluation parameters for care-giving motions, J *Phys Ther Sci* 2007; **19**: 299–306.

[12] Iwakiri K, Sotoyama M, Mori I, Jonai H, Saito S: Shape and thickness of cushion in a standing aid to support a forward bending posture: Effects on posture, muscle activities and subjective discomfort. *Industrial Health* 2004; **42**: 15–23.

[13] Skotte JH: Estimation of low back loading on nurses during patient handling tasks: the importance of bedside reaction force measurement. *J Biomech* 2001; **34**: 273–76.

New Ergonomics Perspective – Yamamoto (Ed.)
© 2015 Taylor & Francis Group, London, ISBN 978-1-138-02751-0

Effects of intervention trampoline exercise on neck strength and physiological responses

Bor-Shong Liu
Department of Industrial Engineering and Management, St. John's University, Tamsui District, New Taipei City, Taiwan

Tung-Chung Chia
Department of Business Administration, Ling Tung University, Taichung City, Taiwan

Hui-Yu Wang
Department of Industrial Engineering and Management, St. John's University, Tamsui District, New Taipei City, Taiwan

Ching-Wen Lien
Nursing Department, Taipei Veterans General Hospital, Beitou District, Taipei City, Taiwan

ABSTRACT: Literatures reviewing showed that the long-term abnormal postures caused operators to be suffered the musculoskeletal disorders of neck areas. The purpose of present study was to provide the trampoline exercise program and examine the effectiveness of an intervention program on neck strength and physiological responses. The trampoline training program consisted of basic trampoline exercises including basic and knee bouncing. The trampoline bed is rectangular 4.28 by 2.14 meters in size fitted into the 5.05 by 2.91 meters frame with around 110 steel springs. Exercises were performed up to subjectively evaluated fatigue, normally in 30–60 sec in one set, and there were similar 30–60 sec recovery times between sets. In the beginning of the training period, the set was performed twice, and it was repeated three times after 2 week. Thus, each participant needed to take the trampoline exercise for 10 minutes thrice a week and total of eighteen trials should be executed during six weeks. The neck muscle strength of flexion, extension, left lateral and right lateral bending were measured in sitting posture by the dynamometer. The heart rate and blood pressure were measured by blood pressure monitoring. Finally, neck strength and physiological responses have been compared to before and after the intervention trampoline exercise. Results of pair-t test showed that the neck muscle strength of flexion, extension, left lateral and right lateral flexion increased for 2.73 kg, 2.62 kg, 1.64 kg and 3.33 kg respectively after intervention the trampoline exercise program. In addition, ranges of motion in neck were also increased. For physiological response, mean incremental heart rate from resting situation was 32.28 bpm after trampoline exercise. The mean systolic blood pressure increased 21.2 mmHg after trampoline exercise. However, mean diastolic blood pressure is not significant difference after trampoline exercise. Trampoline training has been considered as a tool to create a "G environment" for physical training. The trampoline training could be improved general motor skills and to enhance neck strength and muscle balance.

Keywords: musculoskeletal disorders; trampoline exercise; neck strength; cardiac responses

1 INTRODUCTION

Work-related Musculoskeletal Disorders (WMSDs) are caused and/or exacerbated by work-place exertions. One-third or more of registered occupational diseases in the United States, Canada, the Nordic countries, Japan are caused by WMSDs [1]. Over 44 million (one in six) members of the European Union workforce have a long-standing health problem or disability that affects their work capacity [2]. According to the United States (US) Bureau of Labor Statistics (BLS), WMSDs accounted for 29–35% of all occupational injuries and illnesses involving days away from working private industries from 1992 to 2010 [3]. The number of reported work-related MSDs declined from 435,180 in 2003 to 335,390 in 2007 and the direct costs of WMSDs were 1.5 billion dollars for the year 2007. The indirect costs were 1.1 billion dollars for WMSDs for the year 2007 [4]. Neck pain and complaints have become a significant health and economic problem in our society [5, 6]. Work-related musculoskeletal disorders in neck and shoulders are frequently reported among computer operators [7–9]. The incidence of functional disorders of the neck, typically associated with pain and muscular fatigue, is becoming a severe problem in industrial countries, even among young and middle-aged people [10]. Further, review literatures showed that the long-term abnormal postures caused operators to be suffered the musculoskeletal disorders of head and neck areas [11]. There is ample evidence that cervical strength is compromised in various cervical disorders [12]. The sustained muscle contraction required to hold the head in various positions and the fatigue caused by muscular weakness are suspected of being contributing factors in chronic neck pain.

A supervised neck/shoulder exercise regimen was effective in reducing neck pain cases in air force helicopter pilots. This was supported by improvement in neck-flexor function post-intervention in regimen members [13]. Zebis et al. [14] evaluated the effect of implementing strength training at the workplace on non-specific neck and shoulder pain among industrial workers. Participants were randomized to 20 weeks of high-intensity strength training for the neck and shoulders three times a week (n = 282) or a control group receiving advice to stay physically active (n = 255). High-intensity strength training relying on principles of progressive overload can be successfully implemented at industrial workplaces, and results in significant reductions of neck and shoulder pain. Exercises on a mini-trampoline consist of a multi-component approach involving strength and balance training, physical fitness, body stability, muscle coordinative responses, joint movement amplitudes and spatial orientation [15]. Thus, the purpose of present study was to provide the trampoline exercise program and examine the effectiveness of an intervention program on neck strength and physiological responses.

2 METHODS

2.1 *Participants*

Present study recruited 39 participants including 19 males and 20 females. In addition, age was divided into two groups (≤39 years old and ≥40 years old). Average height of male subjects was 172.4 cm (SD = 5.6) and their average weight was 78.9 kg (SD = 15.5). The average height of female subjects was 160.2 cm (SD = 5.6) and their average weight was 56.1 kg (SD = 9.3). All subjects were healthy and reported no musculoskeletal problems that might influence performance detrimentally.

2.2 *Trampoline training program*

The trampoline training program consisted of basic trampoline exercises including basic and knee bouncing. The trampoline bed is rectangular 4.28 by 2.14 meters (14 ft 1 in × 7 ft 0 in) in size fitted into the 5.05 by 2.91 meters (17 ft × 10 ft) frame with around 110 steel springs. Exercises were performed up to subjectively evaluated fatigue, normally

Figure 1. (a) Basic trampoline exercise; (b) Knee bouncing exercise.

in 30–60 sec in one set, and there were similar 30–60 sec recovery times between the sets. In the beginning of the training period, the set was performed twice, and after 2 week, it was repeated three times. Thus, each participant needed to take the trampoline exercise for 10 minutes thrice a week and total of eighteen trials should be executed during six weeks (Fig. 1). For each subject, a resting period of at least 5 min or longer, if required, was provided until heart rate remained a steady rhythm before the actual experiment was started; the resting heart rate was collected afterward (to serve as a baseline measurement). During the resting period, participants were requested to sit, relax, and remain silent. Mean heart rate (bpm) and incremental heart rate (working heart rate minus resting heart rate) were considered indices of physiological workload. The heart rate, systolic blood pressure and diastolic blood pressure were measured by blood pressure monitoring (Hem-7070, Omron, Japan). Further, the neck muscle strength and neck range of motion on flexion, extension, left lateral and right lateral bending were measured in before and after intervention trampoline exercise in sitting posture by the dynamometer (TKK, Japan) and Goniometer (Johnson, USA). Finally, neck strength, ranges of motion and physiological responses have been compared to before and after the intervention trampoline exercise.

2.3 Data analysis

All data were coded and summarized using SPSS Statistics (Version 21.0, IBM, New York, USA). Furthermore, the paired-sample t test was used to compare responses before and after trampoline exercise. Multivariate Analysis of Variance (MANOVA) was utilized to identify significant differences between gender and age after intervention trampoline exercise. The level of significance used for all analyses was $p < 0.05$.

3 RESULTS

3.1 Intervention trampoline exercise

Results of the paired-sample t test were shown in Table 1. The neck strength and range of motion were significant difference after intervention trampoline exercise ($p < 0.001$). The neck strength of lateral bending increased in 1.64 kg and 3.33 kg for left and right lateral bending respectively. Neck strength of flexion and extension increased in 2.73 kg and 2.62 kg respectively after intervention trampoline exercise. For range of motion, left and right lateral bending increased in 9.28 and 10.18 degrees respectively. The 8.1 and 7.54 degrees increased for flexion and extension respectively after intervention trampoline exercise.

Table 1. Comparison of neck strength and range of motion before and after completion of the intervention trampoline exercise.

Variables	After exercise	Before exercise	Paired-difference	t	P value
Left lateral bending (kg)	5.67	4.03	1.64	6.04	0.001
Right lateral bending (kg)	8.34	5.01	3.33	7.42	0.001
Flexion (kg)	7.26	4.53	2.73	6.17	0.001
Extension (kg)	8.87	6.25	2.62	7.78	0.001
Left lateral bending (°)	49.15	39.87	9.28	6.07	0.001
Right lateral bending (°)	48.28	38.10	10.18	7.34	0.001
Flexion (°)	51.05	42.95	8.10	5.10	0.001
Extension (°)	61.67	54.13	7.54	4.44	0.001

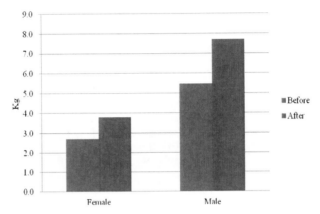

Figure 2. Comparison of neck strength between genders after intervention trampoline exercise.

3.2 Effect of gender and age

Results of MANOVA showed that the gender effects had found (Pillai's trace = 0.55, $F(9, 27) = 3.66$, $p < 0.01$) on 0.95 of statistical power (alpha = 0.05, two-tail). Male subjects have more effectively in neck strength of lateral bending and there is more than 40% increasing after intervention trampoline exercise (Fig. 2). However, there are not significant differences in neck strength and range of motion between age groups.

3.3 Physiological response

For physiological response, mean incremental heart rate from resting state was 32.28 bpm after trampoline exercise. The mean systolic blood pressure increased 21.2 mmHg and mean diastolic blood pressure also increased 1.42 mmHg after trampoline exercise. Results of MANOVA showed that the gender effects had found in Table 3. The mean systolic blood pressure of male subjects was higher than females after trampoline exercise (145.47 mmHg versus 122.7 mmHg). Incremental heart rate and incremental systolic blood pressure were higher in male subjects. For age effect, results of MANOVA showed that younger subjects had higher physiological responses after trampoline exercise.

Effect of exercise times on heart rate showed in Fig 2 and three growing stages have found. In first stage, there is significant decreasing physiological workload after eighth trampoline exercise. The second stage is from eighth to fourteenth trampoline exercise. Final, there is steady state after fourteenth trampoline exercise. Thus, intervention trampoline exercise program should be at least fourteenth exercise for stable outcome.

Table 2. Comparison of incremental neck strength and range of motion after intervention trampoline exercise.

Gender		Left lateral bending (kg)	Right lateral bending (kg)	Flexion (kg)	Extension (kg)	Left lateral bending (°)	Right lateral bending (°)	Flexion (°)	Extension (°)
Female	Mean	1.10***	2.36***	2.10***	2.01***	7.30	9.75	10.10	5.50*
	SD	0.77	2.16	1.81	1.80	8.55	9.54	9.77	11.30
Male	Mean	2.21	4.37	3.39	3.26	11.37	10.63	9.89	9.68
	SD	2.19	3.08	3.43	2.25	10.32	7.87	6.21	9.63
Age									
≥40	Mean	1.92	4.02	3.07	2.56	7.63	6.21	9.74	5.26
	SD	2.01	3.33	2.72	2.27	7.14	8.29	11.99	10.28
≤39	Mean	1.38	2.69	2.40	2.67	10.85	13.95	6.55	9.70
	SD	1.34	2.09	2.83	1.99	11.36	7.37	7.42	10.69

NB: *p < 0.05, **p < 0.01; ***p < 0.001.

Table 3. Comparison of incremental physiological responses after intervention trampoline exercise.

Gender		HR	SBP	DBP	Δ HR	ΔSBP	ΔDBP
Female	Mean	111.66	122.70***	72.32***	29.89***	17.40***	1.13
	SD	17.83	12.86	7.05	15.87	12.38	7.36
Male	Mean	111.63	145.47	75.80	34.85	25.27	1.74
	SD	15.71	16.24	9.83	15.00	14.40	8.13
Age							
≥40	Mean	104.70***	134.42	74.27	27.12***	19.84*	0.55*
	SD	12.83	18.59	8.12	12.08	14.64	6.10
≤39	Mean	118.36	133.39	73.80	37.08	22.46	3.26
	SD	17.47	18.47	9.23	17.01	13.17	8.62

NB: *p < 0.05, **p < 0.01; ***p < 0.001.

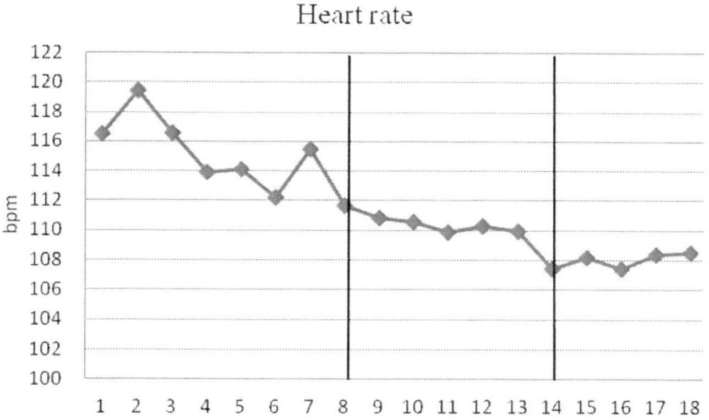

Figure 3. Effect of exercise times on heart rate.

4 DISCUSSION AND CONCLUSION

In a cross-sectional study of 1,065 visual display terminal workers, the 12-month prevalence of pain in the neck, shoulder, hand/wrist, and elbow/lower arm was 55%, 38%, 21%, and 15%, respectively [16]. On the other hand, a cross-sectional study of 14,384 workers in an

Iranian car manufacturing company showed that self-reported 12-month neck pain was prevalent in only 7% and shoulder pain in only 6.1% of the workers [17]. Neck-and-shoulder pain and complaints constitute an important public-health problem. Äng et al. [13] indicated that a supervised neck/shoulder exercise regimen was effective in reducing neck pain cases in air force helicopter pilots. This was supported by improvement in neck-flexor function post-intervention in regimen members. Exercises on a mini-trampoline consist of a multi-component approach involving strength and balance training, physical fitness, body stability, muscle coordinative responses, joint movement amplitudes and spatial orientation [15]. Sovelius et al. [18] showed that trampoline exercise and strength training methods during the 6-week training period were found to be effective in reducing muscle strain during in-flight and cervical loading testing, especially in the cervical muscles. A round trampoline with a diameter of 4.3 m was used. Exercises were performed up to subjectively evaluated fatigue, normally in 30–60 s in one set, and there were similar 30–60 s recovery times between the sets. In the beginning of the training period, the set was performed twice, and after 2 week, it was repeated three times. Aragão et al. [19] investigated whether a 14-week mini-trampoline training contributes to improvements of dynamic stability performance in elderly subjects. They reported more improvement of plantar flexor muscle strength (~10%) and the ability to regain balance during forward falls. This regaining of balance was associated with a higher rate of hip moment generation after perturbation. Present study also reported more improvement of neck strength (~40%) and range of motion in lateral bending (~20%) and flexion/extension (~10%). Trampoline training has been considered as a tool to create a "G environment" for physical training.

The trampoline training could be improved general motor skills and to enhance neck strength and muscle balance. Giagazoglou et al. [20] also indicated that trampoline intervention resulted in significant improvements of participants' performance in all motor and balance tests. In conclusion, trampoline training can be an effective intervention for improving functional outcomes and can be recommended as an alternative mode of physical activity programming for improving balance and motor performance. In addition, Jenson et al. [21] reported that physiological responses and performance were examined during and after a simulated trampoline competition. Trampoline-specific activities were quantified by video-analysis. Countermovement Jump (CMJ) and 20 Maximal Trampoline Jump (20-MTJ) performances were assessed. Heart Rate (HR) and quadriceps muscle Temperature (Tm) were recorded and venous blood was drawn. A trampoline gymnastic competition includes a high number of repeated explosive and energy demanding jumps, which impairs jump performance during and 24 h post-competition.

The trampoline training could enhance neck strength and neck range of motion. However, numerous studies from a variety of countries have shown that trampoline use can be a potentially dangerous activity, e.g., falling off or striking injuries [22] or musculoskeletal disorders. In addition, the safety and injuries prevention should be investigated in further study.

ACKNOWLEDGEMENT

This study was supported by a grant from the Ministry of Science and Technology (Taiwan) and Project No. NSC100-2682-129-001-MY3.

REFERENCES

[1] Punnett L, Wegman DH. Work-related musculoskeletal disorders: the epidemiologic evidence and the debate. Journal of Electromyography and Kinesiology 2004; 14: 13–23.
[2] Bevan S, McGee R, Quadrello T. Fit for work. Musculoskeletal disorders in the European workforce. 2009.

[3] American Federation of Labor and Congress of Industrial Organizations (AFL-CIO), Report on Death on the Job, the Toll of Neglect: a National and State-bystate Profile of Worker Safety and Health in the United States. 2012.

[4] Bhattacharya A. Cost of occupational Musculoskeletal Disorder (MSDs) in the United States. International Journal of Industrial Ergonomics 2014; 44: 448–454.

[5] Bovim G, Schrader H, Sand T. Neck pain in the general population. Spine 1994; 19: 1307–1309.

[6] Fejer R, Kyvik KO, Hartvigsen J. The prevalence of neck pain in the world population: a systematic critical review of the literature. European Spine Journal 2006; 15(6): 834–848.

[7] Gerr, F., et al. "A prospective study of computer users: I. Study design and incidence of musculoskeletal symptoms and disorders," American journal of industrial medicine, 41(4): 221–235.

[8] Brandt, L.P., Andersen, J.H., Lassen, C.F., Kryger, A., Overgaard, E., Vilstrup, I., & Mikkelsen, S. (2004). Neck and shoulder symptoms and disorders among Danish computer workers. Scandinavian journal of work, environment & health, 30(5): 399–409.

[9] Andersen JH, Harhoff M, Grimstrup S, Vilstrup I, Lassen CF, Brandt LP, Mikkelsen S. Computer mouse use predicts acute pain but not prolonged or chronic pain in the neck and shoulder. Occupational and environmental medicine 2008; 65(2): 126–131.

[10] Garcés G, Medina D, Milutinovic L, Garavote P, Guerado E. Normative database of isometric cervical strength in a healthy population. Medicine and Science in Sports & Exercise. 2002; 34: 464–470.

[11] Arvidsson I, Hansson GÅ, Erik Mathiassen S, Skerfving S. Neck postures in air traffic controllers with and without neck/shoulder disorders. Applied Ergonomics 2008; 39(2): 255–260.

[12] Prushansky T, Gepstein R, Gordon C, Dvir Z. Cervical muscles weakness in chronic whiplash patients. Clinical Biomechanics 2005; 20(8): 794–798.

[13] Äng BO, Monnier A, Harms-Ringdahl K. Neck/shoulder exercise for neck pain in air force helicopter pilots: a randomized controlled trial. Spine 2009; 34(16): E544-E551.

[14] Zebis MK, Andersen LL, Pedersen MT, Mortensen P, Andersen CH, Pedersen MM. Implementation of neck/shoulder exercises for pain relief among industrial workers: a randomized controlled trial. BMC musculoskeletal disorders 2011; 12(1): 205.

[15] Miklitsch C, Krewer C, Freivogel S, Steube D. Effects of a predefined mini-trampoline training programme on balance, mobility and activities of daily living after stroke: a randomized controlled pilot study. Clinical Rehabilitation 2013; 27(10): 939–947.

[16] Klussmann, A, Gebhardt H, Liebers F, Rieger MA. Musculoskeletal symptoms of the upper extremities and the neck: a cross-sectional study on prevalence and symptom-predicting factors at Visual Display Terminal (VDT) workstations. BMC Musculoskeletal Disorder 2008; 9: 96.

[17] Alipour A, Ghaffari M, Shariati B, Jensen I, Vingard E. Occupational neck and shoulder pain among automobile manufacturing workers in Iran. American Journal of Industrial Medicin 2008; 51: 372–379.

[18] Sovelius R, Oksa J, Rintala H, Huhtala H, Ylinen J, Siitonen S. Trampoline exercise vs strength training to reduce neck strain in fighter pilots. Aviation, Space, and Environmental Medicine 2006; 77: 20–25.

[19] Aragão FA, Karamanidis K, Vaz MA, Arampatzis A. Mini-trampoline exercise related to mechanisms of dynamic stability improves the ability to regain balance in elderly. Journal of Electromyography Kinesiology 2011; 21: 512–518.

[20] Giagazoglou P, Kokaridas D, Sidiropoulou M, Patsiaouras A, Karra C, Neofotistou K. Effects of a trampoline exercise intervention on motor performance and balance ability of children with intellectual disabilities. Research in Developmental Disabilities 2013; 34: 2701–2707.

[21] Jensen P, Scott S, Krustrup P, Mohr M. Physiological responses and performance in a simulated trampoline gymnastics competition in elite male gymnasts. Journal of Sports Sciences 2013; 31(16): 1761–1769.

[22] Eager D, Scarrott C, Nixon J, Alexander K. Survey of injury sources for a trampoline with equipment hazards designed out. Journal of Paediatrics and Child Health 2012; 48(7): 577–581.

New Ergonomics Perspective – Yamamoto (Ed.)
© 2015 Taylor & Francis Group, London, ISBN 978-1-138-02751-0

The effects of wearing encapsulating protective clothes, workload, and environment temperature on heat stress

Peng-Cheng Sung & Po-Sheng Hsu
Department of Industrial Engineering and Management, Chaoyang University of Technology, Taichung City, Taiwan

Shu-Zon Lou
School of Occupational Therapy, Chung Shan Medical University, Taichung, Taiwan

ABSTRACT: Heat stress is the net heat burden on the body from the combination of the body heat generated from working, environmental sources and clothing. It is also a well-recognized safety and health hazard especially for workers that have to wear Encapsulating Protective Clothing (EPC) at work. Currently, the exposure surveillance of heat stress for workers wearing protective suit is still limited to a general assessment of the macro-environment (ambient environment) climate using the WBGT index. An underestimate of the heat stress based on macro-environment approach may result, and subsequently increase the worker's risk for heat-induced illnesses and injuries. An overestimation of the heat stress and the resultant control measures may lead to a loss of productivity. Therefore, it is more meaningful to evaluate the heat stress based on measurements taken in the micro-environment in the protective suit. According, this study assessed both the climate condition and heat strain using an integrated real-time micro-environment climate condition/heat strain monitor developed in the laboratory to evaluate the potential risk of heat hazards to improve worker safety and health. Besides this integrated real-time monitor, this study also adopted a Polar-rs800cx heart rate monitor and an electronic scale to explore the effects of macro-environment temperature (21 ± 1°C or 29 ± 1°C), EPC wearing condition (with or without), and workload (moderate/heavy/rest) on skin temperature, tympanic temperature, micro-environment temperature, and heart rate of six male subjects. In summary, all the main factors including "macro-environment temperature", "EPC wearing condition", and "workload" showed significant effects on skin temperature, tympanic temperature, heart rate, and micro-environment temperature. Significant differences between the macro- and micro-environments were also found in this study where the temperatures in the micro-environment are significantly higher than those in the macro-environment.

Keywords: Heat stress; Heat strain; Encapsulating protective clothing; Macro-environment climate; Micro-environment climate

1 INTRODUCTION

Heat stress is the net heat burden on the body from the combination of the body heat generated from working, environmental sources (air temperature, humidity, air movement and radiation from the sun or hot surfaces/sources) and clothing [1]. When the core temperature rises above 38°C, a series of heat-related illnesses, or heat stress disorders, can then develop [1, 2]. Heat stress is also a well-recognized safety and health hazard especially for workers that have to wear Encapsulating Protective Clothing (EPC) in their work [3, 4]. Operations conducted in hot weather with high humidity, such as asbestos removal, casting, construction, refining, and hazardous waste site activities, especially those that require workers to wear EPC can result in health effects ranging from transient heat fatigue to serious illness or death.

Currently, the exposure surveillance of heat stress for workers wearing EPC is still limited to a general assessment of the macro-environment (ambient environment) climate using the WBGT index [5]. Only one study [6] has been found assessing the micro-environment climate under impermeable EPC for three different ambient environments, 18.2°C, 22.3°C, and 26.9°C WBGT. The mean (±S.D.) WBGT of the micro-environment climate for each of the three ambient environments were 27.9 (±1.6)°C, 30.3 (±1.6)°C, and 32.6 (±1.5)°C WBGT, respectively. The work-rest paradigm is 30-min:30-min and the time-weighted work rate is 350 W (300 Kcal/h). The results suggest that generally, adding 10°C to the ambient WBGT index reflects the micro-environment for that type of EPC. An underestimate of the heat stress based on macro-environment approach may result, and subsequently increase the worker's risk for heat-induced illnesses and injuries. An overestimation of the heat stress and the resultant control measures may lead to a loss of productivity. It is more meaningful to evaluate the heat stress based on measurements taken in the micro-environment in the protective suit.

Heat Strain is the overall physiological response resulting from heat stress [7]. The physiological responses are dedicated to dissipating excess heat from the body. Workers may also differ in their bodily response (namely heat strain) to the heat stress. This difference in heat tolerance may be due to a variety of factors, such as acclimatization, age, gender, and physical condition on a particular day. Identifying heat-intolerant individual and provide real-time monitoring have been long been recognized as important issues in need of research and development [4, 8, 9]. Previous studies [10–14] have evaluated the physiological strain in workers wearing protective suit in environmentally controlled room. The physiological responses recorded continuously in these experiments including heart rate, arterial pressure, oxygen consumption, sweat rate, and skin, oral, tympanic, and rectal temperatures. However, none of these studies provide WBGT measurements of the micro-environment in the encapsulating protective clothing. Therefore, it is impossible to provide the data necessary for the establishment of relationship between heat stress and climate condition/heat strain in the EPC.

Heat stress may occur year-round in foundries, kitchens, or laundries, or only a few days during the summer in almost any indoor/outdoor work setting [7]. The potential health hazards from work in hot environments also depend strongly on physiological factors that lead to a range of susceptibilities [1]. Therefore, professional judgment is of particular importance in assessing the level of climate condition and physiological heat strain to adequately provide guidance for protecting nearly all healthy workers [7]. According, this study assessed both the climate condition and heat strain of subjects inside the EPC using an integrated real-time micro-environment climate condition/heat strain monitor developed in the laboratory to evaluate the potential risk of heat hazards to improve worker safety and health.

2 METHODS

2.1 Participants and encapsulating protective clothes

Six healthy unacclimatized Taiwanese males voluntarily participated in this experiment. The mean age, height, and weight of the subjects were 23.5 ± 1.8 years, 171.5 ± 7.1 cm, and 72.6 ± 19.8 Kg. MSA PLASTIKLOS (MSA the Safety Company) Level C encapsulating chemical protective garments used for chemical handling and transportation, industrial hazardous waste handling, and HAZMAT handling etc., was adopted for this proposed study.

2.2 Real-time micro-environment climate condition/heat strain monitoring system

Figure 1 shows the schematic diagram of the real-time personal micro-environment climate condition/heat strain monitoring system. This monitoring system consists of two sets of sensors and a Polar S810 heart rate monitor. The first sensor set included a relative

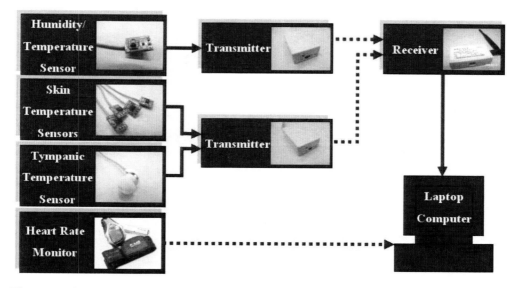

Figure 1. The schematic diagram of the real-time personal climate condition/heat strain monitoring system.

humidity/temperature sensor used to evaluate the climate condition (micro-environment WBGT) in the EPC. The second sensor set consists of four skin temperature sensors and one tympanic temperature sensor to evaluate the heat strain of the worker that donned EPC. Two sets of transmitters and receivers manufactured in the Chaoyang Ergonomics Laboratory were used to wirelessly transmit the heat stress/heat strain to a laptop computer [15]. The dotted lines in Figure 1 are used to indicate that the data are transmitted wireless while the solid lines are used to indicate that the devices are directed linked. A Visual Basic based program had also been developed in house to process and analyze the data.

2.3 *Simulated tasks*

Walking on Octane PRO 450 elliptical trainer were performed by each subject to achieve Time Weighted Average (TWA) metabolic rates of 250 Kcal/h and 350 Kcal/h representing the moderate and heavy workloads categorized by the ACGIH [1]. For each task demand, each subject completed two 60-minute walk/recovery trials separated at least 24 hours. During each 60-minute trial, a subject completed two walk/recovery cycles where each cycle is made up of 15 minutes of walk followed by 15 minutes of rest representing 50% work and 50% rest regimen.

2.4 *Statistical analysis*

The independent variables in this experiment are workload (rest/moderate/heavy), macro-environment temperature ($21 \pm 1°C$ or $29 \pm 1°C$), and EPC wearing condition (with or without). The order of presentation of the workload and EPC were randomized. The performance measures are heart rate, skin and tympanic temperatures, micro-environment WBGT and sweat loss. Descriptive statistics were computed for the performance measures and anthropometry data. Repeated measures ANOVA was used to determine whether there were significant differences between independent variables on dependent variables. All data were analyzed for statistical significance at $p \leq 0.05$ using the SPSS 17 (SPSS Inc., Chicago, Illinois) statistical software.

Table 1. Summary of the repeated-measured ANOVA results.

Source of variance	Skin temperature		Tympanic temperature		Heart rate		Micro-environment temperature	
	F	Sig	F	Sig	F	Sig	F	Sig
Macro-environmental temperature	110.451	0.000	58.538	0.001	22.957	0.005	823.411	0.000
EPC	31.074	0.003	13.927	0.014	28.109	0.003	206.345	0.000
Workload	13.070	0.002	16.722	0.001	137.139	0.000	39.486	0.000

3 RESULTS

The repeated-measured ANOVA results (see Table 1) indicated that all the main factors (macro-environment temperature, EPC wearing condition, and workload) showed significant effect on skin temperature, tympanic temperature, heart rate, and micro-environment WBGT. No statistically significant effects for all the main factors on sweat loss were found.

Higher macro-environment temperature increased skin and tympanic temperatures by 1.00°C and 0.73°C respectively than lower macro-environment temperature. Higher macro-environment temperature also generated 6.85 beats per minute (bpm) more heart rate than the lower macro-environment temperature. In addition, higher macro-environment temperature produced 6.70°C higher micro-environment WBGT than the lower macro-environment temperature.

Wearing EPC increased skin and tympanic temperatures by 0.47°C and 0.11°C. Wearing EPC also increased heart rate by 10.3 bpm. In addition, 2.35°C greater micro-environment WBGT was found when wearing EPC.

Moderate and heavy task demands generated significantly higher skin temperature (0.55°C and 0.55°C greater) than the rest condition ($p < 0.05$). Moderate and heavy task demands also generated significantly higher tympanic temperature (0.23°C and 0.31°C greater) than the rest condition ($p < 0.05$). In addition, moderate and heavy task demands produced significantly higher heart rate (29.6 bpm and 37.8 bpm greater) than the rest condition ($p < 0.001$). Heavy workload also significantly increased heart rate by 8.2 bpm than moderate workload. In addition, moderate and heavy task demands generated significantly higher micro-environment WBGT (1.94°C and 2.39°C greater) than the rest condition ($p < 0.05$). No statistically significant difference was found between moderate and heavy workloads on skin temperature, tympanic temperature, and micro-environment WBGT.

4 DISCUSSION AND CONCLUSION

Bishop et al. (2000) [6] measured the WBGT under impermeable EPC when subjects worked for 30 min followed by 30 min rest for up to 4 hours at 350 W in ambient environment of 18.2°C, 22.3°C, and 26.9°C WBGT. Their results showed that the clothing adjustments for micro-environment WBGT were 9.7°C, 8.0°C, and 5.7°C, respectively. For this research, the measured micro-environment WBGT under EPC in 21 ± 1°C and 29 ± 1°C macro-environment were 26.1°C and 32.5°C, respectively. The clothing adjustments WBGT yielding in this study, 5.1°C and 3.5°C for ambient WBGT of 21 ± 1°C and 29 ± 1°C, were lower but comparable to Bishop et al.'s (2000) [6] findings.

Chad and Brown (1995) [16] indicated that strenuous task (lifting) increased heart rate and sweat loss due to increased blood flow to the periphery compared to relatively sedentary light task (typing). Their results also showed that both tasks experienced similar thermal stress where no significant differences were observed in core and skin temperatures. Similar results were also found in this study where no significant differences were found between moderate and heavy workloads on skin and tympanic temperatures. This research also showed similar result that heavy workload increased heart rate compared to moderate workload and rest condition. However, disparity exists in these two studies over the effect on sweat loss. The reason may be due to different work and recovery time (157 min versus 60 min) duration.

Logan and Bernard (1999) [17] investigated the relationship between heat stress (macro-environment climate) and heat strain of workers in aluminum smelters and found no relationship between oral temperature and heat stress level. In contrast, this study found significant heat stress effect on tympanic temperature, skin temperature, and heart rate where higher macro-environment WBGT increased the heat strain of the subjects.

The results of this present investigation have shown that adopting the micro-environment WBGT accompanied with the heat strain data collected inside the EPC in the field will better the risk surveillance for assisting or removing of the workers to prevent heat-related disorders than using the ambient WBGT alone. In the future, heat acclimatized subjects could be recruited and tested with different work-rest schedules for longer duration to further explore the effects of independent factors included in this study on dependent variables.

ACKNOWLEDGEMENTS

We thank the National Science Council of Taiwan Grants NSC 102-2221-E-324-026 for supporting this work.

REFERENCES

[1] ACHIG. Threshold limit values for physical agents in the work environment. *American Conference of Governmental Industrial Hygienists* 2008. Cincinnati, Ohio.
[2] World Health Organization (WHO). Health factors involved in working under conditions of heat stress. *WHO Technical Report Series 142* 1969. Geneva: WHO.
[3] NIOSH. Occupational safety and health guidance manual for hazardous waste site activities. *National Institute for Occupational Safety and Health* 1985. Washington DC, USA.
[4] NIOSH. Occupational exposure to hot environments, Revised Criteria. *National Institute for Occupational Safety and Health* 1986. Washington DC, USA.
[5] ISO 7243. Hot Environments-Estimation of the heat stress on working man, based on the WBGT-index (wet bulb globe temperature). Geneva: *International Standards Organization* 2003.
[6] Bishop B, Gu DL, Clapp A. Climate under impermeable protective clothing. *Int J Ind Ergonom* 2000;**25(3)**: 233–8.
[7] OSHA. MNOSHA heat stress guide. *Minnesota Department of Labor and Industry, Occupational Safety and Health Division* 2012. St. Paul, MN, USA. http://www.doli.state.mn.us/OSHA/PDF/heat_stress_guide.pdf.
[8] Bernard TE and Kenney WL. Rational for a personal monitor for heat strain. *Am Ind Hyg Assoc J* 1994;**55(6)**: 505–14.
[9] Reneau PD and Bishop PA. Validation of a personal heat stress monitor, *Am Ind Hyg Assoc J* 1996;**57**: 650–67.
[10] Mclellan TM and Selkirk GA. Heat stress while wearing long pants or shorts under firefighting protective clothing. *Ergonomics* 2004;**47(1)**: 75–90.
[11] Turpin-Legendre E and Meyer JP. Comparison of physiological and subjective strain in workers wearing two different protective coveralls for asbestos abatement tasks. *Appl Ergon* 2003;**34(6)**: 551–6.
[12] Marszalek A, Smolander J, Soltynski K, Sobolewski A. Physiological strain of wearing aluminized protective clothing at rest in young, middle-aged, and older men. *Int J Ind Ergonom* 1999;**25(2)**: 195–202.
[13] Malcolm S, Armstrong R, Michaliades M, Green R. A thermal assessment of army wet weather jackets. *Int J Ind Ergonom* 2000;**26(3)**: 417–24.

[14] Cadarette B, Cheuvront SN, Kolka MA, Stephenson LA, Montain SJ, Sawka MN. Intermittent microclimate cooling during exercise-heat stress in US army chemical protective clothing. *Ergonomics* 2006;**49(2)**: 209–19.

[15] Liu YP, Chen HC, Sung PC. Wireless logger for biosignals. *Int J App Sci Eng* 2010;**8(1)**: 27–37.

[16] Chad KE and Brown JM. Climatic stress in the workplace: its effect on thermoregulatory responses and muscle fatigue in female workers. *Appl Ergon* 1995;**26(1)**: 29–34.

[17] Logan PW and Bernard TE. Heat Stress and Strain in an Aluminum Smelter. *Am Ind Hyg Assoc J* 1999; 60(5): 659–65.

New Ergonomics Perspective – Yamamoto (Ed.)
© 2015 Taylor & Francis Group, London, ISBN 978-1-138-02751-0

Ergonomic issues related with risk assessment in construction sites in Korea

Hyeon-Kyo Lim
Department of Safety Engineering, Chungbuk National University, Cheongju, Chungbuk, Korea

Seong-Rok Chang
Department of Safety Engineering, Pukyong National University, Busan, Korea

Kwang-Won Rhie
Department of Safety and Health Engineering, Hoseo University, Asan, Chungbam, Korea

ABSTRACT: Risk Assessment in the industrial fields became a mandatory activity by the Korean government recently. However, risk assessment can be conducted rather easily in manufacturing plants because most workers commit repetitive activities everyday so that many of hazard factors can be realized by a worker of discretion, if any. However, in the construction sites, it is not so easy as in the manufacturing plants because of characteristics of the construction sites.

Therefore, this study was carried out to elicit problems related with risk assessments and to consult and monitor the management improvement procedures in construction sites from the viewpoint of industrial ergonomists. The authors visited a large-scaled chemical plant construction site a few days a week for three months by turns. During the visit period, the authors reviewed risk assessment procedures for every management level, and participated in risk assessment meetings. Interviews with construction workers as well as their managers including officials from the head office were conducted.

The results revealed that though most workers as well as managers realized the necessity of prior risk assessment, the quality of risk assessment was not so high as expected. The major problem was the lack of skill to elicit human factors that may result in industrial accidents systematically, and the second one was the insensitive risk assessment system they adopted for risk assessment. And the managerial system that the construction company runs also had the problem which was negligent of feedback, cumulation, and propagation of acquired information. In consequence, the way how to improve the quality and effect of risk assessment was debated in this study.

Keywords: risk assessment; hazard analysis; human error; accident prevention; safety management

1 INTRODUCTION

Risk assessment in the industrial fields became a mandatory activity by the Korean government recently. However, risk assessment can be conducted rather easily in manufacturing plants because most workers commit repetitive activities everyday so that may of hazard factors can be realized by a worker of discretion, if any. However, in the construction sites, it is not so easy as in the manufacturing plants because of characteristics of the construction sites.

Therefore, this study was carried out to elicit problems related with risk assessments and to consult and monitor the management improvement procedures in a construction site from the viewpoint of industrial ergonomists. In this poster, the focus will concentrate only on industrial ergonomic problems. There may exist limitations due to a single site, a single company though, it would be enough to explain the current situation of the Korean construction site since the company concerned was one of the representative, outstanding, and large companies in Korea.

2 SCHEME OF STUDY

On the site, a large-scaled chemical plant was supposed to be constructed within about two years. Visit period admitted for study was only three months. Therefore, empirical study and field investigation were concentrated on the early stage of construction, or pile driving operations which were planned for more than three months.

The authors visited the site a few days a week regularly for three months by turns. During the visit period, the authors reviewed risk assessment procedures for every management level, and participated in hazard analysis—risk assessment meetings. Meanwhile, interviews with construction workers as well as their managers including officials from the head office were conducted also.

The study was carried out in the sequence as shown in the Fig. 1.

3 METHOD

3.1 Step 1. Safety management system analysis

Safety management system was analyzed through personal interviews and document surveys, and results were summarized as Fig. 2. Most workers were proud that they were members of an outstanding construction company. However, the effect of safety management with repeated risk assessment meetings was questionable [1, 2]. Diverse and repeated activities could be shown in Fig. 3.

3.2 Step 2. Review of documents

All the documents related with safety management and accident prevention were reviewed by consulting members.

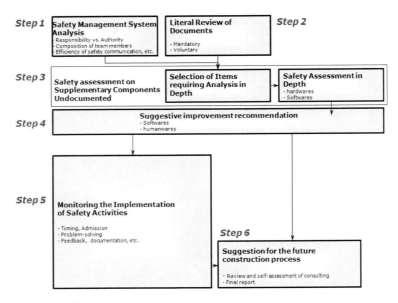

Figure 1.   Major work flow in the present study.

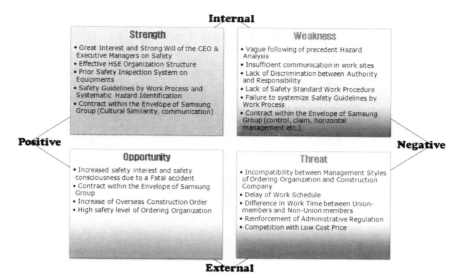

Figure 2.   Result of SWOT analysis.

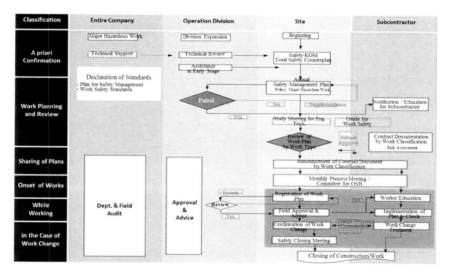

Figure 3.   Major safety management activity flow conducted in the company for construction process.

(a) Safety Management System

- Document management state was excellent due to the experience of a fatal accident occurred the year before. Thus, risk assessment was carried out well. However, oversimplification in technical application undermines discriminability between hazard factors. Risk assessment mainly concentrated on hardwares so that comprehensive analysis including softwares and humanwares could not be expected [3].
- Notification of hazards and derivation of countermeasures were carried out though, follow-up and confirmation of completion were neither specific nor sufficient.
- Too many items were told in meetings and site patrols so that major safety focus could be diffused.
- Progressive safety management could be expected because constitution of safety management staffs satisfies only legal requirements.

157

- Failure to discriminate the concepts of authority and responsibility of safety managers could disturb voluntary progressive management activities.
- PCM result was being used without exact definition of "Safety Standard Work".
- Communication between project team members was not sufficient so that appropriate information was not shared.

(b) Construction Plan & Construction-related Manual

- Safety manuals for work activities were systematically furnished in accordance with those in company scale. Prior safety analysis on construction process were carried out properly through PCM (Pre-Construction Meeting) before subcontractors begin to work.
- Application of risk assessment was not strict in the site. Especially, confirmation of countermeasures against hazard factors were incomplete. For instance, it was observed that piling technique in the site was SDA (Separation Doughnet Auger) whereas safety plan prepared was that for PHC.

3.3 *Step 3. Safety assessment on supplementary hardwares and softwares undocumented*

Specification of facilities, devices, and machines were reviewed and confirmed again through on-site investigation because the construction site had a fatal accident due to inconsistency in equipment specification the year before. Major focus concentrated was inconsistency not only in hardwares but also softwares and humanwares that might result in accidents.

The results showed that accident prevention activities were somewhat inefficient compared with diversity and amount of hazard assessment and safety management activities. For example, safety staffs had little knowledge about "Safety Management with Visual Confirmation Ques". As a consequence, it was concluded that education for effective hazard assessment was urgent and should be continued for upskilling.

3.4 *Step 4. Suggestive improvement recommendation*

With the base of analysis result from the previous stages, improvement methods were developed by the authors. Major points were as following;

- Sensitivity enhancement of risk assessment and practical application of risk assessment results
- Level-up of effectiveness of safety management activities
- Effective prevention of unsafe acts of workers.

All suggestions were got into shape of a final report and submitted to the company, which provoke interests of the head office in the company so that senior managers arranged a direct interview with the authors on the construction site. Fig. 4 shows major improvement points suggested by the authors in the present study.

3.5 *Step 5. Monitoring the implementation of safety activities*

In order to make confirm whether practical management activities were carried out on sites with the base of risk assessment, regular monitoring was conducted for additional two months, two days a week.

Throughout the monitoring activities, discrepancy between paper work and practical construction process was pointed out repeatedly, and managed to give several cues against industrial accidents which were not accepted by the safety managers sincerely though, but practically occurred thereafter.

3.6 *Step 6. Suggestion for the future construction process*

After completion of the field study, more study to enhance communication efficiency and human relations between engineers and industrial workers were recommended by the authors,

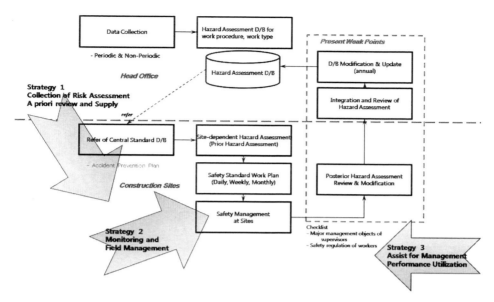

Figure 4.   Improvement points suggested by the authors.

in the shape of meetings, questionnaires, or interviews. All the suggestions and related materials were submitted for the following construction process.

Thereafter the construction process were completed safe. The presented study could be recorded the first ergonomic approach in construction area in Korea.

4   DISCUSSION

There were lots management problems compared with safety management activities in manufacturing area, most of which were resulted from the characteristics of construction engineering. In manufacturing companies, most workers spend time in the same site. On the contrary, in construction site, variation of workers are serious so that stable and consistent safety management activities are difficult to imagine. In that sense, work activity including management activity can apt to lose consistency.

One of the major problems observed was inefficiency of risk assessment application. Most workers were eager to commit risk assessment, unfortunately however, they had little ergonomic knowledge to elicit ergonomic hazards from work sites so that they failed at their first step, or hazard identification stage. Though the head office supplied standard hazard information, on the contrary, it might get worse the situation, since most workers usually copied the supplied material line by line without contemplating hazard effect of the factors concerned. It may be the result of supervision policy of the head office, but it can disturb the creative efforts for hazard identification and management.

To promote applicability of risk assessment and to elicit hazard factors systematically, work procedures should be arranged in sequence, which is an awkward work [4, 5]. In spite of that, efforts to standardize and describe work procedure in a standard format should be continued. Without them, it can not be expected to elicit hazard factors including human factors systematically [6]. Subsequent techniques such as Ergo-HAZOP should be supplied and educated in advance [7].

Another point to be considered is that work requirement for risk assessment should be maintained as minimum. In case of the site considered, risk assessment was a part construction process. It was conducted at most three times a day, at different management level. It means that most safety managers participate in risk assessment every day, which gave themselves an illusion that they do their best for accident prevention. However, the result can be different from their expectation because of risk assessment efficiency.

Disposal without cumulation of risk assessment was also a problem. Risk assessment was committed from the first stage of the construction operations without referring the previous materials partly because construction technique was different, and partly because construction object was different. Construction work procedures may be different site by site, construction technique to technique. But, if result of risk assessment will be discarded without cumulation, whenever construction site changes, it would be waste of man-powers. All the information and risk assessment results should be accumulated by the head office in the company scale, and be utilized as a reference in the later construction operations. It is the managerial tip to enhance safety management levels.

5   CONCLUSION

During the study, not a few problems were observed. Among the problems, the authors wanted industrial ergonomists to pay attention to the following points.

In any case, irregular work would be difficult and more apt to result in accidents. Therefore, an effort to make "Standard Work Procedures" as much as possible is quite necessary to take a long-term view. Unfortunately, however, safety managers in the construction site concerned in the present study had a somewhat different view. Nominal standard work procedure were scattered over lots of the safety documents, in which there was no consistency to penetrate them.

And, enhancement of risk assessment skill was quite big problem than expected. In case of the company concerned, risk assessment materials are prepared by subcontractors. Even some parts of them were purchased by a professional safety agency. Diverse and a lot of safety management activities were carried out energetically under the titles of "Hazard Analysis" or "Risk Assessment", where hazard points were mainly referred from standard templates, at most, supplied by the head office, not from the brains of analysts.

Once hazard assessments are carried out, they should fed back to risk assessment data base for the future utilization. However, safety management system concerned had a rule to cumulate only accident cases which are known as rare events. For the efficiency of management activities as well as accident prevention, safety managers should equip themselves with ergonomic skills.

REFERENCES

[1] Korea Occupational Safety and Health Agency, *10 Major Accident Types in Each Industry: Accident Analysis in Detailed Industry Classification*, 2010.
[2] Occupational Safety and Health Research Institute, *Accident Report on Industrial Accident Cause Investigation*, Korea Occupational Safety and Health Agency, 2008–2011.
[3] U.S. DOD, *System Safety Program Requirements*, DC US Department of Defense, MIL-STD-882D.
[4] Bahr, N.J., *System Safety Engineering and Risk Assessment: A Practical Approach*, Taylor & Francis, NY, 1997.
[5] Bateman, M., *Tolley's Practical Risk Assessment Handbook*, 5th ed., Elsevier, Oxford, UK, 2006.
[6] Salmon, P., Stanton, N.A., Walker G., *Human Factors Design Methods Review*, Defence Technology Centre, 2003.
[7] Kim, D.G., Lim, H.K., "Development of Ergo-HAZOP Technique for Identification and Prevention of Human Errors in Conventional Accidents," *Journal of KOSOS*, Vol.28, NO.8, pp. 46–51, 2013.

New Ergonomics Perspective – Yamamoto (Ed.)
© 2015 Taylor & Francis Group, London, ISBN 978-1-138-02751-0

A study on work-life balance and QOL during pregnancy

Chinatsu Mizuno
Mejiro University, Iwatsuki-ku, Saitama-shi, Saitama, Japan

Toshihiro Furukawa
Tokyo University of Science, Shinjuku-ku, Tokyo, Japan

Miyoko Kume
Seitoku University, Matsudo-shi, Chiba, Japan

ABSTRACT: In this study, in reference to "Work Family Conflict", a concept proposed by Kahn in which role conflict is discussed as interactions between the work role and the family role, the work family conflict for working pregnant women is considered as "work maternity conflict". The determinants of conflict that interact with job satisfaction and maternity life satisfaction from the aspects of physical and mental health of individuals in pregnancy were analyzed to determine measures that support the balance between work and family life. A web survey was conducted with working women living in Japan who were pregnant for the first time and the data obtained from individuals were analyzed statistically. Those who were living with a spouse were included and young women who were pregnant for the first time were excluded. it is believed that administration and operation of required supports based on recognition of situation of supports and its' sufficiency level for working pregnant women during their pregnancy may reduce conflicts regarding work and maternity life providing an opportunity to further achieve improvements in QOL of women in their maternity cycle.

Keywords: pregnant women; working women; work-life balance; work-family conflict; QOL

1   PROBLEMS AND OBJECTIVES

Harmonization of balance between work and life is an important concept for those who have professions in order to live by their own values in health. In recent years, such trend as to balance work and life has been steadily on the rise in Japan. In particular, health management of working pregnant women is significantly important from perspectives of health of their own as well as growth and development of unborn baby. In spite of a number of reports ever made about impacts on working pregnant women and their delivery, their suggestions are not uniform.[1-2] It may be more important than ever to establish a society where pregnant women are able to live their irreplaceable pregnant duration safely and at ease in ways they like while maintaining their physical and mental health. In this study, in reference to "Work Family Conflict",[3] a concept proposed by Kahn in which role conflict is discussed as interactions between the work role and the family role, the work family conflict for working pregnant women is considered as "work maternity conflict". The determinants of conflict that interact with job satisfaction and maternity life satisfaction from the aspects of physical and mental health of individuals in pregnancy were analyzed to determine measures that support the balance between work and family life.

2 ORGANIZATION OF CONCEPT AND HYPOTHESIS

The term "maternity life" used in the present study refers to a life in general during pregnancy including physical, mental and social aspects. In addition, the term "Work—Maternity Conflict" here (to be abbreviated as WMC) refers to a conflict interacting with personal satisfaction degree of work and maternity life as well as physical and mental health in general.

2.1 Hypothetic models

There is one hypotheses to be examined in the study. It is how "availability of instrumental and mental supports at workplace and home" and "availability of social support" affects Work—Maternity Conflict. Assuming based on conventional studies that conflicts are reduced in general as long as supports are readily available, a hypothesis is led that "Work—Maternity Conflict may be reduced only if supports are available." Fig. 1 shows a simplified relationship.

3 METHODS

In relation to the hypotheses described above, empirical examinations are performed according to the procedures set out below.

3.1 Study method

A web survey was conducted with working women living in Japan who were pregnant for the first time and the data obtained from individuals were analyzed statistically. Those who were living with a spouse were included and young women who were pregnant for the first time were excluded.

3.2 Variables

Variables to be used in analyses are shown in detail as follows:

1. Work—Maternity Conflict
   Subjects of the study are pregnant women living with their spouses who are employed. Overview of the subjects has revealed such situation that they play multiple roles in work and family life and as pregnant women. Therefore, we have prepared Work—Maternity Conflict Scale in reference to Work—Family Conflict Scale in Japanese version prepared by Watai with Work—Family Conflict described above in mind.

   It has been determined to examine Work—Maternity Conflict in consideration of three forms based on "time", "physical and mental strain" and "activities" as well as two directions, "from work to maternity life" and "maternity life to work." As for response to each question item, four-point scale was adopted which ranged from "exactly yes = 4 points" to "not at all = 1 point."

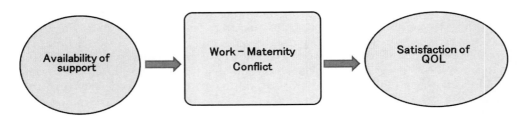

Figure 1. Hypothetic model.

2. Work—Maternity Conflict Coping

We surveyed Work—Maternity Conflict Coping, which is defined from the situation of coping process as cognitive and action-oriented efforts to manage demands from roles in both work and family in facing with Work—Maternity Conflict, from a perspective of utilization of supports. Specifically, 11 items were surveyed in total, such as two items of instrumental support, two items of mental support, utilization of social resources, utilization of support measures provided by employers and utilization of public system during pregnancy. For each item, four-point scale was adopted which ranged from "significantly supported (utilized) = 4 points" to "not supported at all (not utilized) = 1 point." Four-point scale was adopted also for supports from family members which allocated points by number of people who supported, i.e. "three persons or more = 4 points", "two persons = 3 points", "1 person = 2 points" and "no person = 1 point" and subjects were requested to respond them.

3. Quality of Life (QOL)

As for QOL, subjects were asked to respond to 7 items of satisfaction degree of work (working condition, contents of job, motivation, etc.), 2 items of that of family life (relationship with spouse, contentment in family life), 2 items of physical health degree (such as necessity for efforts to keep physical health) and 2 items of mental health degree (such as necessity for efforts to keep mental health) which were measured by four-point scale ranging from "very good = 4 points" to "not good at all = 1 point."

3.3 *Analysis*

First, an average score for all the questions related to work maternity conflict was calculated and two groups were created based on the mean value, a high-score group and a low-score group. The basic statistics of the core attributes of each group were calculated and compared between the high-score group and the low-score group. The average value was 30.78 points (standard deviation of 8.91), so 31 points was determined as the cut-off value. The student's t-test and the Pearson's χ^2 test were used for analysis. Next, factor analysis by the principal factor method, a promax rotation was performed for 12 questions about work maternity conflict, 14 questions about support and 13 questions about QOL. The statistical software AMOS 8.0 for Windows was used for path analysis. A maximum likelihood method was used for estimates. A two-sided significance level of 5% was used for tests.

4 RESULTS

4.1 *Attributes*

1. Whole picture of respondents

We asked 54,131 subjects to respond to our survey who satisfied our requirements by web research monitor. With 471 of valid responses (0.89%) obtained. Larger number of subjects with educational background of four-year universities or higher and working hours of 40 or longer were observed in high-score group.

As for WMC, a factor analysis (major factor method, promax rotation) was performed by asking questions of 12 items in reference to Work—Family Conflict Scale. Three factors were extracted based on degree of eigenvalue attenuation and interpretability. Cumulative contribution ratio of the three factors was 65.74%. The first factor was named as "conflict regarding work" consisting of three items including "having difficulty to take time for carrying out responsibility as a pregnant woman due to much time to be spent for work", and the second factor as "physical and mental conflict" consisting of six items including "reduced capability of work due to tension and anxiety" and "unable to enjoy free time at home due to stresses at workplace" and the third factor as "conflict regarding time for maternity life" consisting of three items including "much time to be spent for carrying out responsibility as a pregnant woman by the sacrifice of work." Both groups showed similar tendency for items regarding time for maternity life except for the item of "less time for career progression due to time required for pregnant duration" for which certain difference was observed between the two groups (p = 0.002).

Similarly, unifactoriality was recognized by factor analyses for each item of question asking about utilization of supports; i.e. two items about instrumental support, two items about mental support and nine items about social support. Utilization of supports was higher in low-score group regarding number of people who provided supports at home and easiness to have consultation at workplace, to apply for prenatal care and health guidance, to commute in different time schedule, to take day off before and after delivery and to apply for job displacement for light labor.

Further, unifactoriality was also recognized by factor analyses for each item of question asking about QOL; i.e. seven items about satisfaction degree of work, two items about satisfaction degree of family life, two items about physical health degree and two items about mental health degree. As for items about satisfaction degree of work, satisfaction degree of pace of work and work conditions was higher in low-score group. In addition, low-score group showed higher satisfaction in relationship with spouse about satisfaction degree of family life with more impacts observed in physical health degree than high-score group. Mental health degree was similar in both groups.

2. Discussion of impacts of support utilization, Work—Maternity Conflict and QOL on satisfaction degree

First of all, how support utilization has impacts on subscales of Work—Maternity Conflict is confirmed. Similar trends and results were obtained in both groups. In particular, higher utilization of social supports was observed in high-score group in relation to Work—Maternity Conflict. Both groups showed negative correlation with mental support. Further, both groups showed significantly high positive correlation with each subscale of Work—Maternity Conflict.

5 CONSIDERATION

In other words, it has been proved that utilization of support by working pregnant women has such an associated structure that affects conflicts regarding work and maternity life and that the conflicts are related to satisfaction degree of QOL. With similar structure in both groups, it indicates that utilization of instrumental support at home, mental support at workplace and social support typically provided during pregnancy may have tendency to affect reduction of conflicts. From those described above, it is believed that administration and operation of required supports based on recognition of situation of supports and its' sufficiency level for working pregnant women during their pregnancy may reduce conflicts regarding work and maternity life providing an opportunity to further achieve improvements in QOL of women in their maternity cycle. Specifically, it may contribute to improvement of balance.

REFERENCES

[1] Senzaki keiko, *A study on self-employed health care, especially maternal occupation*, Bosei-eisei 1979; Vol.20(1): p135–141.
[2] Sakudou Masamichi, *A Study on the work of pregnancy pregnant women*, Bosei-eisei 1991; Vol.32(2): p168–175.
[3] Kahn RI, *Organization stress,* New York: Wiley; 1964.
[4] Watai Izumi, *Development of a Japanese Version of the Work-Family Conflict Scale(WFCS), and Examination of its Validity and Reliability, San Ei Shi* 2006; Vol.48: p71–81.

New Ergonomics Perspective – Yamamoto (Ed.)
© 2015 Taylor & Francis Group, London, ISBN 978-1-138-02751-0

A study on work-life balance for single mothers

Chinatsu Mizuno
Mejiro University, Iwatsuki-ku, Saitama-shi, Saitama, Japan

ABSTRACT: This study used the Work-Family Conflicts (WFC), which are conflicts between interacting work and family roles proposed by Kahn, and analyzed the determinants of work-life conflicts in single mothers to discover a system that would support both work and family life. We commissioned an Internet research company to perform this questionnaire survey. In particular, for mothers in single-parent families, it was suggested that more attention must be paid to strain-based conflict, where the stress generated in one role has an influence over the other role, rather than to "time-based" conflict or "behavior-based" conflict caused by differences in the required behaviors in the family and work. The "support function" provided by the flexibility of family roles overcomes these conflicts.

Keywords: single mothers; working women; work-life balance; work-family cinflict; support function

1 PROBLEMS AND OBJECTIVES

Since the Equal Employment Opportunity Law went into effect in 1987, the rate of female workers in Japan has continued to increase, and it has been pointed out that more women want to keep their jobs after having babies. Compared to other developed countries, however, there is still a strong division of roles by gender as the social norm in Japan. Women who have children tend to bear a heavier load than men in managing housework, child rearing and work, which is pointed as one of the causes of the declining birthrate. Also, under recent changes in family forms, such as increases in the number of divorces and single people, it is becoming more important to develop systems that promote the independence of single-parent families and the healthy growth of their children.

In addition, since it is especially necessary for single parents to effectively manage both work and child rearing, we need to understand, examine, and evaluate their current status; the ideal work environment and system that they require; and how much of these are currently available to make the workplace more friendly to single-parent families, in addition to traditional families.

This study used the Work-Family Conflicts (WFC), which are conflicts between interacting work and family roles proposed by Kahn,[1] and analyzed the determinants of work-life conflicts in single-parent families to discover a system that would support both work and family life. The study paid particular attention to the working mothers of single-parent families and explored what factors in the conflict of work-life balance have an influence on the flexibility of family functions controlled by the family roles that form the foundation of life.

2 METHODS

We commissioned an Internet research company (Point On, Inc.,) to perform this questionnaire survey. A preliminary survey was sent to 59,386 contract members who were extracted from those that readily agreed to participate in this survey. In total, 345 responses with no missing values (187 of the traditional families and 158 of the single-parent families) were used for analysis (response rate: 38.9%). This survey was conducted with the registered members of a mobile research company who readily agreed to participate in the survey. The participants were free to respond or withdraw at any time. The privacy information of participants has been protected under the privacy policy of Point On, Inc.

2.1 Survey details

Basic attributes (sex, age), family conditions (family form, number of children), work conditions (work hours) and 18 items of the Japanese version of the Work-Family Conflict (WFC) scale created by Watai[2] were used. The subscales consist of the combinations of two directions, Work Interference with Family (WIF) and Family Interference with Work (FIW) and their three forms based on "time," "strain," and "behavior." Watai developed the Japanese version of this scale. The responses were counted by a 5-grade evaluation of 1 (Not at all true) to 5 (Absolutely true).

For the flexibility of family roles, "Members of our family alternatively take charge of housework according to need," one of the adaptability score items for family members, was used from the family function evaluation scale of Kusata & Okado et al.[3] The responses were counted by a 5-grade evaluation of 1 (Never) to 5 (Always). These evaluation scales have demonstrated sufficient validity and reliability.

3 RESULTS

3.1 Subjects

An attribute of subjects; as follows. (Table 1).

3.2 Analysis of WFC scale

In order to determine the adaptability of participants to the WFC scale, an exploratory factor analysis was performed and subscales were determined. Means and standard deviations of the 18 items of the WFC scale were computed. Eight items in which a floor effect was observed were excluded from the subsequent analysis. Then, the remaining 10 items were analyzed with the principal factor method. Taking into account the changes of fixed values (10.17, 4.75, 3.52, 2.01, ...) and the interpretability of factors, a three-factor structure was considered appropriate. Therefore, three factors were again assumed and a factor analysis with the principal factor method and Promax rotation was performed. This analysis yielded sufficient factor loadings. Table 2 shows the final factor pattern and inter-factor correlations after promax rotation. These 10 items accounted for 56.7% of the total variance.

The first factor consisted of four items, which showed high loadings in problem-solving behavior and effective behavior in both behavior-based work-to-family and family-to-work interference directions. This was named "behavior-based bi-directional conflict between work and family." The second factor consisted of three items, which showed high loadings in time-based work-to-family conflict. This was named "time-based work-to-family conflict."

Table 1. The attribute of subjects.

	Traditional families	Single-parent families
The number of households	158	187
Parental age (age)	37.2 ± 4.5	39.6 ± 1.1
Age of a child (age)	8.5 ± 2.6	9.5 ± 1.1
The number of children	1.5 ± 0.8	1.7 ± 1.1
Working time (hour)	9.1 ± 4.6	8.7 ± 3.3

Table 2. WFC scale factor analysis results (factor pattern after promax rotation).

	Items	Factor I	Factor II	Factor III
A18	Problem-solving behavior in family is not helpful in workplace.	**0.89**	0.00	−0.07
A16	Behavior good in family is not effective in workplace.	**0.85**	−0.07	0.07
A15	Behavior effective in workplace is not useful in being a good parent or a spouse.	**0.66**	0.06	−0.03
A13	Problem-solving behavior I use at work is not effective in solving problems in the family.	**0.55**	0.04	0.06
A1	Work takes more time than I expect, which I want to spend with my family.	−0.01	**0.88**	−0.03
A2	I have to spend more time on work, and it is difficult to secure time to fulfill my family responsibilities and do housework.	0.01	**0.87**	0.04
A3	I need to spend much time to fulfil my duties at work, and I do not have time to do things with my family.	0.01	**0.83**	0.08
A8	I am mentally exhausted when I come home from work, and I often have problems with my family.	−0.01	−0.07	**1.02**
A7	I am totally exhausted when I come home from work, and I sometimes do not have time to do things with my family or fulfill my family responsibilities.	−0.02	0.10	**0.83**
A9	Due to stress at work, I sometimes even cannot do the things that I like to do.	0.05	0.09	**0.67**

Factor correlation matrix	I	II	III
I	–	0.42	0.41
II		–	**0.67**
III			–

The third factor consisted of three items, which showed high loadings in content related to "physical stress." This was named "strain-based work-to-family conflict."(Table 2).

3.3 Correlation

In factor analysis of the WFC scale, the means of items that showed high loadings in respective factors were calculated and scores for behavior-based bi-directional conflict between work and family (mean 2.4, SD 1.1), time-based work-to-family conflict (mean 2.4, SD 1.1) and strain-based work-to-family conflict (mean 2.2, SD 1.1) were determined. The α factors calculated to determine the internal consistency were sufficient: α = 0.82 in behavior-based bi-directional conflict between work and family, α = 0.91 in time-based work-to-family conflict and α = 0.90 in strain-based work-to-family conflict. There was no significant difference between traditional families and single-parent families for all of these scores. The mean value of the flexibility in family roles (hereinafter "flexibility") was 2.91 and SD was 1.02. Similarly, there was no significant difference between traditional families and single-parent families for these scores. Table 3 shows the intercorrelations of the WFC scale and flexibility in traditional families and single-parent families together and Table 4 shows the intercorrelations in traditional families and single-parent families separately. When traditional families and single-parent families were analyzed together, there was a significant positive correlation in all items, except for strain-based work-to-family conflict and flexibility. However, when traditional families and single-parent families were analyzed separately, the correlation pattern was different. In traditional families, nearly all items related to flexibility were not correlated, but in single-parent families, all items showed a significant positive correlation. (Tables 3, 4).

Table 3. Intercorrelations of WFC scale and flexibility (traditional families and single-parent families together).

	Time-based work-to-family conflict	Strain-based work-to-family conflict	Behavior-based bi-directional conflict between work and family	Flexibility
Behavior-based bi-directional conflict between work and family	–	0.65**	0.39**	0.11**
Time-based work-to-family conflict		–	0.38**	0.046
Strain-based work-to-family conflict			–	0.17*
Flexibility				–

$*p < 0.05, **p < 0.01$.

Table 4. Intercorrelations of WFC scale and flexibility (traditional families and single-parent families separately).

	Time-based work-to-family conflict	Strain-based work-to-family conflict	Behavior-based bi-directional conflict between work and family	Flexibility
Behavior-based bi-directional conflict between work and family	–	0.67**	0.43**	0.06
Time-based work-to-family conflict	0.59**	–	0.41**	−0.02
Strain-based work-to-family conflict	0.28**	0.29**	–	0.04
Flexibility	0.26**	0.20*	0.33**	–

$*p < 0.05, **p < 0.01$.
Top right corner: Traditional Families, The lower left: Single-parent Families.

4 DISCUSSION

This study explored what work-family conflict factors are influenced by the flexibility in family roles that form the foundation of life in single-parent families. In traditional families, the flexibility in family roles did not have a direct influence on work-family conflict. However, in single-parent families, there were "time-based," "behavior-based," and "strain-based" work-to-family conflicts. From this result, it was clear that family life was significantly influenced by work. There were also "behavior-based" family-to-work conflicts, suggestive of some influence on work from problems or events within the family. In line with the previous studies that claimed that high work-family conflict increases the degree of depression, this study showed "strain-based" work-to-family conflict in single-parent families. This supports the characteristics of single-parent families where a parent has to hold more roles compared to traditional families. In particular, for mothers in single-parent families, it was suggested that more attention must be paid to strain-based conflict, where the stress generated in one role has an influence over the other role, rather than to "time-based" conflict or "behavior-based" conflict caused by differences in the required behaviors in the family and work. This study showed that the "support function" provided by the flexibility of family roles overcomes these conflicts. Therefore, it is necessary to study the social support systems that are suitable to the needs of users. In this study, it was suggested that the development of "flexible" social systems are required to increase the work-life balance of single mothers and early implementation of such systems would reduce work-family conflicts. It is necessary to make efforts to develop an environment that harmonizes "life" and the "way of work" from a "flexible" and multilateral viewpoint that allows the optimum work-life balance for each parent.

REFERENCES

[1] Kahn RI, *Organization stress,* New York: Wiley; 1964.
[2] Watai Izumi, *Development of a Japanese Version of the Work-Family Conflict Scale (WFCS), and Examination of its.*
[3] Kusata & Okado, *Underwriting method for the family relationship*, Psychological test studies; 1993, p573–581.

New Ergonomics Perspective – Yamamoto (Ed.)
© 2015 Taylor & Francis Group, London, ISBN 978-1-138-02751-0

Evaluation of the energy-saving performance of heat-resistant paint in winter

Takashi Oda & Kimihiro Yamanaka
Tokyo Metropolitan University, Hino, Japan

Mitsuyuki Kawakami
Kanagawa University, Yokohama, Japan

ABSTRACT: In recent years, the importance of various forms of energy saving for residential and work spaces has increased as a result of apprehension related to global warming and anxiety about electric power supply after the Great East Japan Earthquake. Measures related to cooling and heating loads have become particularly important, and there is a trend toward improving building insulation to reduce the influence of the outside environment and to allow using cooling and heating equipment in an efficient manner. With regard to building insulation performance, in the past, insulation materials were often used to direct the transfer of heat toward the inside of the structure. Recently, there has been a growing expectation that the application of heat-resistant or heat-insulating paint, which can improve energy-saving performance, can serve as a new energy-saving technique that differs from the conventional mechanism of installing insulation materials. However, the different mechanism and properties of this technique mean that conventional insulation performance evaluation methods cannot be used to evaluate the energy-saving performance. As a consequence, each manufacturer is presently using its own proprietary standards to evaluate energy savings. For end users selecting a product, the lack of common indicators of energy-saving performance obstructs the promotion of effective energy saving, and therefore a common method of quantitative evaluation is desirable. Furthermore, although there have been numerous reports on the energy-saving performance of heat-resistant paint in summer, few such evaluations are available for winter. In the present study, we aimed to develop an empirical quantitative evaluation of the energy-saving performance of heat-resistant paint in winter. Specifically, heat-resistant paint and conventional wall paint were applied to steel boxes placed in a room kept at constant temperature and humidity, after which heat sources were placed in the boxes and the energy-saving performance of each type of paint was evaluated from the change in temperature at the box walls and inside the boxes. The experimental results showed that the heat-resistant paint reduced the amount of heat escaping through the walls, and thus it can be expected to reduce heat loss. Furthermore, in the case of the heat-resistant paint, the amount of heat passing through the walls was 16% less than with the conventional paint.

Keywords: Energy-saving; Heat-resistant paint; Comfortable work area

1 INTRODUCTION

Among energy-saving paints, many are so-called high-reflectance paints, which when applied to the roof and outer walls of a building can help to regulate heat on the outside of the building by reflecting sunlight with high efficiency (1). The standards JIS K 5602 "Determination of reflectance of solar radiation by paint film" and JIS K 5675 "High solar reflectance paint for roof" detail a quantitative method of evaluating high-reflectance paints (2,3).

The method described in these standards determines solar reflectance from the measured values of spectral reflectivity by using the spectral radiation distribution of standard sunlight. In contrast to the situation for outdoor paints, there are energy-saving paints that are used on interior surfaces not exposed to solar radiation as a means of increasing cooling and heating efficiency, but there has been no agreement on a common evaluation standard by which to evaluate the energy-saving effect of an interior paint (4).

Against this back ground, the ceramic insulative paints with the greatest market share among energy-saving paints for indoor surfaces and two varieties of conventional energy-saving paint were compared, and the conduction heat flux values were directly detected in the present study. From these results, a more unified verification of effectiveness is proposed that estimates the heat transfer inhibition effect of energy-saving paints intended for indoor surfaces, with the ultimate aim of a quantitative evaluation of effectiveness. In addition, the effects of high-reflectance paints were compared.

2   EXPERIMENTAL METHOD

Iron boxes (sides: 500 mm; thickness: 1.6 mm) were used in the present study. Iron was chosen as the base material because it is a thermally stable raw material with low thermal resistance to minimize the the impact of individual differences. The ceramic insulative paint, the conventional energy-saving paint, and the high-reflectance paint were each applied to the inside of a separate box to create three different samples. The specifications of the three paints are as shown in table 1.

In the experiment, the conduction heat flux from the inside to the outside of the box and the structural temperature at the point on both the inside and the outside of the box ceiling were measured while heat was generated inside each iron box by a small fan heater (FH-120, FUKADAC); the setup is shown in figure 1. The heat flux was measured by a heat flux sensor (MF-180, Eko Instruments), and the temperature inside and outside the box was measured by a thermocouple. The experiment was performed in a room with constant temperature and humidity and the ambient temperature set at 22 °C. Measurements were taken starting at 30 min from the start of operation of the heater and the thermal behavior was studied at 10 min intervals from then. In addition, it was verified before the start of each trial that the box and its inside temperature were the same temperature as the soundings (22 °C).

Table 1.   Specifications of the three paints.

	Conventional energy-saving paint	Ceramic insulative paint	High-reflectance paint
Diluent	Water	Water	Water
Coating method	Spray	Spray	Spray
Number of times recoating	2	2	2
Standard application amount	0.2~0.4 kg/m²/times	0.2~0.23 kg/m²/times	0.22 kg/m²/times
Specific gravity	1.24	0.78	1.24

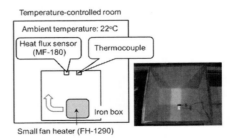

Figure 1.   Experimental setup.

Figure 1 shows the installation point of the heat flux sensor for the measurement of the conduction heat flux (inside the box). The ceiling, which is not directly exposed to the warm air from the small heater, was used as the heat flux sensor installation point, and the sensor was affixed with double-sided tape. The conduction heat flux to the outside was measured by a logger (3635–04, Hioki E. E. Corporation) at a sampling rate of 1/min while the heater was operating inside the box.

Similarly, figure 1 also shows an iron box specimen and the installation points of the thermojunction temperature sensors (inside the box). The thermojunction temperature sensors were affixed to the inner and outer surfaces of the box ceiling with aluminum tape. The changes in the inside and outside surface temperature of the box structure were measured by a logger (309, Centre) at a sampling rate of 1/min while the heater was operating inside the box.

3   RESULT OF CONDUCTION HEAT FLUX

Figure 2 shows the conduction heat flux over time. The horizontal axis in the figure shows the elapsed time, and the vertical axis shows the conduction heat flux from inside to outside the box. The three curves indicate the conduction heat flux of the three boxes. According-ing to the figure, the trend in conduction heat flux appears to differ by paint between the first half and second half of the measurements. The measurement time has been split into three segments (1–10 min, 11–20 min, and 21–30 min) to characterize these differences. Figure 3 shows the results of consolidating the conduction heat flux for each paint in each time period. The horizontal axis shows the time period, and the vertical axis shows the average conduction heat flux from inside to outside the box of each paint in each time period. The three curves show the results for each box. As can be seen in the figure, the conduction heat flux of the box to which ceramic insulative paint was applied tended to be smaller than that of other boxes at every time period. Table 2 shows the results of taking the values of the figure as characteristic values and conducting a two-way analysis of variance with time period (A) and variety of paint (B) as factors. As seen in the table, both the measurement time period and the variety of paint were significant ($p < 0.05$ and $p < 0.01$, respectively), but a pairwise comparison by the Holm method showed that there were no significant differences between conduction heat fluxes for either the measurement time period or the variety of paint.

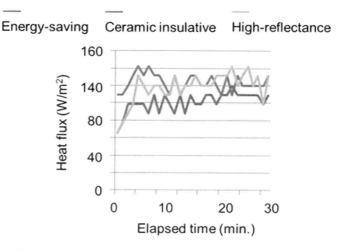

Figure 2.   Results of measurement the conduction heat flux.

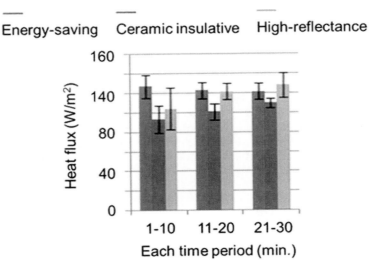

Figure 3. Results of the conduction heat in each time period.

Table 2. ANOVA table for conduction heat flux (**:p < 0.01, *:p < 0.05).

Source of variation	SS	df	MS	F-value
Time period	2066.88	2.00	1033.44	4.23*
Variety of paint	8267.54	2.00	4133.77	16.92**
Interaction	2545.68	4.00	636.42	2.61
Within	19784.25	81.00	244.25	
Total	32664.35	89.00		

4 RESULT OF STRUCTURAL TEMPERATURE (INNER WALL AND OUTER WALL)

Figures 4 shows the changes in inner wall and outer wall temperatures for the boxes. The horizontal axes in the figures show the elapsed time and the vertical axes show the wall surface temperature. The two curves in each figure show the temperature changes for the inner wall and outer wall.

As can be seen in the figure, it was found that the inner wall temperature tended to be higher, from immediately after commencing operation of the heater, in the box to which ceramic insulative paint was applied, and a temperature gradient occurred between the inside and outside of the box. In contrast, the inner wall and outer wall temperatures of the other boxes changed almost identically.

Figure 5 shows the inner–outer wall temperature difference for each box at each time period. The horizontal axis in the figure shows the paint and the vertical axis shows the inner–outer wall temperature difference. The three curves show each measurement time period. Table 3 shows the results of taking the values of the figure as characteristic values and conducting a two-way analysis of variance with time period (A) and variety of paint (B) as factors. As noted in the table, both the measurement time period and the variety of paint were significant (p < 0.05 and p < 0.01, respectively). Furthermore, pairwise comparison by the Holm method showed that the inner–outer wall temperature difference was significantly larger for the ceramic paint.

These results indicate that heat loss from the box with ceramic insulative paint was the lowest, which is consistent with the variety of paint being an influencing factor in the results of the conduction heat flux, as shown in table 2.

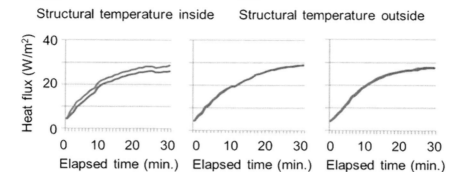

Structural temperature inside   Structural temperature outside

Figure 4.   Changes in inner wall and outer wall temperatures.

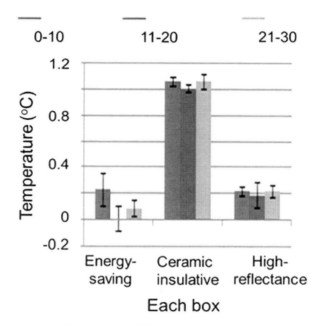

Figure 5.   The inner–outer wall temperature difference.

Table 3.   ANOVA table for inner–outer wall.

Source of variation	SS	df	MS	F-value
Time period	0.16	2.00	0.08	3.71*
Variety of paint	15.76	2.00	7.88	388.64**
Interaction	0.12	4.00	0.03	1.49
Within	1.62	81.00	0.02	
Total	17.66	89.00		

5   CONCLUSION

In the present study of the heat transfer inhibition effect of energy-saving paints, the effects of three varieties of paint—an ordinary paint, a ceramic insulative paint, and a high-reflectance paint—were experimentally verified for iron boxes. The results obtained can be summarized as follows.

1. The variety of paint was shown to be an influencing factor on conduction heat flux, which is an index that represents the loss of heat to the outside of the box.
2. The inside–outside temperature gradient of the box was steeper for the box with ceramic paint than for the other boxes, indicating that heat loss was the smallest for the box with ceramic paint.

REFERENCES

[1] Japan Paint Manufacturers Association, "Guideline of manufacture and environment 2011", http://www. toryo.or.jp/jp/anzen/reflect/reflect-info2.pdf, (30/6/2014 access).
[2] Japanese Industrial Standards, "Determination of reflectance of solar radiation by paint film", JIS K5602, 2008.
[3] Japanese Industrial Standards, "High solar reflectance paint for roof" detail a quantitative method of evaluating high-reflectance paints", JIS K 5675, 2011.
[4] Ministry of Land, Infrastructure, Transport and Tourism, "Energy-saving standard of architectural structure", http://www.mlit.go.jp/common/000996591.pdf, (01/7/2014 access).

New Ergonomics Perspective – Yamamoto (Ed.)
© 2015 Taylor & Francis Group, London, ISBN 978-1-138-02751-0

Key points of procedures of Multiple Role Map program toward Japanese nurses: Differentiation between individual and group approaches

Yasuyuki Yamada
Graduate School of Health and Sports Science, Juntendo University, Inzai-Shi, Chiba, Japan

Yasuyuki Hochi
Faculty of Health and Sports Science, St. Catherine University, Matsuyama City, Ehime, Japan

Motoki Mizuno
Graduate School of Health and Sports Science, Juntendo University, Inzai-Shi, Chiba, Japan

ABSTRACT: *Objectives*: In general, nurses could be regarded as hard workers engaged in multiple roles such as nurse, administrator, parent and partner. Therefore, researchers have discussed and examined the appropriate work-life balance. Our original tool, the Multiple Role Map (MRM) program, also contributed to this task. The MRM program enabled us to collect narrative data about the amount and number of multiple roles, role expectation, role personality, Negative Spillover (NSP), Positive Spillover (PSP), Compensation (COM) and Segmentation (SEG). Through the MRM program, nurses could not only estimate both of merits and demerits of engaging in the multiple roles, but also choose some practical coping methods. However, it had some issues in the procedures. For example, it required participants about sixty to ninety minutes to describe and complete the MRM form (A3 paper size). Furthermore, participants had to understand about basic role concepts of the NSP, PSP, COM and SEG according to the guideline. To accomplish more effective data collection in MRM study, this study clarified the key points of the study procedures.

Methods: This study compared the response rate, recovery rate and the quality of the data between two study procedures, individual and group approaches. The subjects of the individual approach were 20 Japanese hospital nurses. We got informed consent and explained the guideline of MRM after their working time individually. Completed MRM form was collected through the placement method. The subjects of group approach were 29 Japanese hospital nurses participated in the nursing seminar that featured the theme of team building. We got presentation time for the aim of informed consent and explanation of the MRM guideline. The mailing method was adopted for the data collection.

Results: As the results of recruit performance, both approaches accomplished 100.0 percent response rate. The recovery rate of the individual approach was 100.0 percent and that of the group approach was 35.7 percent. Mean number of characters in completed MRM from, regarded as index of quality of data, were 463.0 (SD = ±207.8) in individual approach and 620.6 (SD = ±344.2) in group one.

Conclusions: The simple and brief recruit procedures in the group approach with mailing method were attractive for researchers. However, it had a risk of lower recovery rate. When we select group approach, we should make much effort to reduce non-respondent bias and self-selection bias.

Keywords: Work-life balance; Nurses; Role conflict; Multiple roles; Spillover

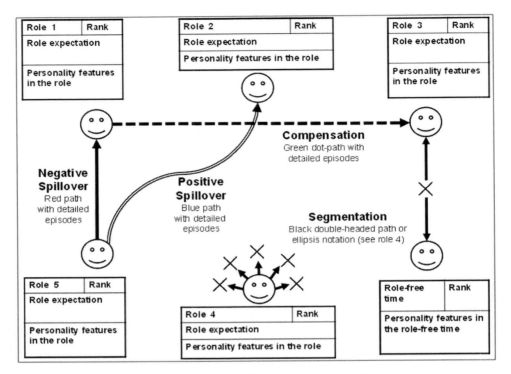

Figure 1.   MRM form (A3 paper size) and guideline for the description [3].

1   INTRODUCTION

Working female nurses can be regarded as hard workers engaging in multiple roles such as nurse, administrator, friend, wife, mother and daughter [1]. Hence, how to support their good work-life balance was meaningful task. To make a contribution to this task, we applied our original tool, Multiple Role Map (MRM) program, to working female nurses in Japan. The MRM program requires participants to fill in the name, priority, role expectation and role personality in each role on the MRM from [2]. Additionally, participants describe the Positive Spillover (PSP)[1] effects, Negative Spillover (NSP)[2] effects, Compensation (COM)[3] effects and state of Segmentation (SEG)[4] on the interface between two roles (Figure 1). Through the experience of the MRM program, participants get wide perspective to monitor their work and daily-life conditions and to select effective coping strategies. Through the analysis of completed MRM from, researchers are able to discuss about the practical support for the intervention.

On the other hand, researchers need to know that some participants hesitate to do the MRM program because of the work load. For example, participants have to understand some academic concepts, such as the NSP, PSP, COM and SEG, based on the guideline [3].

---

1. PSP: Experiences in which psychological and physiological states spill over from one role to another role in relation to one's skills, behavior, positive mood, sense of accomplishment and support [4–5].
2. NSP: Experiences in which skills, behavior patterns (i.e. type of activities), strain, emotions, beliefs and attitudes spill over negatively from one role to another role [6].
3. COM: Experiences in which dissatisfaction in one role leads to trying to find more satisfaction in other roles [7].
4. SEG: Segmentation indicates irrelevance between two roles, in which psychological and physiological states remain independent [7].

Table 1. Differentiations of procedures and performance between individual and group approaches.

		Individual approach	Group approach
Procedures	Informed consent	One-to-one	One-to-group
	Data collection method	Leaving method	Mailing method
	Investigation period	1 week	1 day
Recruit performance	Number of participants	20	28
	Response rate	100.0%	100.0%
Data collection performance	Recovery data	20	10
	Recovery rate	100.0%	35.7%
Quality of data	Number of multiple roles	Mean = 6.2 SD = ±2.8	Mean = 5.6 SD = ±0.7
	Number of PSP episodes	Mean = 4.8 SD = ±2.5	Mean = 6.6 SD = ±3.3
	Number of NSP episodes	Mean = 2.7 SD = ±1.6	Mean = 5.3 SD = ±4.4
	Number of COM episodes	Mean = 5.4 SD = ±0.8	Mean = 5.0 SD = ±2.4
	Number of characters	Mean = 463.0 SD = ±207.8	Mean = 620.6 SD = ±344.2

In the preliminary study, participants spent sixty to ninety minutes to complete the MRM form [1]. Since these factors have potential to disturb the performance of scientific approach, key points of the effective study procedures should be discussed. Therefore, this study aimed to compare the performances of recruit, data collection and quality of data between two different study procedures, individual and group approaches (Table 1).

2 METHODS

2.1 *Participants and procedures*

The subjects of the individual approach were 20 Japanese hospital nurses. We got informed consent and explained the guideline of MRM after their working time individually. Completed MRM form was collected through the placement method. The subjects of the group approach were 29 Japanese hospital nurses participated in the nursing seminar that featured the theme of team building. We got presentation time for the aim of informed consent and explanation of the MRM guideline. The mailing method was adopted for the data collection in group approach.

2.2 *Multiple Role Map program*

In the MRM program, participants described the information about features of multiple roles in the MRM sheet (A3 paper size) based on defined instructions (Fig. 1) [3]. Through the MRM program, we could collect descriptive data about the contents of multiple roles, role expectation, role personality, priority and episodes about the PSP, NSP, COM and SEG. In this study, we asked participants to think about six role conditions, nurse, manager, mother (or daughter), marital/live-in partner, friend and private time (role-free time), these were frequency reported in our preliminary study [1].

2.3 *Measurements*

This study evaluated the recruit performances by the respondent rate. The performance of data collection was examined by the recovery rate. Quality of data was evaluated by amount of information in the MRM form included in the number of PSP, NSP and COM episodes and characters.

3 RESULTS

As the results, both approaches accomplished 100.0 percent response rates. The recovery rate in the individual approach was 100.0 percent and that in the group approach was 35.7 percent. The amount of information in the MRM from in individual approach was lower than group approach. Mean number of characters in completed MRM from was 463.0 (SD = ±207.8) in individual approach and 620.6 (SD = ±344.2) in group approach.

4 DISCUSSION

In the individual approach with placement method, we spent much time and effort to the recruitment and informed consent. Thanks to this procedure, we could establish a rapport with participants and accomplish 100.0 percent response and recovery rates. Hence, we suggest effective recruitment and informed consent as one of the key points of the MRM study.

On the other hand, through the group approach with mailing method, we collected data from just 35.7% of participants. Remarkably, collected data was comparatively high quality. In this case, we have to estimate effects from non-respondent bias and self-selection bias. Therefore, when we select group approach, we should make much effort to reduce these biases.

REFERENCES

[1] Yamada Y, Ebara T, Kinooka Y, Mizuno M, Hirosawa M, Kamijima. Narrative evidence of the work-family positive spillover in Japanese midwives: A descriptive study using the Multiple Role Map program. *News letter of Human Ergology* 2012; 98.

[2] Yamada Y, Hochi Y, Mizuno M, Kawata Y, Oki K, Hirosawa M. Development of the Multiple Roles Map Program for the Enhancement of Self-understanding. In: Lin DM, Chen H, editors. *Ergonomics for All*, London: CRC press Taylor & Francis Group; 2010, pp. 369–73.

[3] Yamada Y, Mizuno M, Ebara T, Hirosawa M. Merits and demerits of engaging in athletic, academic and part-time job roles among university student-athletes in Japan. *J Hum Ergol* 2011;40(1–2):141–50.

[4] Grzywacz JG, Marks NF. Reconceptualizing the work-family interface: an ecological perspective on the correlates of positive and negative spillover between work and family. *J Occup Health Psycol* 2000; 5(1):111–26.

[5] Hanson GC, Hammer LB, Colton CL. Development and validation of a multidimensional scale of perceived work—family positive spillover. *J Occup Health Psychol* 2006;11(3):249–65.

[6] Geurts SAE, Demerouti E. Work/non-work interface: A review of theories and findings. In: Schabracq MJ, Winnubst JAM, Cooper CL, editors. *The Handbook of Work and Health Psychology*, New York: John Wiley & Sons, 2003, pp.279–312.

[7] Roehlinget P, Moen P, Batt R. Spillover. In: Moen P, editor. *It's about Time: Couples and Careers*, Ithaca, New York: Cornell University Press, 2003, pp.101–21.

New Ergonomics Perspective – Yamamoto (Ed.)
© 2015 Taylor & Francis Group, London, ISBN 978-1-138-02751-0

Factors that increase women's health awareness during the childcare period

Motoko Kosaka

Kansai University of International Studies, Hyogo, Japan

ABSTRACT: ***Purpose***: Many women who are caring for children usually also care for the health of their families, which include their children and husbands. Therefore, this study sought to identify factors that facilitate women's health awareness during the childcare period and thereby develop an intervention program.

Method: An anonymous self-reported questionnaire was conducted on mothers taking their children for health examinations from December 2009 to January 2010. The survey slip enclosed a questionnaire when sending a notification of medical consultation recommendation for the health examination and was collected at the time of the examination. The survey has set criteria for 5 items as "Priority of health behavior", "Awareness of health conduct", "Purpose in life", "Active coping behavior", "Emotional supporting network", and "Preventive health behavior" with involvement of basic attribute (i.e. age, family constitution, educational background, work), health conditions, lifestyles, life factor (i.e. Sleep, meal, exercise, smoking, drinking), and preventive health behavior. Factor analysis with promax rotation was used to analyze items related to living conditions. Multiple regression analysis with stepwise method was conducted on the 5 items related to preventive health behavior (objective variables) and basic attributes/life factors/extracted factors as explanatory variables (significance probability: 5% level). Factor analysis with promax rotation was also conducted with a preventive health behavior standard.

Result: The response rate was 264/406 (65.0%).

Two factors were extracted, which were termed "subjective symptom" and "decision-making" from the factor analysis. When we conducted a multiple regression analysis adding those factors into explanatory variable, "Decision-making," and "subjective symptom" were extracted from the multiple regression analysis. As a result of having performed a factor analysis of a preventive health action, "Eat aggressively" and "avoid eating" were extracted from the factor analysis of preventive health action.

Conclusion: "Subjective symptoms" and "decision-making" are related to the preventive health behavior of child-rearing women. Meal choices that prioritize the family's health, movements that improve subjective symptoms, and intervention of the production of a friend are effective in increasing health consciousness.

Keywords: Adult female; Health behavior; Public health; Occupational health; Family health

1 BACKGROUND

As personal health is affected by the social environment such as the home environment, school, an area, and the workplace, it is necessary to maintain such an environment to protect the health of the nation in Great Society. In the home environment, the time that a father spends on housework or childcare is usually shorter than that of a mother. The mother is usually responsible for the health of the family, which includes the child and the husband.

Therefore, the child and father are affected by the lifestyle and health choices of the mother. However, research on interventions regarding the health choices of mothers is lacking. Such interventions are necessary to promote health at home. Therefore, I aimed to examine factors that promote healthy choices in women caring for children and from there develop an intervention program.

2 METHOD

The subjects were 264 mothers whose babies and toddlers underwent medical checks in B-Ward of A-City. In December 2009 through January 2010, I implemented an anonymous self-reported survey. The questionnaires were sent to the mothers via regular post, together with information about the medical checks for babies and toddlers. We collected the completed surveys at the medical check places.

The contents of the survey included basic attributes, eight items concerning health status and living habits, and "the seven health habits of Breslow." There were also five items concerning preventive health activities such as "priority of health behavior," "awareness of health conduct," "purpose in life," "active coping behavior," "emotional support network," and the scale of "preventive health behavior." I stated the purpose of the research and the non-usage of data other than for research purposes in the request, and that the answers were anonymous. Completed questionnaires were considered to indicate consent to participate in this study.

Factor analysis with promax rotation was used to analyze the eight items of health conditions and living habits, and multiple linear regression with the stepwise procedure was used to analyze the five items on preventive health behavior as objective variables, basic attributes, life factors, and the extracted factors as explanatory variables (the criterion for statistical significance was set at $p < 0.05$). Concerning the scale of preventive health action, I implemented "the factor analysis using promax rotation" with SPSS Ver 17.0 for statistic process.

3 RESULTS

3.1 *Basic attributes*

The response rate was 264/406 (65.0%).

The average age of the respondents is 32.5 (standard variation: 4.561), with those in their 20 s and 30 s constituting over 95% of the sample. Concerning the number of children per participant, 33.3% of mothers had only one child, 47.3% had two children, and 19.3% had three or more children. Family units consisting of a husband, a wife, and children accounted for 90.2%. Fulltime housewives made up 73.5% of the sample. Among them, 84.8% perceived themselves as healthy, and 91.3% indicated themselves as being disease-free. On the other hand, 89.4% reported having some symptoms. Fifty-eight percent did not attend medical checks.

3.2 *Health behavior of women in the childcare period*

Factor analysis was conducted on the eight items on health conditions and lifestyle habits as main factors in order to identify the potential factors influencing living conditions. I extracted two factors from the scree plot and implemented a promax rotation. The communality was >0.16, and the load was >0.35 (Table 1). "The existence or non-existence of stress" and "the existence or non-existence of symptoms" had high factor loadings and were named "subjective symptom" because they are aware with some symptoms for the first factor. For the second factor, "the existence and non-existence of an unbalanced diet" and "medical check" had high loadings. This factor was termed "decision-making" because there were several paths to reach those specific goals.

3.3 Factors related to the implementations of preventive health behavior

The dependent variables consisted of five items related to preventive health actions, while basic attributes, life factor, subjective symptom, and decision-making were independent variables. Multiple linear regression analysis with stepwise procedure (significance probability at five percent standard) was conducted on these variables (Table 2).

When we conducted a multiple regression analysis adding those factors into explanatory variable, "Decision-making," and "subjective symptom" were extracted from the multiple regression analysis.

"Preventive health behavior" had a mean of 10.46 and standard variation of 3.683. In order to identify the potential factors of preventive health activities, a promax rotation by the chief source method was conducted on the 21 items of preventive health behavior. The results revealed nine items that did not have sufficient loadings, and were therefore excluded; the promax rotation with the chief source method was then conducted again. Thereafter, factor analysis was repeated while deleting items with factor loadings <0.35. Table 3 presents the

Table 1. Factor analysis of living conditions.

	Factor 1	Factor 2	Commonality
Stress	**0.603**	−0.001	0.364
Presence or absence of symptoms	**0.444**	0.035	0.189
Presence or absence of disease	0.182	−0.011	0.034
Weight fluctuation	0.062	0.024	0.003
Medical examination	0.204	**0.453**	0.192
Unbalanced diet	−0.067	**0.421**	0.199
Nutritional balance	0.004	0.344	0.118
Health outlook	−0.240	0.316	0.202
Factor correlation			−0.159

Table 2. Factors involved the execution of preventive health behaviour.

Item	Category score	Standard partial regression coefficients as the dependent variable each factor β				
		Priority of health behaviour	Awareness of health conduct	Purpose in life	Active coping behaviour	Emotional supporting network
Generation	(1:teens-20s, 2:30s–40s)					−0.134
The number of children	(3:3 1:1 or more people)	−0.128				
Family structure	(1: nuclear family, 2: extended family)			−0.241		
Occupations	(1: full-time, 2 part, etc., and 3: none)	0.188				
Sleeping hours	(High score: 1–3)				0.288	
Motion	(High score: 1–3)					−0.125
Weight control	(High score: 1–3)		−0.134			
Subjective symptoms	(High score: 2–5)			0.179	0.145	0.177
Decision-making	(High score: 3–9)	−0.247			−0.161	−0.172
Decision coefficient (R²)		0.085	0.018	0.094	0.148	0.084
						p < 0.05

185

Table 3. Factor analysis of preventive health behavior.

Item	Factor 1	Factor 2	Commonality
Often eating seaweeds such as kelp, seaweed, and laver	**0.809**	–0.118	0.659
Often eating vegetables with dense colors such as spinach and carrot	**0.448**	0.108	0.216
Processed soy products, such as tofu, are eaten.	**0.411**	0.141	0.194
Restricting the consumption of sweet foods and drinks as much as possible	–0.048	**0.615**	0.378
Restricting salt intake	0.093	**0.566**	0.334
Food containing animal fat, such as butter lard, is cut down.	0.076	**0.455**	0.216
Factor correlation			0.043

final model and factor correlations from the promax rotations. With two factors before rotation, the ratio to explain the full dispersion of six items was 33.3%.

The first factor was composed of three items such as "often eating seaweeds such as kelp, seaweed, and laver" and "often eating vegetables with dense colors such as spinach and carrot." The contents include eating aggressively in a conscious manner. I termed the factor "eat aggressively." The second factor had a high loading with items indicating restriction of eating certain foods due to health reasons such as "restricting the consumption of sweet foods and drinks as much as possible" and "restricting salt intake." I termed this factor as "avoid eating."

4 DISCUSSION

4.1 Features of women in the child care

Over 90% of the participants eat breakfast, indicating that most of the sample is conscious of the concept of nutritional balance. Analysis of the scale of preventive health behavior revealed two factors: "eat aggressively" and "avoid eating." According to the pre-study, with children's growth, the number of people conscious of nutritional balance is increasing. In addition, a previous report indicated that some women make it a rule to lead healthy lives and diets from the moment they become pregnant. We can therefore assume that women who are caring for children are highly conscious of meals. Many full-time housewives of nuclear families, who are assumed to be focusing on childcare and household duties, participated in the present survey.

Habits and impulses are of daily habits which will be inherited to the families. The present results indicate that interventions through education and instructions from the beginning of pregnancy, with meals as catalysts, would probably be effective.

4.2 Implementation factors for preventive health behavior

Decision-making and subjective symptoms were revealed to be key factors in the implementation of preventive health behavior. Many of the participants reported experiencing some stresses and/or symptoms despite perceiving themselves as disease-free and healthy. Furthermore, we found that those without subjective symptoms engaged in active coping behavior and received social support. Thus, it might be helpful to build the positive responding method and social support systems from the perspective of subjective symptoms.

Decision-making regarding health was higher when "priority of health behavior," "active coping behavior," and "emotional network" were high. According to Munakata, there are four conditions for taking self-care actions: 1. find health issues and activity targets, 2. strong motivation and less feeling of being burdened, 3. self-confidence, and 4. networks that support and evaluate self-decisions [1]. The results of the present study therefore seem reasonable with these conditions in mind, and I realize that the programs meeting four conditions are effective. Many women who are caring for children are from nuclear families, which increases the chances of them being isolated, because they are leading lives focusing on child care. Upon these matters, the existence of fellow people building health while child care makes psychologically and physically them healthy.

5 CONCLUSION

From the health awareness of women caring for children and their lifestyle habits, we obtained specific and usable supporting measures such as positive responding actions concerning meals, subjective symptoms, and networking. Future studies should develop more achievable programs. In addition, we need to verify the propagation effects on family members.

REFERENCE

[1] Munakata T. Supporting healthy self-decision: SAT health counselling. Minzoku Eisei 2002;17:1–15 (in Japanese).

New Ergonomics Perspective – Yamamoto (Ed.)
© 2015 Taylor & Francis Group, London, ISBN 978-1-138-02751-0

Survey of problems while wearing a hay fever prevention mask in 2009 and 2012

Mika Morishima & Yuki Shimizu
Faculty of Textile Science and Technology, Shinshu University, Ueda, Japan

Koya Kishida
School of Psychology, Chukyo University, Nagoya, Japan

Takashi Uozumi
Department of Computer Science and Systems Engineering, Muroran Institute of Technology, Muroran, Japan

Masayoshi Kamijo
Faculty of Textile Science and Technology, Shinshu University, Ueda, Japan

ABSTRACT: Our long-term goal is to develop a high-performance hay fever mask that is comfortable to wear. This study comprised two surveys that were conducted in 2009 and 2012. Both surveys involved distributing questionnaires to university students in five or six areas of Japan. There were 1,519 positive responses in 2009 and 3042 in 2012. In both surveys, masks were the most frequent hay fever countermeasure used by people of both genders. However, for both genders and surveys, most mask users encountered problems, such as "humidity," "misting of eyeglasses," and "breathing difficulty," which were influenced by the thermal properties, hygroscopic properties, and air flow properties of the masks. After investigating the correlations between response variables, correlations were observed between symptoms, countermeasures, and problems for both genders and surveys. Thus, improving these properties should result in less discomfort while wearing masks. In addition, resolving these problems should also improve other problem awareness.

Keywords: Maks; Hay fever; Survey; Problem awareness

1 INTRODUCTION

Our long-term goal is to develop a high-performance hay fever mask that is comfortable to wear. In the daily life of Japanese people, a hygiene mask is often worn to protect against influenza, to prevent common colds, for etiquette, protection against dryness, and for protection against pollens, among other reasons. During periods of high levels of pollens from cedar, Japanese cypress, Japanese alder, and other sources, Hay Fever (HF) sufferers wear masks to protect themselves against these pollens. In addition, wearing a mask is recommended in the Japanese guidelines for allergic rhinitis [1]. However, mask users are aware of problems while wearing these masks.

As a basic study to develop a hay fever protection mask, a survey of problems that HF sufferers faced while wearing masks was conducted in 2009 [2]. In addition, another survey using the same questionnaire was conducted 3 years later; the combined results were then compared. Moreover, problem awareness based on the characteristics of individual HF sufferers that were possibly correlated with symptoms, countermeasures used, and problems while wearing a mask were assessed by co-occurrence analysis. We anticipated that by improving the physical properties of masks related to common factors of problem awareness in these two surveys would result in less discomfort while wearing a mask. Moreover, although the correlations among other variables varied between these survey years, we anticipated that resolving these problems would result in improving other problem awareness.

2 SURVEY METHODS

Two surveys were conducted by distributing questionnaires to university students in the Tohoku, Kanto, Chubu, Kinki, and Kyushu areas of Japan in October 2009 and February 2010. Another survey was conducted in November 2012 and February 2013 that included these areas in addition to the Chugoku area.

Figure 1 shows the flowchart for survey data analysis. The questions and responses were written in Japanese. The questions were: 1. "Do you have hay fever symptoms?"(yes or no); If yes, continue to question 2. 2. "What symptoms do you have?"(open-ended response); 3. "What measures do you use?" [Multiple choice: Mask (C1), Eyeglasses (C2), Gargling (C3), Over-the-counter drugs (C4), Prescription drugs (C5), Injection (C6), Other (C7), None (C8)]; If you selected mask, continue to question 4. 4. "Do you have any problems when wearing a mask?" (yes or no); 5. "What problems do you experience with the mask?" [Multiple choice: Humidity (P1), Fogged eyeglasses (P2), Difficulty breathing (P3), Awkwardness (P4), Make-up coming off (P5), Ear pain (P6), Feeling hot (P7), Distracted by mask (P8), Poor fit (P9), Other (P10)].

For question 1, we did not inquire about the types of pollen that caused respondents' HF and we did not ask how HF was diagnosed. For question 2, there was a wide variety of responses. Therefore, using a text mining approach, the recorded data were categorized based on a list of words that were used at a high frequency and then combined. To assess the problems encountered while wearing a mask from a holistic perspective, co-occurrence relationships were determined based on individual responses. Pilot studies were conducted using this questionnaire with smaller groups of university students between 2006 and 2009. The options given for questions 3 and 5 were selected based on the results of these preliminary studies.

3 RESULTS AND DISCUSSION

3.1 *HF sufferers*

From the combined results for question (1), there were 1,519 positive responses in 2009 (715 from men and 804 from women) and 3042 positive responses in 2012 (1475 from men and 1567 from women). In the second survey, 41.1% of men and 42.5% of women were reportedly HF sufferers. On comparing the results of these years, we observed that there was no significant difference with regard to the proportions of men and women who were HF sufferers (Men, $P = 0.040 > 0.01$; Women, $P = 0.477 > 0.01$). The incidences of HF in our surveys were slightly higher than those reported in another study[3]. However, our survey results were not based on medical diagnoses. In addition, we did not determine the kinds of pollen allergens that affected our respondents. Both surveys were conducted from October or November to February, which were periods when HF symptoms were less likely to be problematic. Therefore, our results were holistic results derived from respondents' memory and cognizance.

3.2 *HF countermeasures*

Table 1 shows the results for the HF countermeasures reported by men and women in the second survey and the statistical comparison results between the two surveys and between genders (P values based on Chi-square tests). Masks were used most frequently by both genders. On comparing the results for genders in the second survey, we observed that more women who were HF sufferers used masks and took prescription drugs compared with men. These

Table 1. Percentages of HF countermeasures and P values by Chi-square tests.

| Countermeasure | Men | | Women | | P-value |
	Percentage in 2nd survey(%) (N = 585)	P value 1st survey and 2nd survey	Percentage in 2nd survey(%) (N = 631)	P value 1st survey and 2nd survey	Men and women in 2nd survey
C1; Mask	57.1	0.001**	78.8	0.000**	0.000**
C2; Eyeglasses	10.6	0.315	7.8	0.000**	0.087
C3; Gargling	22.9	0.727	21.9	0.991	0.665
C4; Over-the-counter drugs	30.6	0.671	27.7	0.367	0.272
C5; Prescription drugs	29.7	0.463	40.9	0.980	0.000**
C6; Injection	2.4	0.129	1.3	0.365	0.141
C7; Other	5.0	0.310	5.5	0.000**	0.646
C8; None	17.9	0.744	8.7	0.000**	0.000**

**; <1% significance level.

results were the same as the combined results of our first survey. In the responses for other countermeasures used, these included yogurt (0.6%), eye washing (0.5%), traditional Chinese medicine (0.3%), and some others, which were almost the same as those in the first survey.

On comparing the results of the two surveys, we observed that the increase in mask use was 13.0% for men and 23.9% for women. This depended on wider use of different kinds of masks with different sizes, shapes, and colors. This appeared to be a result of habituation with wearing a mask after the widespread outbreak of H1N1 influenza. On comparison, we observed that there was a decrease of 8.0% of eyeglass wearers and 8.0% of HF sufferers who did not use countermeasures among women.

3.3 HF Symptoms

For question 2, 4082 open-ended responses were provided. Therefore, we focused on words that were used at high frequencies, which generated a list of 62 words. The responses were categorized based on this list. Table 2 shows the combined results for symptoms for mask users, users of countermeasures other than a mask, and no use of countermeasures. There were no significant differences in symptoms among these groups (Men, $P = 0.783$; Women, $P = 0.583$). On comparing the results between the two surveys, we observed that there were no significant differences among each group. Based on these results, there were no significant correlations between wearing a mask and symptoms for men and women, which were the same results as those in the first survey.

3.4 Problem awareness in mask users

From the combined results for question 2, despite the popularity of wearing a mask, 75.1% of men (n = 334) and 84.7% of women (n = 497) reported problems while wearing a mask. These percentages were significantly different between men and women ($P = 0.001$). On comparing the results for different survey years, we observed that there were no significant differences for each gender (Men, $P = 0.975$; Women, $P = 0.704$).

Table 3 shows the percentages of respondents that reported each problem and the P values based on Chi-square tests. In the second survey, problem awareness of humidity, misting of eyeglasses, and breathing difficulties had higher percentages for both genders. These problems were because of the thermal properties, hygroscopic properties, and air flow properties at the time of wearing a mask. Moreover, a higher percentage of women reported problems such as makeup coming off, ear pain, and poor fit. By comparison, a higher percentage of men reported distraction as a major problem. Comparing the major problems reported in

Table 2. Percentages of men and women reporting each symptom.

Symptoms	Men (%)			Women (%)		
	Mask users (N = 334)	Users of counter-measures other than a mask (N = 146)	No use of counter-measures (N = 105)	Mask users (N = 497)	Users of counter-measures other than a mask (N = 84)	No use of counter-measures (N = 50)
Itchy eyes	81.1	80.1	70.5	83.9	81.0	74.0
Runny nose	85.0	78.1	77.1	87.1	81.0	70.0
Sneezing	58.4	53.4	48.6	70.6	57.1	52.0
Sniffle	59.6	60.3	48.6	56.9	45.2	38.0
Itchy throat	9.9	6.8	8.6	21.7	14.3	16.0
Sore throat	10.5	7.5	5.7	12.5	11.9	0.0
Painful eyes	9.6	9.6	1.9	9.5	6.0	2.0
P value 1st survey and 2nd survey	0.853	0.310	0.848	0.946	0.987	0.445

Table 3. Percentages of problems and P-values by Chi-square tests.

Problems	Men		Women		P-value
	Percentage in 2nd survey(%) (N = 251)	P-value 1st survey and 2nd survey	Percentage in 2nd survey(%) (N = 421)	P-value 1st survey and 2nd survey	Men and women in 2nd survey
P1; Humidity	59.4	0.234	67.7	0.261	0.029
P2; Misting of eyeglasses	58.2	0.585	48.7	0.677	0.017
P3; Breathing difficulties	46.6	0.977	47.5	0.062	0.823
P4; Awkwardness	9.2	0.014	11.6	0.000**	0.316
P5; Makeup coming off	1.6	0.792	47.7	0.019	0.000**
P6; Ear pain	29.1	0.254	45.8	0.001**	0.000**
P7; Feeling hot	24.7	0.482	27.8	0.517	0.381
P8; Distraction	36.7	0.531	23.0	0.000**	0.000**
P9; Poor fit	8.8	0.010	21.9	0.731	0.000**
P10; Other	7.6	0.897	6.4	0.250	0.566

**; 1% significance level.

the second survey with those in the first, problems reported by men were nearly the same. The problems of awkwardness and distraction decreased by 14.0% and 14.8%, respectively, among women. In contrast, the problem of ear pain increased by 14.7% among women.

3.5 Co-occurrence analysis of problems for individual responses

A holistic perspective of problem awareness was considered using co-occurrence analysis [4]. As an index of the degree of similarity for individual responses among respondents, Jaccard coefficients, J, were determined using equation (1). Moreover, by focusing on the highest Jaccard coefficient in each combination, more combinations were determined.

$$J = \frac{X_1 \cap X_2}{X_1 \cup X_2} \qquad (1)$$

Considering $J > 0.1$ in the second survey, we found the following. For men (n = 251): (P1) mask, runny nose ($J = 0.810$), itchy eyes (0.693), sniffle (0.402), (P1) humidity (0.394), sneezing (0.307), (P2) misting of eyeglasses (0.227), (P3) breathing difficulty (0.159), (C4) over-the-counter drug use (0.143), and (P8) distraction (0.108); for women(n = 421): (P1) mask, runny nose (0.843), itchy eyes (0.717), sneezing (0.530), (P1) humidity (0.416), sniffle (0.228), (P2)misting of eyeglasses (0.171), and (P5) makeup coming off (0.112). In the first survey for men (n = 84): (P1) mask, itchy eyes (0.881), runny nose (0.750), sneezing (0.512), sniffle (0.321), (P1) humidity (0.226), and (P8) distraction (0.131); and for women (n = 164): (P1) mask, itchy eyes (0.902), runny nose (0.799), sneezing (0.567), sniffle (0.396), (P1) humidity (0.274), (P3)breathing difficulties (0.189), and (P5) makeup coming off (0.110).

From these results, the common problem awareness for different survey years and genders were humidity along with four symptoms. In addition, common problem awareness that were observed between the survey years were distraction for men and make up coming off for women. When problems of humidity were improved based on the physical properties, it was indicated that discomfort while wearing a mask became less frequent. Moreover, we anticipated that the other problems would also be resolved.

4 CONCLUSIONS

From the combined results of surveys conducted in 2009 and 2012, problem awareness included "humidity," "misting of eyeglasses," and "breathing difficulty" with high percentages in both years and genders. Subsequently, a holistic perspective of problem awareness was assessed by co-occurrence analysis. From the correlation results among symptoms, countermeasures, and problems, it was indicated that the common problem awareness was humidity in both years and genders. Improving these problems based on physical properties of masks resulted in less discomfort while wearing them. Moreover, we anticipate that the other problems will also be resolved.

REFERENCES

[1] K. Okubo, Y. Kurono, S. Fujieda, S. Ogino, E. Uchio, H. Odajima, *et al.*, "Japanese guideline for allergic rhinitis," *Allergol Int*, vol. 60, pp. 171–89, Mar 2011.
[2] M. Morishima, K. Kishida, T. Uozumi, and M. Kamijo, "Experiences and Problems with Hygiene Masks Reported by Japanese Hay Fever Sufferers," *International Journal of Clothing Science and Technology*, vol. 26, p. in press, 2014.
[3] T.D. Richards S, Roberts H, Harries U., "How many people think they have hay fever, and what they do about it," *Br J Gen Pract.*, vol. 42, pp. 284–286 1992.
[4] L.W. Nees Jan van Eck, "How to Normalize Co-Occurrence Data? An Analysis of Some Well-Known Similarity Measures," *Journal of the American Society for Information Science and Technology*, vol. 60, pp. 1635–51, 2009.

New Ergonomics Perspective – Yamamoto (Ed.)
© 2015 Taylor & Francis Group, London, ISBN 978-1-138-02751-0

Investigation on factors affecting nursing performance in Intensive Care Unit in a hospital

Eric Min-yang Wang & Novie Canggang

Department of Industrial Engineering and Engineering Management,
National Tsing Hua University, Hsinchu, Taiwan

ABSTRACT: ICU is defined as a place that provides health care service for critically ill patients who needs exigent treatment to strengthen health care with support from surveillance equipment for patients. ICU nurses become one of the important roles in providing health care in ICU. ICU nurses have to contact with death, critical situations of patients, facing large of pressure in performing their job in ICU. But they always demanded to provide the best performance of care that result in higher quality care in monitoring patients for long hours. Nurse working performance in this study defined as an action and situation where nurse as a health-care provider conducting their duties in providing health-care services for guards patient's life. This study was covered three parts of purposes. First is to determine the work-related factors that perceived by ICU nurses which could affect their nursing in performing their duties. Second is to identify the impact of the work-related factors to the nurse. Third is to provide suggestions in order to improving the design effort and quality care in ICU. The result showed that there were 9 factors which could affect nurse while they progress their duties in ICU including social interaction, ICU occurrence, organizational issues, discussions, decision making standard, motivation, education and training, verification and source of stress. The impact of the identify factors include all the action that related to their working process such as mood, speed, effectiveness, motivation, workload, spirit, etc. Job satisfaction for level seven was also found for 36% participants in this study. This study suggest that there were need some adjustment in staffing, working shift and hours, work scheduling and others for helping participants cope and improve in their quality of care.

Keywords: ICU; Nurse working performance; health-care; patient safety

1   INTRODUCTION

In the health care industry, Intensive Care Unit's (ICU) nurses are taking as an important role for patient who needs an emerge care to survive through some medical care and equipment. Each working places and working activities might conduced in stress and high workload that influence to each working performances. Nursing workload result from work demands that related to patient safety and quality care. ICU was frequently occurring in some incidents as well as a high complexity in care providing process (Ballangrud, Hedelin & Hall-Lord, 2012). In previous has defined nursing performance required to carry out all nursing activities in the critical care regarding quality, patient safety and value of care for patients (Kurtzman, Dawson, & Johnson, 2008). In this study, nurse working performance defined as an action or situation where nurse as a health-care provider conducting their duties for guard patient's life. ICU nurses are mainly involved in performing routine and repetitive activities; thus errors are common happened while they are performing duties in ICU. Job satisfaction also correlated to job performance of hospital professionals (Lynn McFarlane Shore, 1989) and indicated job satisfaction become a better predictor of job performance. Besides that, errors were found to have relation with nurse working performance that impact to patient safety. In condition of error will affect to nurse performance and directly impact to patient's mortality that associated with decreased of job satisfaction and increased emotional fatigue (Mrayyan & Hamaideh, 2009). Some negative impacts are probably associated with a greater risk of incidents. It was considered that negative impacts could result in negative outcomes that not only in patient safety but also impact to care quality, care provider satisfaction, and QWL and even in economic outcomes (Carayon & Gurses, 2005). Thus, it is a necessary to identifying the factors that faced by ICU nurses which could affect nurse while progressing duties in ICU.

Meanwhile in Taiwan, the number of patient bed were keep increasing while the number of nurses also increasing from year to year (Ministry of Health in Taiwan, 2013). Nurses take a role as one of the important health care providers who in touch with patients frequently. With balance number between the patients and the nurses, none of the patients will be ignored. But this fact could not assure the quality of the service provided by health care providers have given best quality performance of patients caring. Health-care providers' quality service has to be improved continuously to give best of services to patients. This study was carried out to investigate the work-related factor affect nurse in performing their duties. The goal is to identify the work-related factors and find out the impact of identified factors. Suggestion also provided in this study. Hopefully, improving the design effort to prevent negative work-related factors in critical care environment will be realized.

2   METHODS

Qualitative and quantitative design was conducted in this study. Participants are nurses working in ICU of one subject hospital in North Taiwan. This research has been approved by Institutional Review Board (IRB) of subject hospital and Research Ethics Contact (REC) of National Tsing Hua University. There were three phases' methods. The first phase is individual interview, the second phase is construction questionnaire survey questions sheet based on the result of individual interview in the first phase. The third phase is questionnaire survey. Individual interview are involved in deeper understanding how the peoples give meaning and clarify their experiences (Hasson, McKenna, & Keeney, 2013). Questionnaire survey are used to gather an exploratory data from a large group of participants in a cost-effective manner (Kilpatrick et al., 2013). Both of individual interview and questionnaire survey was conducted in open-ended questions (see Appendix A, Appendix B). By combining these two methods there will be a deeper depth and strength of the data collection in this study. Participants only are in touch in first phase and third phase.

In the first phase, individual interview are used to obtain the variables of work-related factors based on the nurses' in ICU experiences. Individual interview content were related to urgent decision making; critical decision making; simple decision making; job satisfaction; heaviest workload and least workload. Participants were group into three group based on how long they have been worked in ICUs in their entire life. First group are the participants who have been worked in ICU environment for less than three years (not include full three years); The second group are the participants who have been worked in ICU environment for three years (include full three years) to less than seven years (not include full seven years); and the third group are the participants who have been worked in ICU for seven years (include full seven years) or more. Then will be randomly choosing four nurses (two are from surgical department nurses and the others two are from medical department nurses) from each group to do the first phase individual interview.

In the second phase, based on the result of individual interview in phase 1, the variables of work-related factors perceived by ICU nurses who affect to nursing working performance were identified. These identified work-related factors' variables were presented as the questionnaire survey questions for third phase in this study.

In the third phase, the participants who did not do the individual interview in first phase are all required to do questionnaire survey. Questionnaire survey utilized 4- Point Likert Scale. Respondents were asked to seal the questionnaire sheets in an envelope and placed in a bigger pocket provided by the researcher.

3 RESULTS

There were 45 participants conducting in this study. All the participants were female. For individual interview, average age was 29.3 years old, average working years were 5.5 years old and average working hours per day was 8.6 hours/day. For questionnaire survey, average age was 30.2 years old, average working years was 4.2 years and average working hours per day was 8.6 hours per day. Table 1 shows twenty six identified variables in phase 1. Each variable were designed as one question in questionnaire survey. Adding two more question regarding jobs satisfaction level and reason by participants.

3.1 *Identified factors*

Questionnaire Survey results were analysed using SPSS version 20 for factor analysis with Bartlett test and KMO. The Output of the analysed data shows in Table 2. Value of KMO was found less than 0.5. The reason of this scene is because the number of sample were too less in this study since in subject hospital, total nurses in surgical ICU and medical ICU are 45 participants.

In factor analysis found that there were 9 representative factors, each indicated the eigenvalues greater than one will formed a factor shows as Table 3. The result of the identified factors comes with each variable shows as Table 4.

Table 1. Variables affect nurse while progress duties in ICU.

Variables	
V1	Doctor's instruction
V2	Order from doctor
V3	Discussion with other health-care team
V4	Get along with other health-care providers
V5	Past experiences
V6	ICU courses every month
V7	Training every three years
V8	Information found by own-self
V9	Guidance by senior
V10	SOP in ICU
V11	Hospital law and regulation
V12	Hospital management
V13	Patient's family pressure
V14	Patient's family argument
V15	Get along with patient's family
V16	Patient's agreement
V17	Patient's condition
V18	High turnover rate
V19	Nursing shortage
V20	Paper work
V21	Exam in ICU
V22	Complicated procedures
V23	Salary
V24	Working hours
V25	Working shift
V26	Become leader

Table 2. Output data for KMO and Bartlett's test.

Kaiser-Meyer-Olkin measure of sampling adequacy		0.347
Bartlett's Test of Sphericity	Approx. Chi Square	524.268
	df	325
	Sig.	0.000

Table 3. Numbers of factors formed.

Component	Extraction sums of squared loadings		
	Total	% of variance	Cumulative %
1	5.889	22.690	22.690
2	3.573	13.742	36.432
3	2.454	9.438	45.869
4	1.965	7.559	53.428
5	1.673	6.434	59.862
6	1.604	6.169	66.032
7	1.363	5.241	71.272
8	1.178	4.530	75.802
9	1.059	4.075	79.877

Table 4. Factors affect nurse in ICU.

Variables	
1. Social interaction	• Get along with other health-care providers • Get along with patient's family • Patient's agreement • Patient's condition
2. ICU occurrence	• Patient's family pressure • Patient's family argument • High turnover rate • Nursing shortage
3. Organizational issues	• Hospital management • Paper work • Working hours • Working shift
4. Discussions	• Discussion with doctor • Discussion with other health-care team
5. Decision making standard	• Past experiences • SOP in ICU • Hospital law and regulation
6. Motivation	• Information found by own self • Examination in ICU • Salary
7. Education and training	• ICU courses every month • Training every three years
8. Verification	• Order from doctor • Guidance by senior
9. Source of stress	• Complicated procedures • Become leader

First factor social interactions found by get along well and having good relationship with other health-care team and patient's family, it could help participants in smooth work and mood whole day. Other elements are participants decide next treatment care based on patient's agreement and condition. Although there were some patients can't express what they want, but patient's still can response to participants by movement such as nod their head.

Second factor ICU occurrence often perceived by participants. Patient's family pressure came from in meets patient's family needs. When participants remind and tutor to patient's family, it could lead to the argument. Another, it's common known that high turnover rate and nursing shortage were common problems often occur in ICU. These elements could increase the demands to the current participants and could enlarge participant's workload.

Third factor organizational issues could effect on workload and concentration. Participants were asked to prepare for the paper work in their work-off time, result in less their rest time, some of the participants admitted they used during work time to do the paper work and result as interruption in perform their work and affect them managing their time not only working time but also their rest time.

Fourth factor discussions with other-health care team; they shared what they know and what they don't know from the discussions. It could help a lot to each participant in performing their duties inside ICU.

Fifth factor decision making standard, participants were thought of these elements in solving the first step in a situation. These variables were useful could help them in perform their duties. Participants who have less experience inside ICU were admitted having problem in performing decision making in their duties inside ICU.

Sixth factor motivation, Participants were doing own—homework by observing the newest cases and technology information related to ICU work. ICU department also provided exam every month to help participants in adding their work knowledge. Salary in ICU also found as one of the motivation for participants.

Seventh factor education and training resulted participants learned a lot from the courses and training in new cases, new patient's condition, new technology, new way to using equipment and some tutorial to unexpected situation inside ICU.

Eighth factor verification, participants admitted they will not directly do what doctor's order for them. They will first evaluate whether the order were right to progress or not. In evaluating, participants will consider some information guide by senior. If the order were not right, they were try to communicate and remind to doctor.

Ninth factor source of stress, procedures inside ICU are complicated and confusing. Becoming a leader in one shift affect to workload and concentration level in ICU because leader is not only has to take of care their own patient but also to control and monitor other participants' movement in one shift.

3.2 Job satisfaction level

There were 36% of participants having level 7 as their job satisfaction level in ICU. The lowest level was found in level 3 which only 3% of the participants having this job satisfaction level. Reason of participants gave their job satisfaction level was provided in Table 5. The percentages show how many participants mentioned the elements for satisfied and unsatisfied of their job satisfaction level. The unit it per cent and each reason could be mentioned by the participants more than once. There were some elements were not mentioned in previous literature such as working shift, self-improvement, patient's family needs, long working experience years, team work and accomplishment.

4 DISCUSSION

This study found that nurse having a least workload condition in performing routine activities that could probably increase their working speed spirit. This result was different from previous literature of Sawatzky (1996) mentioned in doing routine activities rank highly in stressors

Table 5. Job satisfaction level reasons.

Satisfied	Unsatisfied
Salary and benefit 22%	Working shift 15%
Team-work among nurse 16%	Need-self-improvement 15%
Adapt well in ICU environment 12%	Pressure 15%
Accomplishment 10%	Communication 12%
Working hours 8%	Nursing shortage 10%
Interest in this job 8%	Health-care team attitudes 10%
Nice supervisors 8%	Over loading 10%
Long working experienced 6%	Meet patient's family needs 5%
Respect from patient after care 4%	Lack of respect 5%
Lots of room for onward 2%	Working hours 2%

identified in the research of stress in critical care nurses. Possible reason is because the type of critical care of each research. However, between these still has a similar result mentioned that stress occur in facing the unstable patient. Although from this study found that nurse performing their duties with stress and workload within, they having higher pride and accomplishment than others' department nurse. This could motivate ICU nurse in decreasing their intention to leave from ICU. Younger working experiences nurse perceived workload more than the others who having long working year's experiences. It results decrease nurse retention level working in ICU. This fact was not considered in the beginning of this study. By considering and imply this fact in this study field, the result of this study will be more contributed by having result in factor affect least working year experience nurse and long year working experience nurse. Different working time and shift could be applied in order to avoid intention of leave by ICU nurse. In addition, individual meeting between nurse manager and nurses should be carry on to know nurses thought and psychological stress perceived by ICU nurse.

Social interaction for factor one found for the most dominant factor, they always respect and follow patient's agreement and condition before giving treatment care. Source of stress for factor nine found for the less dominant factor, they found less in sharing complicated procedures in ICU since they have been recognize and learn in their previous education; since not every participants could be arranged as a leader in one shift, thus only few participants shared regarding becoming leader factor.

Working shift in ICU subject hospital found to be messy. Participants have difficulties in managing their working time and rest time. Although there are so many research found this difficulties, this fact still happened in some medical environment. Subject hospital was suggested to manage this difficulty by giving at least one rest day for the nurse before having different working shift especially for the nurse who will have late-night shift for her next shift. Late due off work also decrease the level of job especially the nurses who had late night shift. It is a crucial issues in having improvement late due off work issues in ICU. By having a punctual schedule of turn over duties, late due off work issues could be eliminating. Beside the medical education, time management education also should be implied to nurse in ICU to having ability in managing their own time.

Data in this study was restricted of some limitations to make further result regarding the link of participant general information and questionnaire survey result. Also, volume in this study was weak result from the limitation of sample size. Analysis of the factors was weak of reliability and validity. By having link to participant's general information to the identified factors in this study, the level of contribution in this study will be volume up and probably increasing nursing performance.

5 CONCLUSION

This study was drive to investigate the factors that affect the ICU nurses as participants while the progress their duties inside ICU. It found that there were 9 factors identified in this study which are social interaction, ICU occurrence, organizational issues, discussions, decision making standard, motivation, education and training, verification and source of stress. The result in this study could apply in implication of hospital regulations, smoothing work progress and for improvement of systematic in ICU. There were also some suggestions provide in this study. From human factors and ergonomics perspective, considering human and environment to improve the quality care provided in ICU, suggestions provided as follow: by managing well staffing by nurse manager in more paying attention and understanding in what nurse in shift complaint about, focus in working shift and working hours adjustment in order having a good management of interval between working time and rest time of the ICU nurses, increasing attention from nurse manager, supporting flexible work scheduling, helping in cope with stress by having some relax ritual activities for each nurse, minimizing time spent on non-nursing tasks for the nurse on their work-off day and self-emotion control were could improve the emotional fatigue happened on the nurse to the design effort to prevent the negative work-related factors in ICU in a hospital.

6 FUTURE STUDY

This study found an interesting fact on junior nurses, their intention to leave the current job in ICU was quite strong. It was recommended to having deeper research to junior nurses. Other things are, to having deeper and continuously research from this study's limitations mentioned in the previous. By enlarging the scope of the participants in several departments could gain the enough volume for validity and reliability in this field. There are also some reasons in job satisfaction level that not mentioned in the previous literature, thus recommend for a future studies concern in this study field.

6.1 Study limitations

This research was progressed in one of the regional hospital in North Taiwan. Since every hospital has their own authorization and to tend to be different due to the differences conditions, systems, procedures, practices, policies thus the result of this study might have a possibility in mismatch to other hospitals. In other words, the results, conclusions and suggestion in this study can be uncertain in apply to the other hospitals. Because of IRB issues mentioned in not using any data information from participants, thus some part of this study having difficulties in making further result discussions.

ACKNOWLEDGEMENTS

This study was assisted by subject hospital ICU department. We would like to give deepest appreciation to subject hospital and National Tsing Hua University Research Ethics Contact (REC). We would also like to thank to all ICU department's nurses as participants in this study.

REFERENCES

Al-khasawneh, A.L., & Futa, S. (2013). The Relationship between Job Stress and Nurses Performance in the Jordanian Hospitals. *Asian Journal of Businness Management*, 5(2), 267–275.

Amin, S.G. (2011). A study to determine the influence of workload on nursing personnel. *ProQuest Dissertations and Theses*.

Arevalo, J.J., Rietjens, J.A., Swart, S.J., Perez, R.S., & van der Heide, A. (2013). Day-to-day care in palliative sedation: survey of nurses' experiences with decision-making and performance. *Int J Nurs Stud*, 50(5), 613–621. doi: 10.1016/j.ijnurstu.2012.10.004.

Ballangrud, R., Hedelin, B., & Hall-Lord, M.L. (2012). Nurses' perceptions of patient safety climate in intensive care units: a cross-sectional study. *Intensive Critical Care Nurs*, 28(6), 344–354. doi: 10.1016/j.iccn.2012.01.001.

Bakalis, N., Bowman, G.S., & Porock, D. (2003). Decision making in Greek and English registered nurses in coronary care units. *Int J Nurs Stud*, 40(7), 749–760. doi: 10.1016/s0020-7489(03)00014-2.

Beckmann, V., Baldwin, I., Hart, G.K., Runciman, W.B. (1996). Human error in Health Care. *The Australian Incident Monitoring Studyin Intensive Care (AIMS-ICU): An Analysis of the First Year of Reporting. Anaesthesia and Intensive Care 24 (1996): 320–329.*

Carayon, P. (2010). Human factors in patient safety as an innovation. *Appl Ergon, 41*(5), 657–665. doi: 10.1016/j.apergo.2009.12.011.

Carayon, P., & Gurses, A.P. (2005). A human factors engineering conceptual framework of nursing workload and patient safety in intensive care units. *Intensive Critical Care Nurs, 21*(5), 284–301.1–12 doi: 10.1016/j.iccn.2004.12.003.

Carayon, P., Wetterneck, T.B., Rivera-Rodriguez, A.J., Hundt, A.S., Hoonakker, P., Holden, R., Gurses, A.P. (2013). Human factors systems approach to healthcare quality and patient safety. *Appl Ergon.* doi: 10.1016/j.apergo.2013.04.023.

Carlson, E., Rämgård, M., Bolmsjö, I., & Bengtsson, M. (2013). Registered nurses' perceptions of their professional work in nursing homes and home-based care: A focus group study. *Int J Nurs Stud.* 1–25. doi: 10.1016/j.ijnurstu.2013.10.002.

Carmel, S. (2006). Health care practices, professions and perspectives: a case study in intensive care. *Soc Sci Med, 62*(8), 2079–2090. doi: 10.1016/j.socscimed.2005.08.062.

Chang, H.-L., Lin, C.-F. (2008). Factors Related to Newly-Under Graduated Nurses Turnover Rates.

Chiu, M.-C., Wang, M.-J., Lu, C.-W., Pan, S.-M., Kumashiro, M. Ilmarinen, J. (2007). Evaluating work ability and quality of life for clinical nurses in Taiwan. *Nurs Outlook, 55*(6), 318–326. doi: 10.1016/j.outlook.2007.07.002.

Chiang, H.-Y. (2006). Barriers to Nurses' Reporting of Medication Administration Errors in Taiwan. *Journal of Nursing Scholarship.* 392–399.

Chiang, Y.-M., & Chang, Y. (2012). Stress, depression, and intention to leave among nurses in different medical units: implications for healthcare management/nursing practice. *Health Policy, 108*(2–3), 149–157. doi: 10.1016/j.healthpol.2012.08.027.

Chu, C.-I., & Hsu, Y.-F. (2011). Hospital nurse job attitudes and performance: the impact of employment status. *J Nurs Res, 19*(1), 53–60. doi: 10.1097/JNR.0b013e31820beba9.

DeLucia, P.R., Ott, T.E., Palmieri, P.A. (2009). Performance in Nursing. *Reviews of Human Factors and Ergonomics, 5*(1), 1–40. doi: 10.1518/155723409x448008.

Donald Hay, M. (1972). The Psychological Stresses. *Psychosomatic Medicine, 34.*

Donchin, Y. (2003). A look into the nature and causes of human errors in the intensive care unit. *Quality and Safety in Health Care, 12*(2), 143–147. doi: 10.1136/qhc.12.2.143.

Duffield, C., Diers, D., O'Brien-Pallas, L., Aisbett, C., Roche, M., King, M., & Aisbett, K. (2011). Nursing staffing, nursing workload, the work environment and patient outcomes. *Appl Nurs Res, 24*(4), 244–255. doi: 10.1016/j.apnr.2009.12.004.

El-Masri, M., & Fox-Wasylyshyn, S.M. (2007). Nurses' roles with families: perceptions of ICU nurses. *Intensive Crit Care Nurs, 23* (1), 43–50. doi: 10.1016/j.iccn.2006.07.003.

Gabrielle, S., Jackson, D., & Mannix, J. (2008). Adjusting to personal and organisational change: Views and experiences of female nurses aged 40–60 years. *Collegian, 15*(3), 85–91. doi: 10.1016/j.colegn.2007.09.001.

Gurses, A.P., & Carayon, P. (2009). Exploring performance obstacles of intensive care nurses. *Appl Ergon, 40*(3), 509–518. doi: 10.1016/j.apergo.2008.09.003.

Hasson, F., McKenna, H.P., & Keeney, S. (2013). A qualitative study exploring the impact of student nurses working part time as a health care assistant. *Nurse Educ Today, 33*(8), 873–879. doi: 10.1016/j.nedt.2012.09.014.

Hunt, S.T. (2009). NursingTurnover: Costs, Causes & Solutions. *SuccesFactors Healthcare.*

Jamal, M. (2011). Job stress, job performance and organizational commitment in a multinational company. *International Journal of Business and Social Science, 2.*

Kao, C.-C. (2011). Multi-aspects of Nursing Man-Power in Taiwan. *Cheng Ching Medical Journal, 7.*

Kilpatrick, K., Dicenso, A., Bryant-Lukosius, D., Ritchie, J.A., Martin-Misener, R., & Carter, N. (2013). Practice patterns and perceived impact of clinical nurse specialist roles in Canada: Results of a national survey. *Int J Nurs Stud, 50*(11), 1524–1536. doi: 10.1016/j.ijnurstu.2013.03.005.

Knezevic, B., Milosevic, M., Golubic, R., Belosevic, L., Russo, A., & Mustajbegovic, J. (2011). Work-related stress and work ability among Croatian university hospital midwives. *Midwifery, 27*(2), 146–153. doi: 10.1016/j.midw.2009.04.002.

Kostagiolas, P., Korfiatis, N., Kourouthanasis, P., & Alexias, G. (2014). Work-related factors influencing doctors search behaviors and trust toward medical information resources. *International Journal of Information Management, 34*(2), 80–88. doi: 10.1016/j.ijinfomgt.2013.11.009.

Kurtzman, E.T., Dawson, E.M., & Johnson, J.E. (2008). The current state of nursing performance measurement, public reporting, and value-based purchasing. *Policy Polit Nurs Pract, 9*(3), 181–191. doi: 10.1177/1527154408323042.

Lin, Y.-H., & Liu, H.-E. (2005). The impact of workplace violence on nurses in South Taiwan. *Int J Nurs Stud, 42*(7), 773–778. doi: 10.1016/j.ijnurstu.2004.11.010.

McCallum, J., Duffy, K., Hastie, E., Ness, V., & Price, L. (2013). Developing nursing students' decision making skills: are early warning scoring systems helpful? *Nurse Educ Pract, 13*(1), 1–3. doi: 10.1016/j.nepr.2012.09.011.

Mrayyan, M.T., & Hamaideh, S.H. (2009). Clinical errors, nursing shortage and moral distress: The situation in Jordan. *Journal of Research in Nursing, 14*(4), 319–330. doi: 10.1177/1744987108089431.

Myny, D., Van Hecke, A., de Bacquer, D., Verhaeghe, S., Gobert, M., Defloor, T., & Van Goubergen, D. (2012). Determining a set of measurable and relevant factors affecting nursing workload in the acute care hospital setting: a cross-sectional study. *Int J Nurs Stud, 49*(4), 427–436. doi: 10.1016/j.ijnurstu.2011.10.005.

Radtke, J.V., Tate, J.A., & Happ, M.B. (2012). Nurses' perceptions of communication training in the ICU. *Intensive Crit Care Nurs, 28*(1), 16–25. doi: 10.1016/j.iccn.2011.11.005.

Rattray, J., Jones, M.C. (2007). Essential elements of questionnaire design and development. *J Clin Nurs, 16*(2), 234–243. doi: 10.1111/j.1365–2702.2006.01573.x.

Rechel, B., Buchan, J., & McKee, M. (2009). The impact of health facilities on healthcare workers' well-being and performance. *Int J Nurs Stud, 46*(7), 1025–1034. doi: 10.1016/j.ijnurstu.2008.12.008.

Sawatzky, J.A. (1996). Stress in critical care nurses: Actual and Perceived. *Heart & Lung, Vol. 25, No. 5.*

Shore, L.A., Martin, H.J. (1989). Job Satisfaction and Commitment. *Human Relations, 42*, 625–638.

Tzeng, H.M. (2004). Nurses' self-assessment of their nursing competencies, job demands and job performance in the Taiwan hospital system. *Int J Nurs Stud, 41*(5), 487–496. doi: 10.1016/j.ijnurstu.2003.12.002.

Willem, A., Buelens, M., de Jonghe, I. (2007). Impact of organizational structure on nurses' job satisfaction: a questionnaire survey. *Int J Nurs Stud, 44*(6), 1011–1020. doi: 10.1016/j.ijnurstu.2006.03.013.

APPENDIX A. INDIVIDUAL INTERVIEW QUESTION

Investigation on Factors Affecting Nursing Performance in Intensive Care Unit in a Hospital

Individual Interview Questions

➢ Please think about your experiences in some situation when you performing your duty in ICU environment,

1. The situation where you need to do an **urgent decision making**
   What are you considering when you are going to make an urgent decision?
   Please briefly explain why are you considering about that?

2. The situation where you need to do a **critical decision making**
   What are you considering when you are going to make a critical decision?
   Please briefly explain why are you considering about that?

3. The situation where you need to do a **simple decision making**
   What are you considering when you are going to make a simple decision?
   Please briefly explain why are you considering about that?

4. About your **job satisfaction**
   Please briefly rate your job satisfaction level from 1–10: _____
   Please briefly explain the reason of your rating.

5. The situation when you felt **heaviest workload.**
   Please briefly explain about the situation
   What is the impact to you while you progress your daily duties?
   What are your suggestions to prevent the situation?

6. The situation when you felt **least workload.**
   Please briefly explain about the situation
   What is the impact to you while you progress your daily duties?

Investigation on Factors Affecting Nursing Performance in Intensive Care Unit in a Hospital

Questionnaire Survey

Please tick one answer that represents how you feel while you progressing your duties in ICU.

- **Part I (Working Effectiveness based on Factors)**
  1 → **doesn't help at all**   3 → **helpful**
  2 → **doesn't help**   4 → **very helpful**

	1	2	3	4
1. Discussions with doctors on the effectiveness of your work.				
2. Order from doctors on the effectiveness of your work.				
3. Discussions with other health-care team on the effectiveness of your work.				
4. Get along with other health-care providers on the effectiveness of your work.				
5. Past experiences that you faced on the effectiveness of your work.				
6. ICU courses every month on the effectiveness of your work.				
7. Training every three years on the effectiveness of your work.				
8. Information that found and learn by yourself after work on the effectiveness of your work.				
9. The guidance from senior on the effectiveness of your work.				
10. SOP in ICU on the effectiveness of your work.				
11. Hospital law and regulation on the effectiveness of your work.				
12. Hospital management on the effectiveness of your work.				

- **Part II (Duties process and Quality Care based on Factors)**
  1 → doesn't affect at all   3 → affect
  2 → doesn't affect   4 → very affect

	1	2	3	4
13. Patient's family pressure on your duties process in ICU.				
14. Patient's family argument on your duties process in ICU.				
15. Get along with patient's family on your duties process in ICU.				
16. Patient's agreement on your duties process in ICU.				
17. Patient's condition on your duties process in ICU.				
18. High turnover rate on your quality care in ICU.				
19. Nursing shortage on your quality care in ICU.				
20. Paper work on your quality care in ICU.				
21. Exam on your quality care in ICU.				
22. Complicated procedures on your quality care in ICU.				
23. Salary on your quality care in ICU				
24. Working hours on your quality care in ICU.				
25. Working shift on your quality care in ICU.				
26. Become leader on your quality care in ICU.				

- **Part III**

27. Please rate your current job satisfaction level:

[ ]1   [ ]2   [ ]3   [ ]4   [ ]5

[ ]6   [ ]7   [ ]8   [ ]9   [ ]10

28. Please state the reason why you give the level of your job satisfaction.

_____

_____

_____

_____

Thank You for Your Participation

New Ergonomics Perspective – Yamamoto (Ed.)
© 2015 Taylor & Francis Group, London, ISBN 978-1-138-02751-0

The relation between occupational stress and supports by occupations

Nanae Shintani

Department of Nursing, School of Health Sciences, Kansai University of International Studies, Shijimi-Cho, Miki, Japan

ABSTRACT: According to Survey on State of Employees' Health in 2012 Japan, there was 60.9% [58.0% in 2007] of workers having some stress from their occupational life or work. Breaking down those stresses, "Personal Relationship at work (41.3% [38.4% in 2007])" was the most.

Moreover, 90.0% [89.7% in 2007] of the people said that there were some people whom they could talk to about the stress. And those people were "Family Members or Friends (86.7%)" and "Supervisors or Colleagues at work (73.5%)" according to Survey on State of Employees' Health in 2012 Japan. And they also responded that "the stress was resolved (33.0%)", "the stress was not resolved but felt better (61.1%)" after they talked about the stress to some people. So "Talking about the stress to some people" was big influence to someone who has the stress from their occupational life or work.

However, there haven't been enough previous studies about influences of stress and support on stress response by occupations.

So this study aims to find the buffer factor of influence from "Supervisors or Colleagues at work" and "Family Members or Friends" support by occupations while the occupational stress process of reaching stress response.

1,204 workers at an electric equipment industry were collected and classified as White-Collar, Gray-Collar and Blue-Collar to clarify the relation between "Dependent-Variable: Stress Response" and "Independent-Variable: Occupational-Stress and Support from appropriate people (Supervisors/Colleagues at work/Family Members/Friends)" by analysis of variance.

The result indicates as follows. When the degree of job demand is low, White-Collar workers who have a good relationship with their family have less physical response to stress. However, when the degree of job demand is high, they have high physical response to stress. No matter if the degree of job demand is high or low, Gray-Collar and Blue-Collar workers who have a good relationship with their family have a low physical response to stress. Since White-Collar workers with a high degree of job demand and a large Work-Load have a good relationship with their family, cannot fulfill their responsibility for their family, they have a high physical response to stress.

Keywords: Occupational Stress; Job demand; Support

1 BACKGROUND

During the change in economic and industrial structure, the ratio of workers who suffer from strong anxiety, distress and stress about their profession and work life is increasing to 61.5% (58.0% at the research in Heisei 19) [1]. When checking the contents, "Human relation problems in the workplace" was 41.3% (do. 38.4%) and the most frequently reported. Regarding the increase and decrease in employees' mental illnesses in the last three years reported by companies, 51.4% of the companies feel nearly flat and 37.6% feel increasing tendency.

When checking the age group at the onset, those in their 40 s increase to 36.2% and those in their 30 s and 40 s become two of the largest age groups [2]. In response, when asked about the measures of the mental health of 252 listed companies across the country, the companies implemented some kind of measures which account for 86.5%, and it is increased by 7.3 points compared to the last research, which is 79.2% in 2008. Additionally, in the same target companies, the number of companies becomes 63.5% which replied that there are employees who are absent from work or take a leave of absence for more than a month because of mental disorder. Even if the number of companies with implemented measures is increasing, high rate of absence or leave from work is still unchanged [3]. As just described, it is found that there are still a lot of companies with increasing tendency even if some companies tend to put a stop on the increasing rate of people with mental disorder.

Furthermore, the ratio of workers who can get a consultation about workplace stress is 90.0% (89.7% at the research in Heisei 19). The most common consultants are, "family and friends", followed by "Bosses and colleagues". The influence of social support is significant because the stress was released (33.0%) or the stress was not released but relieved (61.1%). At the NIOSH occupational stress model, it is said that the support by family and friends is influenced to the increase and decrease in stress reaction [4].

On the other hand, it is clarified that the stress condition varies according to the difference in profession such as white-collar, gray-collar and blue-collar [5].

2 PURPOSE

The objective of this study is to clarify the influence of support by bosses, colleagues and family as the shock-absorbing factor during the process between recognition of stressor and emerging stress reaction by workers. (Figure 1).

3 METHOD

1204 workers who work for an electrical equipment production company were divided into white-collar, gray-collar and blue-collar workers. Then set the dependent variable as stress reaction, and the independent variable as occupational stress and support by bosses, colleagues and family, then clarified the relationship between these factors by variance analysis.

3.1 Study scale

From the NIOSH survey sheet, the occupational stress, support by bosses and colleagues, and stress reaction were extracted and used the four-degree replies between "good" and "not good" for the family support questions: "Is the marital relationship good?", and "Is the parental relationship good?"

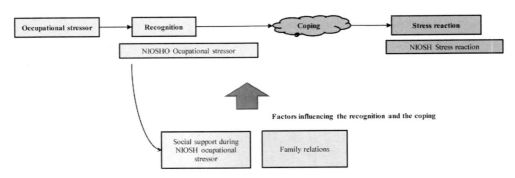

Figure 1. Stress process model.

Table 1. Factor analysis of occupational stressors.

	Support	Degree of demmand	Vague role	Degree of discretion	Utilize of skill	Role conflict
Colleagues at work are reliable when some problems occurred on the work (support 4)	.796	.017	.001	-.072	-.047	-.018
Colleagues at work are willing to listen to personal problems when getting advice (support 6)	.786	.039	-.075	.011	-.040	.070
Immediate superior is willing to listen to personal problems when getting advice (support 5)	.757	-.033	-.019	.002	.013	.066
Colleagues at work are willing to give attention and help to ease your job (support 2)	.696	-.058	-.038	.021	-.017	.019
Immediate superior is reliable when some problems occurred on the work (support 3)	.668	-.035	.076	-.008	-.035	-.061
There is a friendly atmosphere in the group members which I belong to (group 3)	.494	.061	.069	.077	.086	-.089
The group members which I belong to support their opinions each other (group 2)	.407	.038	.048	.080	.141	-.060
Need to do so much works (demand degree 3)	-.011	.910	-.105	.083	-.062	-.030
Can't handle the work because of lack of time (demand degree 2)	-.029	.795	-.182	.040	-.025	-.004
There are cases of extremely increasing workload (demand degree 4)	-.008	.759	.003	-.001	-.034	.040
Need to work quite hard (demand degree 1)	.032	.630	.140	-.100	-.059	-.087
There are jobs which need to concentrate sometimes (demand degree 5)	.009	.456	.183	-.049	.141	-.017
There are cases to process with quick thinking (demand degree 6)	.011	.434	.133	.008	.166	.154
Understood what my responsibility is (ambiguous role 3)	-.067	.090	.890	-.049	-.062	.020
Know exactly what is expected of me (ambiguous role 4)	.011	.036	.788	-.002	.037	.025
Clearly explained what should I do at my job (ambiguous role 5)	.102	-.006	.682	-.010	-.070	-.046
There are clear planed targets and objectives (ambiguous role 1)	.029	-.018	.617	-.021	.018	.064
Feel that I allocated my own work time properly (ambiguous role 2)	-.080	-.151	.536	.166	.000	-.041
I have the freedom to decide the pace of my job (work discretion 3)	-.004	-.036	-.010	.877	-.034	-.026
I have the freedom to decide the quantity of my job (work discretion 2)	.025	-.058	-.099	.815	-.013	.023
I have the freedom to decide the order of my job (work discretion 1)	-.034	.052	.247	.618	-.024	-.022
I have the freedom to decide until I finish my job (work discretion 2)	.044	.093	.004	.564	.057	.075
I have a chance to do what I am good at (utilize skill 2)	-.005	-.024	-.022	.052	.784	-.009
Use skills I got at the previous experience, education and training (utilize skill 3)	.012	.018	.037	-.025	.747	-.022
Use skills I learned at school (utilize skill 1)	-.016	-.011	-.081	-.038	.738	.007
Requested conflicting demands from multiple people (conflict role 3)	.052	-.006	.062	-.047	-.041	.871
Work with a group working in a completely different way (conflict role 2)	.006	-.014	.036	.082	.008	.672
Doing jobs which need not to do (conflict role 4)	-.083	.034	-.098	.011	.018	.577
Credibility factor	.838	.838	.820	.814	.787	.745

Correlation between factors	II	III	IV	V	VI
I . Support	.035	.476	.312	.333	-.299
II . Degree of demand		.138	-.110	.161	.468
III . Vague role			.391	.393	-.262
IV . Degree of discretion				.417	-.091
V . Utilize of skill					.019

3.2 Analytical procedure

1. A factor analysis was conducted about the results of NIOSH stress survey sheet. (Table 1).
2. About the support by bosses and colleagues, it was classified as a supporting factor.
3. Among the occupational stressors other than the supporting factor, a variance analysis was conducted to check the relations including interaction by setting a factor which might be felt as the heaviest burden, family support and support by bosses and colleagues as the independent variable, and stress reaction as the dependent variable.

SPSS Statistics Ver.20 was used for the analysis.

4 ETHICAL CONSIDERATIONS

We got an approval from the ethical committee in Hiroshima university graduate school of integrated science.

5 RESULTS

Among the white-collar workers, the body reaction by stresses was low for better family relations when the degree of demand for jobs was low but the body reaction was high for better family relations when the degree of demand for jobs was high ($F(1, 166) = 7.10, p < 0.01$).

Among the gray-collar workers, the body reaction by stresses was high when the degree of demand for jobs was high ($F(1, 358) = 4.57, p < 0.05$), but the body reaction was low for better family relations ($F(1, 358) = 5.76, p < 0.05$).

In a similar way, among the blue-collar workers, the body reaction by stresses was high when the degree of demand for jobs was high ($F(1, 677) = 33.26, p < 0.001$), but the body reaction was low for better family relations ($F(1, 677) = 9.01, p < 0.01$).

6 DISCUSSION AND CONCLUSION

It was a result that among gray-collar and blue-collar workers, the stresses reaction was low for better family relations and the stresses reaction was high for bad family relations regardless of the demand for jobs.

On the other hand, among white-collar workers, working hours become long if the demand for jobs is high which is different from both gray-collar and blue-collar workers. Therefore, it was considered that the better family relations the higher the stress reaction because they cannot play a role for families. Furthermore, it is considered that the thinking of work-life balance becomes widespread for white-collar workers. It is expected that there will be a lot of people who consider to avoid sacrificing their families as much as possible.

The result of gray-collar and blue-collar workers was common as shown in the NIOSH occupational stress model [4] but the result of white-collar workers was interesting. For white-collar workers who have high demand for jobs and continuing a period of heavy workload, it is found that the better family relations the more negative influence on them.

Managers and supervisors need to consider not prolonging the term of high demand for jobs. If there is no choice but to prolong the term, it is necessary to put measures and thoughts on the work such as setting longer holidays to spend with families or informing when the workload will be reduced.

The finding from this study is that the ways of getting stress are different from job categories. It is indicated that the intervention with taking into account of job categories is necessary when checking the interaction of influence on stress reactions.

REFERENCES

[1] Ministry of Health, Labour and Welfare: Heisei 24 Survey on State of Employees' Health.
[2] A public interest incorporated foundation, Japan Productivity Center 2012.
[3] Private research institute, Institute of Labor Administration 2010.
[4] National Institute for Occupational Safety and Health, Occupational Stress Model.
[5] Toshiko Hirose, Ikue Doi, Kazuo Nitta, The research report about mental health and lifestyle of staff members in Nittazuka Medical Welfare Center, Magazine 2009, Vol. 6, No. 1, Page 1–11.

New Ergonomics Perspective – Yamamoto (Ed.)
© 2015 Taylor & Francis Group, London, ISBN 978-1-138-02751-0

Consideration of the medical-care-support-system for Non-Japanese visitors in Okinawa

Shigeki Tatsukawa

Department of Nursing, School of Health Sciences, Kansai University of International Studies, Shijimi-Cho, Miki, Japan

ABSTRACT: The Japanese government in 2010 showed "The strategy of Health Japan" indicating Medical-Tourism. So there have been lots of studies about medical-care-support-systems for foreign residents in Japan. However, those were only studies from viewpoints of either hospitals or medical professions. To see those issues from a viewpoint of tourism-industry is also necessary. Moreover, considering its systems for Non-Japanese Visitors is very important especially in OKINAWA.

This study aims to find " issues of emergency hospital visit for Non-Japanese Visitors" in Okinawa from a viewpoint of tourism-industry and categorize them into Major and Secondary-items by Affinity Diagram for further consideration about a medical-care-support-system.

18 people from tourism-industry were collected and made 61 labels about "issues of emergency hospital visit for Non-Japanese Visitors" in OKINAWA, which were categorized into 19 Secondary-items ((1) No supporting counter service for emergency sickness or accident, (2) Vague sense of problem for flight repatriation assessment, (3) Risks of hospital visit refusal, (4) Out of trim for different religions (Hospital), (5) Out of trim for different religions (tourism-industry), (6) Confusion about different systems in hospitals between Japan and others, (7) Disable to make international phone-calls from hospitals, (8) Lacking of wheel-chairs in tourism facilities, (9) Issues about visa for prolonged hospitalization, (10) Issues of hospital diet's cultural differences, (11) Lacking of Medical Interpreters, (12) Lacking of Multi-language Display (Hospital), (13) Lacking of Multi-language Display (Tourism facilities), (14) Issues of Language conversation in hospitals, (15) Issues related Insurance, (16) Issues about hospital payment, (17) Lacking support for Non-Japanese during disaster, (18) Lacking knowledge of First-Aid in tourism-industry, and (19) Not enough workers able to give First-Aid treatment in tourism-industry) and 9 Major-items (① No guidelines for Non-Japanese emergency sickness or accident, ② Issues about religious differences, ③ Medical system differences between Japan and others, ④ Issues of equipment shortages, ⑤ Issues by prolonged hospitalization, ⑥ Issues of Language, ⑦ Issues about hospital payment, ⑧ Issues involved in disaster, and ⑨ Issues involved in emergency sickness or accident).

By those Major and Secondary-items, there are several characteristic items such as "Issues of equipment shortages", "Vague sense of problem for flight repatriation assessment", which regional residents or tourism-industry can only see. And perhaps, we couldn't have found the Secondary-item of "Disable to make international phone-calls from hospitals" from viewpoints of hospitals or medical professions. It is necessary to clarify these strategies about the medical-care-support-system for further consideration.

Keywords: Non-Japanese visitors; Okinawa; Emergency hospital visit; Tourism industry; Medical care support system

1 BACKGROUND

1.1 Medical tourism in Japan

The Japanese government in 2010 showed "The strategy of Health Japan" indicating Medical-Tourism to increase the number of Non-Japanese Visitors from other countries. So the Japanese government has forced to build a system of cooperation between hospitals or medical professions and tourism-industry. However, there has been a delay to build its system in Japan comparing with developed countries of Medical-Tourism in Asia such as Singapore and Thailand [1].

1.2 The number of foreign residents and Non-Japanese visitors in Japan

According to Ministry of Justice in Japan 2011, the number of foreign residents in Japan was 2,070,000. And it was 1.63% of all population in Japan [2]. In addition, on 7th September 2013, there was a big announcement for Japan that Tokyo would host Olympic Games in 2020 by International Olympic Committee (IOC). So increasing number of Non-Japanese Visitors to Japan would be expected in near future.

On the other hand, the number of Non-Japanese Visitors having emergency situations of sickness or accident during their stay is predicted to increase. So Japanese hospitals and medical professions worry about the Language barrier and issue of hospital payments the most. Moreover, issue of hospital payment is at the forefront of public attention more than Language barrier in hospital currently [3] [4].

1.3 Current situation of medical-care-support-system for Non-Japanese visitors in Japan

Since the number of Non-Japanese Visitors having sickness or accident during their stay is expected to increase, there are lots of Multiple Language tools prepared in Japan such as handbook of how to visit hospital, medical interview sheet in each hospital and maternal and child health handbook, etc. And almost all of those tools are provided via paper-based or internet-downloading.

However, there should include Non-Japanese Visitors with emergency situations and from viewpoints of "informed-consent", it is necessary for hospitals or medical professions to develop the medical-care-support-system to provide well-understandable explanation about patient sickness and treating plan to avoid difficulty in communicating.

Despite of the necessity to develop the medical-care-support-system avoiding difficulty in communication, hospitals or medical professions in Shizuoka prefecture Japan answered that there was no need to have Medical Interpreters. They also answered there were many experiences of foreign residents unable to communicate in Japanese had visited their hospitals. But they all brought someone (Family, Friends or Colleagues) spoke both their language and Japanese to the hospital. On the other hand, these hospitals and medical professions answered there were difficult experiences of "handling at reception desk, informed-consent of patient sickness and treating plan, educational guidance of patient's daily life to avoid worsen condition of diagnosis, and medical interviews at consultation" [5].

1.4 Non-Japanese visitors in Okinawa

Okinawa Prefecture is made up of the Ryukyu Islands, which at their southern extremity begin at Nansei Island, and lie between Kyushu, the most southwesterly of Japan's four main islands. Okinawa consists of 160 islands of various size scattered across a vast area of ocean. And Okinawa is famous for crystal blue seas, white sand beaches and colorful marine life [6]. According to Okinawa Prefecture, there were estimated 301,4000 of Non-Japanese Visitors. And 80% of those visitors are mostly from Eastern Asian countries (From Taiwan 115,600 people: 38.4%, Hong Kong 54,700 people: 18.1%, China 44,500 people: 14.8%, Korea 26,000 people: 8.6%, USA 6,500 people: 2.2%) [7].

There have been lots of studies about medical-care-support-systems for foreign residents in Japan. However, those were only studies from viewpoints of either hospitals or medical professions. To see those issues from a viewpoint of tourism-industry is also necessary. Moreover, considering its systems for Non-Japanese Visitors is very important especially in OKINAWA.

2 PURPOSE

This study aims to find "issues of emergency hospital visit for Non-Japanese Visitors" in Okinawa from a viewpoint of tourism-industry and categorize them into Major and Secondary-items by Affinity Diagram for further consideration about a medical-care-support-system.

3 METHOD

18 people (2 people from Tourism associations, 5 people from Hotels, 4 people from Non-Japanese supporting associations, 3 people from Travel agencies, 3 people from licensed travel guide, and 1 person from office worker at United States Marine) from tourism-industry were collected and made some labels "issues of emergency hospital visit for Non-Japanese Visitors".

4 researchers with well trained of the Affinity Diagram categorized as follows:

1. Making labels
   The 18 people from tourism-industry write on Post-it within 100 Japanese words about some issues they can imagine when Non-Japanese Visitors have some emergency situations of sickness or accident.
2. Analytical procedure
   The above labels were collected and categorized into Secondary-items and Major-items for qualitative integration by Affinity Diagram.
3. Securement of high reliability and validity
   Analytical decisions of Secondary-items and Major-items were made by agreement of the 4 researchers under a University professor of qualitative study supervising.

4 ETHICAL CONSIDERATIONS

I explained the purpose of this study, how to collect people from tourism-industry, method of analysis, and privacy consideration by an explanation paper to OKINAWA Prefecture. And I obtained all people's consent with a consent form to ensure that it is not possible to identify an individual.

5 RESULTS

The 18 people from tourism-industry were collected and made 61 labels about "issues of emergency hospital visit for Non-Japanese Visitors" in OKINAWA, which were categorized into 19 Secondary-items ((1) No supporting counter service for emergency sickness or accident, (2) Vague sense of problem for flight repatriation assessment, (3) Risks of hospital visit refusal, (4) Out of trim for different religions (Hospital), (5) Out of trim for different religions (tourism-industry), (6) Confusion about deferent systems in hospitals between Japan and others, (7) Disable to make international phone-calls from hospitals, (8) Lacking of wheelchairs in tourism facilities, (9) Issues about visa for prolonged hospitalization, (10) Issues of hospital diet's cultural differences, (11) Lacking of Medical Interpreters, (12) Lacking of

Table 1. Issues of emergency hospital visit for Non-Japanese visitors.

Major-items	Secondary-items
① No guidelines for Non-Japanese emergency sickness or accident	(1) No supporting counter service for emergency sickness or accident
	(2) Vague sense of problem for flight repatriation assessment
	(3) Risks of hospital visit refusal
② Issues about religious differences	(4) Out of trim for different religions (Hospital)
	(5) Out of trim for different religions (tourism-industry)
③ Medical system differences between Japan and others	(6) Confusion about different systems in hospitals between Japan and others
④ Issues of equipment shortages	(7) Disable to make international phone-calls from hospitals
	(8) Lacking of wheel-chairs in tourism facilities
⑤ Issues by prolonged hospitalization	(9) Issues about visa for prolonged hospitalization
	(10) Issues of hospital diet's cultural differences
⑥ Issues of Language	(11) Lacking of Medical Interpreters
	(12) Lacking of Multi-language Display (Hospital)
	(13) Lacking of Multi-language Display (Tourism facilities)
	(14) Issues of Language conversation in hospitals
⑦ Issues about hospital payment	(15) Issues related Insurance
	(16) Issues about hospital payment
⑧ Issues involved in disaster	(17) Lacking support for Non-Japanese during disaster
⑨ Issues involved in emergency sickness or accident	(18) Lacking knowledge of First-Aid in tourism-industry
	(19) Not enough workers able to give First-Aid treatment in tourism-industry

Multi-language Display (Hospital), (13) Lacking of Multi-language Display (Tourism facilities), (14) Issues of Language conversation in hospitals, (15) Issues related Insurance, (16) Issues about hospital payment, (17) Lacking support for Non-Japanese during disaster, (18) Lacking knowledge of First-Aid in tourism-industry, and (19) Not enough workers able to give First-Aid treatment in tourism-industry) and 9 Major-items (① No guidelines for Non-Japanese emergency sickness or accident, ② Issues about religious differences, ③ Medical system differences between Japan and others, ④ Issues of equipment shortages, ⑤ Issues by prolonged hospitalization, ⑥ Issues of Language, ⑦ Issues about hospital payment, ⑧ Issues involved in disaster, and ⑨ Issues involved in emergency sickness or accident) (Table 1).

Some of those 61 labels were related to the tourism-industry or local characteristic ones such as" Unexplainable approximate cost of admission or consultation, Unclear system of hospital visit from the Non-Japanese viewpoint, Unclear usage of ambulance from the Non-Japanese viewpoint, Different system of purchasing medication with doctor's prescriptions, Disable to make international phone-calls from hospitals, Confusion about visa expiring because of the hospitalization, Issue about communication between a patient and a dentist when visitor had a toothache, Lacking of Multi-language Display on the beach if it's safe to swim, Lack of quick response at hotel for Non-Japanese visitors during disaster such as typhoon, There is no place to lend wheel-chairs after 7 p.m. on Kokusai street in NAHA".

6 DISCUSSION AND CONCLUSION

The Major-items of "① No guidelines for Non-Japanese emergency sickness or accident, ② Issues about religious differences, ③ Medical system differences between Japan and others, ⑥ Issues of Language, ⑦ Issues about hospital payment" had already been shown in the study of Maeno, et al, 2010. In Shizuka prefecture, there were difficult experiences of "handling at reception desk, informed-consent of patient sickness and treating plan, educational guidance of patient's daily life to avoid worsen condition of diagnosis, and medical interviews at consultation" [5]. This study in OKINAWA indicated the same issues for Non-Japanese Visitors. But it also included the local and tourism-industry characteristic secondary-items such as (2) Vague sense of problem for flight repatriation assessment, (11) Lacking of Medical Interpreters. The Major items of "④ Issues of equipment shortages, ⑤ Issues by prolonged hospitalization, ⑧ Issues involved in disaster, and ⑨ Issues involved in emergency sickness or accident" were the local and tourism-industry characteristic ones. Especially, ④ Issues of equipment shortages were categorized from the characteristic secondary-items of (7) Disable to make international phone-calls from hospitals. And ⑤ Issues by prolonged hospitalization were from the characteristic secondary-items of (9) Issues about visa for prolonged hospitalization. ⑨ Issues involved in emergency sickness or accident were from (18) Lacking knowledge of First-Aid in tourism-industry, and (19) Not enough workers able to give First-Aid treatment in tourism-industry. By those Major and Secondary-items, there are several characteristic items which regional residents or tourism-industry can only see. And perhaps, we couldn't have found the Secondary-item of "(7) Disable to make international phone-calls from hospitals" from viewpoints of hospitals or medical professions. It is necessary to clarify these strategies about the medical-care-support-system for further consideration.

REFERENCES

[1] D. OGURA. Current Medical Tourism Korea–Medical Korea. CLAIR FORUM/Council of Local Authorities for International Relations, 257, 35~40, 2011.
[2] Legal Affairs Bureau. Statistics of foreign residents. http://www.moj.go.jp/housei/toukei/toukei_ichiran_touroku.html.
[3] I. SAWA. Medical problems at Non-Japanese accomodation facilities. The 1st symposium of Medical Problem and medical interpriters, 31~35, 2008.
[4] K. MINAMITANI. Current situation of Non-Japanese visits at hospitals near inthernational airports. The 1st symposium of Medical Problem and medical interpriters, 23~31, 2008.
[5] M. MAENO, N. ENOMITO, R. MAENO, et al. The Problem of Foreign Language Translation and Supporting for Foreigners in Clinics Accepting Foreigners. University of Shizuoka, Junir college proceeding (24), 13~26, 2010.
[6] Japan National Tourism Organization. Japan:the official guide OKINAWA. URL: http://www.jnto.go.jp/eng/location/regional/okinawa/.
[7] OKINAWA Prefecture, Derpartment of Culture, tourism and sport. Summery of tourism visitors at Okinawa in 2013. URL; http://www.pref.okinawa.jp/site/bunka-sports/kankoseisaku/documents/h23-gaikyou-rekinen.pdf.

New Ergonomics Perspective – Yamamoto (Ed.)
© *2015 Taylor & Francis Group, London, ISBN 978-1-138-02751-0*

Satisfaction feelings and body-part discomfort of university students on classroom furniture

Chalermsiri Theppitak, Kwanruan Pinwanna & Pilunthan Chauchot
School of Occupational Health and Safety, Suranaree University of Technology, Nakhon Ratchasima, Thailand

ABSTRACT: Sitting on unsuitable classroom furniture leads to poor sitting postures and bodily discomfort. Comfortable and suitable classroom furniture can help motivate students to perform better and encourage their learning processes. The pilot survey was conducted to evaluate satisfaction feelings and body-part discomfort of university students on three types of classroom furniture. Sixty students participated in this study were with age average of 21.05 ± 0.53 year-old (range 20–23). Participants were asked to answer questionnaires about satisfaction feelings on three types of classroom furniture (single-seat, paired-seat and row-seat). A body part discomfort scale was used to evaluate student's experience of discomfort at different body parts before class and after class. The results indicated that the satisfaction feelings of students were high for paired-seat type and low for single-seat type. The body-part discomfort of participants increased after class in all types of classroom furniture. Therefore, the current classroom furniture needs to be improved in order to respond to the requirements and body-dimension of students.

Keywords: Classroom furniture; Satisfaction feeling; Discomfort; Body dimension; Seating; University students

1   INTRODUCTION

Sitting related Musculoskeletal Disorders (MSDs) are frequent and several studies have pointed out an association between back pain and prolonged sitting [1]. University students spend several hours in static or awkward sitting posture on classroom furniture at the university [2]. Sitting on unsuitable classroom furniture leads to poor sitting posture and bodily discomfort. There are different types of classroom furniture that are used in university. Most of classroom furniture cannot be adjusted in order to comfort all students. Comfortable and suitable classroom furniture can help motivate students to perform better and encourage their learning process.

The pilot survey was conducted to evaluate satisfaction feelings and body-part discomfort of university students on three types of classroom furniture. The results from this study can be applied for improvement or re-design classroom furniture that should be suitable with body dimension and requirement of users.

2   METHODS

2.1   *Subjects*

The study group consisted of sixteen male and forty-four female undergraduate students at Suranaree University of Technology, with an average age of 21.07 ± 0.55 year-old (range 20–23). The average of height was 162.14 ± 6.46 cm (range 152–179). The average of weight was 52.84 ± 9.24 kg (range 40–78). Participation of subjects in this study was voluntary, with each participant being provided with adequate and appropriate information about the involvement of their participation. All participants were healthy and had no prior report of musculoskeletal disorder or injury before participation in this study.

2.2 *Measurement and procedure*

The subjects were asked to answer questionnaires about satisfaction feelings and body part discomfort after sitting on three types of classroom furniture on separate days and one day per each type of classroom furniture. The three types of classroom furniture were single-seat, paired-seat and row-seat. The single-seat type is a chair with flat surface plank on the right side of the chair to facilitate writing for one user as Fig. 1 (a). The paired-seat type is a set of one table and two separate chairs for two users as Fig. 1 (b). The row-seat is a chair with flat surface plank on the right side of the chair and each chair connect with other chairs for many users as Fig. 1 (c).The satisfaction feeling questionnaire was used to ask participant's feelings/opinions on four topics of classroom furniture. First topic was on characteristic of seat pan such as height, depth, size and softness. Second topic was on characteristic of desk such as height, size, clearance between participant's body and desk, clearance between thigh and surface of desk. Third topic was about a keeping place for bag. Fourth topic was asking which kind of seat type should be the most to be improved?

A body part discomfort scale is a subjective symptom survey tool used to evaluate subject's experience of discomfort at different body parts [3]. Participants were shown the body part diagram (12 parts of body) and asked to indicate where they felt discomfort while they were sitting on the seat before class and after class. They were then asked to rate how severe the discomfort was on a five point scale, with 0 being none and 5 being unbearable. The higher number means that more uncomfortable they felt at that certain part of the body.

All statistical analyses were performed with the SPSS software version 16. The data for satisfaction feeling of subjects were analysed by descriptive statistics. The body part discomfort score before class was compared with that of after class using a paired-sample t-test for each seat type.

3 RESULTS

3.1 *Satisfaction feeling on three classroom furniture types*

The results were shown in table 1 indicated that most subjects satisfied with row-seat in many characteristics of seat such as seat pan height, seat pan depth, seat pan size, and desk height. The classroom furniture type which needs to be improved the most was the single-type. Subjects felt that the single seat was not suitable for them as it had too low seat pan, too small seat

| (a) Single-seat | (b) Paired-seat | (c) Row-seat |

Figure 1. Illustrated three classroom furniture types: (a) Single-seat; (b) Paired-seat and (c) Row-seat.

pan, too hard, and a small desk size. All participants were not satisfied with seat pan softness and a keeping place for bag in all seat types.

3.2 *Body part discomfort*

The body part discomfort scores after class at all body parts were higher than those of before class in all of classroom furniture types. For single-seat, participants showed high score in first four order of discomfort after class at shoulder, lower back, neck and upper back. For paired-seat, participants showed high score in first four order of discomfort after class at lower back, knee, upper back and neck. For row-seat, participants showed high score in first four order of discomfort after class at neck, lower back, shoulder and upper back.

Table 1. The satisfaction feelings of participants on three classroom furniture types.

Items	Single-seat (n)	Paired-seat (n)	Row-seat (n)
1. Characteristics of seat pan			
Seat pan height			
Too high	0	2	0
Suitable	36	55	54
Too low	23	3	6
Seat pan depth			
Too deep	5	4	2
Suitable	40	52	56
Too shallow	12	2	2
Seat pan size			
Too big	1	4	0
Suitable	32	49	54
Too small	26	5	4
Seat pan softness			
Too hard	31	45	44
Suitable	28	14	15
Too soft	0	0	0
2. Characteristics of desk			
Desk height			
Too high	5	14	8
Suitable	43	44	49
Too low	12	2	2
Desk size			
Too big	0	1	1
Suitable	20	50	31
Too small	39	7	24
Clearance between participant's body and desk			
Too far	7	2	9
Suitable	36	54	45
Too close	17	4	5
Clearance between thigh and desk			
Too far	3	4	3
Suitable	38	52	48
Too close	19	4	8
3. Keeping place for bag			
Convenient	11	15	11
Inconvenient	49	45	48
4. The seat type needs to be improved the most	41	10	7

Figure 2.   The body part discomfort scores before class and after class in three kinds of seat type.

4   DISCUSSION

The results from this study indicated that university students were discomfort from using inappropriate classroom furniture type. The body part discomfort at lower back, upper back and neck increased after sitting in the class in all seat types. These results are consistent with a previous study that students experienced pain while sitting in the classroom and back pain occurred after 1 hour of sitting and increasing with the longer duration of the sitting position at school [4]. The majority of participants agreed that the single-seat needs to be improved the most. Participants were unsatisfied with characteristics of a single-seat such as too low seat pan, too small seat pan, too hard seat pan, and small desk size. These inappropriate seat features may result in discomfort feelings at shoulder, lower back, neck and upper back of participants. Some studies had reported a positive relationship between back pain and seat height [5]. Most participants were unsatisfied with a keeping place for bags because no providing of keeping bag place in all seat types that may contribute to body discomfort. A previous study reported that one of factors that influence the incidence of musculoskeletal pain in school children were a heavy school bag [6].

5   CONCLUSION

Our findings suggested that there were some mismatches between body dimension of participants and classroom furniture design. Further studies need to be conducted to measure body dimension of users and classroom furniture dimension for further improvement or redesign classroom furniture suitable for all users.

ACKNOWLEDGEMENTS

The authors are thankful for the research grant from Suranaree University of Technology. We are thankful to all of the subjects in this study.

REFERENCES

[1]   Magnusson LM, Pope HM. A review of the biomechanics and epidemiology of working postures-technique for assessing postures. *J Sound Vib* 1998;**215**(4):965–76.
[2]   Bendix T. Adjustment of the seated work place with special reference to heights and inclinations of seat and table. *Dan Med Bull* 1987;**34**:125–39.
[3]   Corlett EN, Bishop, RP. A technique for assessing postural discomfort. *Ergonomics* 1976;**19**:175–82.
[4]   Troussier B. Comparative study of two different kinds of school furniture among children. *Ergonomics* 1999;**42**:516–26.
[5]   Yeats B. Factors that influence the postural health of school children. *Work* 1997;**9**: 45–5.
[6]   Negrini S, Carabalona R. Backpacks on! Schoolchildren's perceptions of load, associations with back pain and factors determining the load. *Spine* 2002;**27**(2):187–95.

New Ergonomics Perspective – Yamamoto (Ed.)
© 2015 Taylor & Francis Group, London, ISBN 978-1-138-02751-0

Study of the child care support in Japan

Yoshika Suzaki
School of Nursing, Yasuda Women's University, Hiroshima-City, Hiroshima-Prefecture, Japan

Yukiko Minami
Graduate School, Prefectural University of Kumamoto, Japan

Naoko Takayama
School of Nursing, University of Kindai Himeji, Japan

Hiromi Ariyoshi
Faculty of Medicine, Saga Medical School, Saga University, Saga-City, Saga-Prefecture, Japan

ABSTRACT: Municipalities in Japan have a "Childcare Support Center" for the sake of young parents who do not have an opportunity to talk about their children with their own parents. Young mothers take part in activities of these support center to communicate each other, take counsel with the staff about childcare, or are taught how to play with children by the staff.

Although there exit these facilities, the number of counseling on child abuse in Japan is increasing year by year; for example, the figure was 26,569 in 2003, and 40,639 in 2007. About sixty percent of abusers are children's real mothers. While the birthrate and the number of children are both declining, the number of counseling on child abuse is growing up. It may be because childcare itself has become difficult for mothers. This research tries to reveal the actual condition of childcare, especially about how much mothers who are bringing up their children utilize a childcare support center. Subjects of the survey: We surveyed those who are bringing up their children and using childcare facilities in City M. We distributed the question sheets to 200 people at two childcare facilities in City M. One-hundred and twenty-nine respondents (5 males and 124 females) returned the sheet; namely, the return rate was 64.5%. Twenty-nine respondents were in their twenties, 82 in their thirties, and 18 in their forties. Fifty-two respondents (40.3%) have used the childcare service, and 77 (59.7%) have not used. The reasons of the non-participation were "The time of activities is not convenient" (47.8%) or "I do not have to use it" (32.2%). More than half of those who have not used the support centers answered that they have not used because the time was not convenient. On the other hand, more than half (53.8%) of those who have used the support centers keep using it. Among those who have used it four times or more, 20 people (71.4%) are in their thirties. Now that working mothers are increasing in number, operating hours of support centers have to be reconsidered.

Keywords: Childcare Support Center; municipalities in Japan; young parents

1 BACKGROUND

Municipalities in Japan have a "Childcare Support Center" for the sake of young parents who do not have an opportunity to talk about their children with their own parents. Young mothers take part in activities of these support centers to communicate each other, take counsel with the staff about childcare, or are taught how to play with children by the staff. Although there exit these facilities, the number of counseling on child abuse in Japan is increasing year by year; for example, the figure was 26,569 in 2003, and 40,639 in 2007. About 60% of abusers are **children's real mothers**. **While the** birthrate and the number of children are both declining, the number of counseling on child abuse is growing up. It may be because childcare itself has become difficult for mothers. This research tries to reveal the actual condition of childcare, especially about how much mothers who are bringing up their children utilize a childcare support center.

2 METHOD

1. Subjects of the survey
   We surveyed those who are bringing up their children and using childcare facilities in City M.
2. Survey period
   The survey period was from June to September in 2010.
3. Survey method
   We conducted the questionnaire survey on those who were bringing up their children. The questions include whether they participated in activities of a support center, how many times they have participated, why they have not participated if not, what they are conscious of in childcare, etc. The respondents answered them in writing.

3 RESULTS

We distributed the question sheets to 200 people at two childcare facilities in City M. One-hundred and twenty-nine respondents (5 males and 124 females) returned the sheet; namely, the return rate was 64.5%.

Twenty-nine respondents were in their twenties, 82 in their thirties, and 18 in their forties.

Fifty-two respondents (40.3%) have used the childcare service, and 77 (59.7%) have not used. The reasons of the nonparticipation were "The time of activities is not convenient" (47.8%) or "I do not have to use it" (32.2%).

Among the respondents, 13 people have participated just once, 5 have participated twice, 5 have three times, and 28 have four times or more. The benefits of participation were "I made a friend" (26%) and "It served as a mental diversion" (24%). About childcare, 91.4% answered "I would like to consult," and its contents are related to "child's development" (28.7%), "discipline (20.9%), "treatment of poor physical condition" (18.6%), etc.

Table 1. Good to participate.

	n	%
To make friends	13	26.0
Consultation of experts	1	2.0
Talk to the participants	7	14.0
Change of pace	12	24.0
Freedom from child-rearing	3	6.0
Play of children	5	10.0
Prevention homebound	2	4.0

Table 2. Consultation content.

	n	%
Development	37	28.7
Training	27	20.9
Correspondence at the time of the ill-health	24	18.6
Meal	16	12.4
Termination of breast-feeding	12	9.3
Rebellious phase	5	3.8
Other	7	5.4

4 DISCUSSION

According to the result of Longitudinal Survey of Newborns in the 21st Century[1], the employment rate of mothers was 55.4% one year before giving birth. However, the rate declined to 25.1% immediately after giving birth and bounced back up to 60.5% one year later. It means that about half of the mothers have returned to work after their children's growth. More than half of those who have not used the support centers answered that they have not used because the time was not convenient.

On the other hand, more than half (53.8%) of those who have used the support centers keep using it. Among those who have used it four times or more, 20 people (71.4%) are in their thirties. Now that working mothers are increasing in number, operating hours of support centers have to be reconsidered.

According to previous researches,[2] [3] those whom mothers would like to ask for advice about childcare include "spouse," "mother," "friend," and "doctor" in this order. Nakanishi et al. (2004) [4] indicates that 46.4% of their respondents made friends with other participants after participating a childcare circle. To be concrete, 35.7% of them go out or have a walk together, 30.4% exchange emails, and 28.6% consult with each other about childcare. Especially, 60.7% of the respondents who have participated three times or less made friends. Our survey shows that those whom they would like to ask for advice about childcare include "friend"(83), "spouse"(78), and "mother"(76). These results suggest that mothers bringing up children prefer friends as advisor about childcare because both of them live in similar situations.

Comprehensive Survey of Living Conditions[5] shows that 18.5% of all households are three-generation family while our survey shows that 107 (82.9%) are nuclear family and 19 (14.7%) are three-generation family. These results mean that nuclear family increases in local communities. It may be that "friends" have become closer than "relatives" because of the expansion of nuclear family. A Childcare Support Center should utilize the manpower in the local communities to give advice to mothers who tend to isolate themselves. Mothers who are bringing up children hardly experience various types of childcare in these days because they have few opportunities to communicate with others' children. We suppose that this situation causes mothers to seriously worry about children and consequently leads to child abuse or nervous breakdown due to childcare.

Local communities have to be empowered not only by promoting mutual communication between specialists and mothers or between mothers themselves but also by involving elderly people with childcare. We believe that establishment of support environment also leads to the security of manpower in the support center.

5 CONCLUSION

We have to consider measures to create an energetic local community where the whole community supports childcare so that mothers can bring up their children without anxiety.

Since this research did not survey the situation of local community itself, we have to study effective measures concerning that viewpoint.

REFERENCES

[1] Longitudinal Survey of Newborns in the 21st Century: Household Statistics Office Vital, Health and Social Statistics Division, Ministry of Health, Labour and Welfare, 2010.
[2] A Study on the Needs for Supporting Child Rearing in A Prefecture (Part 1): Locality Comparison of Child–Rearing Anxiety and Coping with Stress, Bull. Shikoku Univ. A40: 1–12, 2013.
[3] Masanao Ito: Journal of health and welfare statistics, Ministry of Health, Labour and Welfare, 2010.
[4] Miki Nakanishi, Michiko Iwado: The relationships among child-rearing mothers and difficulties in raising Children, Human Life Science and Faculty of Human LifeScience,Vol. 3, 2004.
[5] Comprehensive Survey of Living Conditions: Household Statistics Office Vital, Health and Social Statistics Division, Ministry of Health, Labour and Welfare, 2013.
[6] Yasuhiro Ogawa: A Case Study of a "community welfare": How they got "ibasho" for children, Hokusei Gakuen University Graduate School Social Welfare Review 9, 39–47, 2006.
[7] Kimie Shibahara: A Study of the Support Activities in Changes and the Community of Childcare Problem, Human Welfare and Research, 6, 27–46, 2004.

New Ergonomics Perspective – Yamamoto (Ed.)
© *2015 Taylor & Francis Group, London, ISBN 978-1-138-02751-0*

A comparison of the grip setting of the IV pole while walking: Presence or absence, 50% or 60% of a user's height setting

Reiko Hachigasaki
School of Nursing, Toho University, Tokyo, Japan
Graduate School, St. Luke's International University, Tokyo, Japan

Michiko Hishinuma
St. Luke's International University, Tokyo, Japan

Sakae Yamamoto
Tokyo University of Science, Tokyo, Japan

ABSTRACT: *Objective*: The objective of this study was to examine the presence or absence of a grip and its suitable height based on kinematic analysis and subjective assessment of walking while using an Intravenous pole (IV pole), in pursuit of safer and more comfortable utilization of IV poles.

Background: More patients are regularly using an IV pole as its use is promoted to encourage earlier post-surgery ambulation and shorten hospital stays, prompting concern over an impact on accidental falls. Grips have been added to improve pole maneuverability and usability, but a set grip placement is not yet established.

Method: From June–August 2009, data was collected on 8 men and 33 women (41 subjects) in good health, aged 60–70 (average age: 66.3 ± 2.3), who enacted walking with an IV pole. Four situations were presented—using a grip placed at 50% of the subject's height, using a grip at 60%, using a pole with a grip, and using one without. Kinematic analysis and subjective assessment compared walking speed, stride, pace, arm swing, anterior inclination of the body trunk, angle of the elbow manipulating the IV pole, and distance between pole and body.

Results: Comparison between poles with a grip placed at 50% of height and without a grip showed the anterior/posterior and left/right distance between pole and body trunk were greater with a grip ($p < 0.01$). At 60%, while stride was longer without a grip ($p < 0.05$), the anterior/posterior distance between pole and body was greater with a grip. In the subjective assessment of the grip placed at 60% of height, more subjects stated that having a grip was better. Furthermore, a comparison of 50% and 60% grip placement showed the anterior/posterior distance between pole and body was greater with a grip placed at 60% of height ($p < 0.01$). Aside from subjective assessment, many subjects stated that when using the 60% grip placement they "didn't worry about their footing." With the 50% grip placement, many said they could "use it in place of a cane" and "lean on it," indicating that a grip placed at 50% of the user's height facilitated putting weight on the IV pole.

Keywords: IV pole; grip; gait; walk; kinematic analysis

1 INTRODUCTION

More patients are regularly using an Intravenous pole (IV pole) as its use is promoted to encourage earlier post-surgery ambulation and shorten hospital stays, prompting concern over an impact on accidental falls [1]–[3]. IV poles are tools used on a regular basis by patients receiving fluid therapy, and must be utilized in safety and comfort. Grips have been added to improve pole maneuverability and usability. Changes in gate when walking with a pole, the establishment of suitable pole height, and grip height have already been clarified [4]. However, a set grip placement has not yet been established.

2 OBJECTIVE

The objective of this study was to examine the presence or absence of a grip and its suitable height based on kinematic analysis and subjective assessment of walking while using an IV pole, in pursuit of safer and more comfortable utilization of IV poles.

3 METHOD

3.1 *Participants*

There were 41 participants, 8 men and 33 women in good health, aged 60–70 (average age: 66.3, SD 2.3). All were under 164.5 cm in height and had not had experience regularly walking with an IV pole in the past 6 months.

3.2 *Experiment plan*

The experiment was implemented in a room offering an even walking path along a straight line of 11 m consisting of a 3 m preparatory segment, a 5 m normal-walking segment, and a 3 m segment for stopping. An IV bottle (500 ml) was hung from a 5-legged IV pole (IV pole KC-508, Paramount Co., Ltd.) equipped with a removable grip to recreate the administration of IV into a peripheral vein. An U-shaped grip was attached in a horizontal position for the setting that utilized a grip. Based on previous research [5], the height of the IV pole was set at 110% of the user's height. The grip on the IV pole was placed at 50% of height (grip height was 50% of the user's height) or 60%. In addition, the experiment was conducted both with and without a grip for a total of four settings. Because the grip could not be attached to the IV pole at 70% of the user's height, that setting was excluded from the study. Markers were placed on a total of 20 points on the user's body, 10 each on the left and right. Those points were the acromion, elbow joint, wrist, spinous process of the anterior iliac, greater trochanter, lateral malleolus, tip of the toe, 5th toe of the MP joint, and the calcaneal region.

3.3 *Procedures*

Two SONY DCR-DVD 505 were used to film the participants walking back and forth along the path. The side camera for the participants was set up approx. 4 m away from the walking path of the approx. 3 m span that always included 1 lap of walking, while the front camera was set up at the turnaround point along the 11 m walking path. Still frames were taken from video filmed with a 30 Hz interlace system, and image tilt was corrected using Adobe® Photoshop® Elements 8. Each variable was measured in the still frames using the image analysis software, ImageJ.

The side maneuvering the IV pole was labeled the "pole side" and the opposite side was labeled the "free side" (see Fig. 1). Walking speed, stride (pole side, free side), pace, arm swing angle (free side), anterior inclination of the body trunk, elbow angle (pole side), and anterior/posterior and left/right distance between pole and body trunk were measured. Stride was considered the anterior/posterior distance of the left/right heel. Arm swing was considered the maximum angle created by the line linking the acromion and the upper end of the radius on the free side not grasping the pole. The elbow angle on the IV pole side was

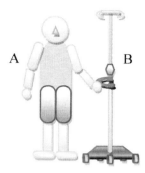

Figure 1. A: Free side, B: Pole side.

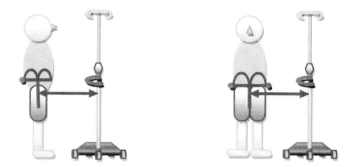

Figure 2. (a) Pole—body anterior/posterior distance, (b) Pole—body left/right distance.

the angle created by 3 points, the acromion, elbow joint, and wrist. The distance between the body trunk and IV pole was measured as the length of the line connecting the centerline of the body and the IV pole (see Fig. 2). The subjective assessment included 6 items. The items regarding maneuverability of the IV pole were a. Ease of maneuverability, b. Stability of IV pole, and c. Ease of walking. Items regarding physical burden were d. Sense of burden on upper limbs, e. Sense of burden on lumbar region, and f. Sense of burden on lower limbs. Furthermore, the users were asked for other impressions not reflected in the subjective assessment.

3.4 Data analysis

The t-test was implemented in the data analysis to make comparisons between the presence or absence of a grip at 50% height, the presence or absence of a grip at 60% height, and the presence of a grip at 50% and 60% height. Significance level was set at 5%. About the subjective assessment of grip presence/absence and grip height, the percentage indicating a positive judgment was calculated, and content regarding each IV pole setting was extracted from other impressions and compared.

3.5 Ethical considerations

The objective of the study was explained to the participants and their signatures were obtained as a declaration their consent. Their state of health was confirmed before and after the study implementation. This study was carried out with the approval of the ethical review board at the university to which the researcher belongs.

4 RESULTS

4.1 *Comparison of gait according to different grip placement*

Table 1. Comparison of grip presence/absence at 50% of height (n = 41).

	50% of height with grip	50% of height without grip	*p*-value
Walking speed (m/sec)	1.07 (0.20)	1.07 (0.19)	0.635
Stride: Pole side (cm)	55.68 (5.10)	55.79 (5.87)	0.822
Stride: Free side (cm)	55.56 (5.20)	56.27 (5.93)	0.090
Pace (steps/sec)	1.94 (0.42)	1.92 (0.42)	0.583
Arm Swing Angle: Free side (deg)	15.29 (10.46)	14.07 (11.43)	0.232
Elbow Angle: Pole side (deg)	136.46 (13.19)	134.09 (1.011)	0.188
Anterior inclination of the body trunk (deg)	182.08 (3.98)	180.73 (5.32)	0.076
Pole—body anterior/posterior distance (cm)	31.49 (10.60)	23.77 (9.84)	**0.000
Pole—body left/right distance (cm)	43.56 (5.09)	41.10 (5.88)	*0.014

mean (SD).

Table 2. Comparison of grip presence/absence at 60% of height (n = 41).

	60% of height with grip	60% of height without grip	*p*-value
Walking speed (m/sec)	1.09 (0.21)	1.08 (0.19)	0.623
Stride: Pole side (cm)	56.18 (5.99)	57.22 (5.85)	*0.014
Stride: Free side (cm)	56.36 (6.12)	57.32 (6.35)	*0.030
Pace (steps/sec)	1.93 (0.42)	1.90 (0.41)	0.301
Arm Swing Angle: Free side (deg)	15.92 (13.25)	16.58 (11.62)	0.637
Elbow Angle: Pole side (deg)	108.58 (22.50)	106.82 (13.96)	0.543
Anterior inclination of the body trunk (deg)	183.27 (3.72)	183.09 (3.33)	0.708
Pole—body anterior/posterior distance (cm)	40.63 (12.05)	27.99 (8.00)	**0.000
Pole—body left/right distance (cm)	42.18 (8.38)	42.91 (4.66)	0.535

mean (SD).

Table 3. Comparison of 50% and 60% of height with grip (n = 41).

	50% of height with grip	60% of height with grip	*p*-value
Walking speed (m/sec)	1.07 (0.20)	1.09 (0.21)	0.287
Stride: Pole side (cm)	55.68 (5.10)	56.18 (5.99)	0.198
Stride: Free side (cm)	55.56 (5.20)	56.36 (6.12)	0.062
Pace (steps/sec)	1.94 (0.42)	1.93 (0.42)	0.697
Arm Swing Angle: Free side (deg)	15.29 (10.46)	15.92 (13.25)	0.570
Elbow Angle: Pole side (deg)	136.46 (13.19)	108.58 (22.50)	**0.000
Anterior inclination of the body trunk (deg)	182.08 (3.98)	183.27 (3.72)	**0.008
Pole—body anterior/posterior distance (cm)	31.49 (10.60)	40.63 (12.05)	**0.000
Pole—body left/right distance (cm)	43.56 (5.09)	42.18 (8.38)	0.220

mean (SD).

Table 4. Subjective assessments by grip presence/absence for 60% of height (n = 41).

	60% of height with grip	60% of height without grip	No difference
a. Ease of maneuverability	33 (80.5)	6 (14.6)	2 (4.9)
b. Stability of IV pole	25 (61.0)	5 (12.2)	11 (26.8)
c. Ease of walking	34 (82.9)	5 (12.2)	2 (4.9)
d. Sense of burden on upper limbs	2 (4.9)	29 (70.7)	10 (24.4)
e. Sense of burden on lumbar region	0 (0.0)	7 (17.1)	34 (82.9)
f. Sense of burden on lower limbs	0 (0.0)	8 (19.5)	33 (80.5)

The number of answerers (%).

4.2 *Subjective assessment by participants*

Among other major impressions not reflected in the subjective assessments a–f (repeated comments), there were negative impressions of the 60% height placement without a grip, including comments such as "I have to hold on tight and push" (12 people) and "The IV pole moves close to my body" (3 people). On the other hand, there were positive impressions of the 60% height placement with a grip, including comments such as "It's easy because I place my hand on the grip" (12 people), "I don't worry about my footing" (8 people), and "Hand positioning is easy" (5 people). In regard to the 50% height placement with a grip, positive comments included "Hand positioning is easy" (6 people) and "I can lean on it" (4 people), while negative comments included "I have to hold on tight and push" (5 people), "It feels heavy" (3 people), and "I get tired" (3 people).

5 DISCUSSION

5.1 *Grip presence/absence*

Placing a grip on the IV pole adds the length of the grip, and the pole can be distanced from the body. Although stride is shortened when walking with an IV pole compared to when walking normally [4], it is surmised that without a grip, the pole cannot be sufficiently distanced from the body and it becomes easier for the feet to make contact with the pole's legs.

In the absence of a grip, a vertical grasp is used to hold on to the pole, with the thumb and index finger at the top. With a grip, a horizontal grasp is used with the palm facing down, since the grip is horizontally attached. In the basic position of the hand, there is the resting position seen when sleeping and under anesthesia, and the functioning position that makes it easy to perform various movements with the hand [5]. Without a grip, hand position is similar to the resting position, and with a grip, hand position is similar to a functioning position. In addition, the rotation of the pole when in use has been indicated by Taga et al. [6]. Without a grip, upper arm strength is needed to prevent rotation and control the IV pole. Therefore, the presence of a grip facilitates movement and eases maneuverability, and thus is a setting that makes it easier to walk.

5.2 *Height of grip placement*

When the placement for maneuvering the IV stand is at 50% of user height, the hand is fixed in a low position with the elbow extended at 130°, and the pole is pulled into the body. In the range of arm movement, there is the normal range where precise motion can be executed that allows ample motor skills in which significant strength can be exercised with the elbow joint serving as a fulcrum, and there is the maximum range of motion where speed and a broad scope of motion can be carried out with the shoulder joint serving as a fulcrum [7]. The 60% grip height is suitable to the normal range because it is comparable to the height of the olecranon inferior margin, which is the height of the elbow from the floor when the elbow is bent at a right angle.

Since the body bends slightly forward with the grip placement at 50% height compared to at 60%, the pole becomes used in place of a cane and can be leaned on. Thus, it is suspected that the height makes it easier to put weight on the IV pole. When that occurs, the center-of-gravity line created by the left/right legs and IV pole shifts forward of the supportive base surface that should be formed by both legs. Therefore, when the IV pole is tilted, upper body balance is lost and the risk of falling increases. Currently, IV poles are not included in JIS standards and there are no prescribed experimental criteria for insuring the safety of maneuverability. In addition, manufacturers prohibit placing weight on IV poles. In terms of safety, it would not be desirable to place the grip at 50% of the user's height.

5.3 *Limitations*

This study does not present results that authentically represent hospital patients, given that the participants were healthy elderly adults, and the study does not reflect a hospital environment because a laboratory was used.

6 CONCLUSION

Placing a grip at 60% of the user's height makes walking with an IV pole safer and more comfortable.

REFERENCES

[1] Kawamura H, *A Complete Book of Error Maps Based on 11,000 Cases of Medical Incidents (in Japanese)*. Tokyo: Igakusyoin; 2003.

[2] Shindo E, Examining IV Stand Casters (in Japanese). *The 23rd Annual Conference of the Japan ergonomics Society Kanto Branch Proceedings*, 1993; pp112–113.

[3] Senba H, Kondo T, Preventing Accidental Falls at Hospitals: Walking with IV poles (in Japanese), *Journal of the Eastern Japan Association of Orthopaedics and Traumatology*, 2004; 481: 16–3.

[4] Hachigasaki R, The influence IV pole height and grip height on gait of healty people ages 60–70 (in Japanese), *Japanese Journal of Nursing Art and Science*, 2012; pp38–47:11–2.

[5] Nakamura R, Saito H, *Fundamental kinesiology (in Japanese)*. sixth ed. Tokyo: Ishiyaku Publishers, Inc; 2003.

[6] Taga M, Terui R, Kamishima S et al., A Usability Assessment of IV Poles (Report 1): An examination of patient and nurse impressions from use (in Japanese). *SCU Journal of Design & Nursing* 2008; 2–1: pp.33–38.

[7] Koitabashi K, Adjustment of the sickbed environment: Bed-making for recumbency patients (in Japanese), Okawara C, Sakai K, editors. *Nursing Ergonomics that Supports Health Care Work*. Tokyo: Ishiyaku Publishers, Inc; 2002.

New Ergonomics Perspective – Yamamoto (Ed.)
© *2015 Taylor & Francis Group, London, ISBN 978-1-138-02751-0*

Resilience, psychological stressors, and stress responses in Japanese university athletes

Yujiro Kawata

School of Child Psychology, Tokyo Future University, Adachi-Ku, Tokyo, Japan
School of Health and Sports Science, Juntendo University, Inzai-Shi, Chiba, Japan

Masataka Hirosawa

School of Health and Sports Science, Juntendo University, Inzai-Shi, Chiba, Japan
Graduate School of Health and Sports Science, Juntendo University, Inzai-Shi, Chiba, Japan

Akari Kamimura

Graduate School of Health and Sports Science, Juntendo University, Inzai-Shi, Chiba, Japan

Kai Yamada

School of Health and Sports Science, Juntendo University, Inzai-Shi, Chiba, Japan
Faculty of Economics, Hosei University, Chiyoda-ku, Tokyo, Japan

Takanori Kato & Kazusa Oki

Graduate School of Health and Sports Science, Juntendo University, Inzai-Shi, Chiba, Japan

Sawako Wakui

School of Health and Sports Science, Juntendo University, Inzai-Shi, Chiba, Japan
Graduate School of Health and Sports Science, Juntendo University, Inzai-Shi, Chiba, Japan

Shino Izutsu

Faculty of Sports and Health Sciences, Japan Women's College of Physical Education, Setagaya-ku, Tokyo, Japan

Motoki Mizuno

School of Health and Sports Science, Juntendo University, Inzai-Shi, Chiba, Japan
Graduate School of Health and Sports Science, Juntendo University, Inzai-Shi, Chiba, Japan

ABSTRACT: For competitive university athletes, success in their specialized sports is great achievement and a highlight of their life in sports; however, athletes often have difficulties in performing at their best due to psychological stressors and stress responses. Thus, athletes and coaches need to learn to cope with stressors appropriately, and resilience may be a factor in their success. According to Davydov et al., resilience is defined as the successful adaptation and swift recovery after experiencing severe adversity, and has an important role in maintaining homeostasis in stressful condition. Therefore, the aim of this study was to examine the relationship between resilience and cognition in the context of stressors and stress responses among Japanese university athletes. We collected the data from 511 Japanese university athletes (358 male, 153 female, $M = 20.0$ years of age, $SD = 1.5$ years) who had participated in national level competitions. We collected the athletes' demographic information (sex, age, grade, main sports event, competitive level, and role on a team). The Adolescent Resilience Scale (ARS) was used to measure resilience; the Daily and Competitive Stressor Scale (DCSS) was used to measure psychological stressors; the General Health Questionnaire-30 (GHQ-30) and the Self-Depression Scale (SDS) were used to measure psychological stress responses. Written informed consent was obtained from all participants. We analyzed the relationships between ARS, DCSS, GHQ-30, and SDS using correlation analyses. There were significant positive correlations between the total score of the DCSS and the total score of the GHQ-30 ($r = 0.46$, $p < 0.001$), and the total score of the SDS ($r = 0.27$, $p < 0.001$). There were significant inverse correlations between the total score of the ARS and the total score of the DCSS ($r = -0.20$, $p < 0.001$), and the subscale scores of the ARS and the subscale scores of the DCSS. There were also significant inverse correlations between the total score of the ARS and the total score of the GHQ-30 ($r = -0.35$, $p < 0.001$), and the subscale scores of the ARS and the subscale scores of the GHQ-30. Moreover, there was significant inverse correlation between the total score of the ARS and the total score of the SDS ($r = -0.45$, $p < 0.001$). We concluded that resilience might have an impact on the stress-related cognition and stress responses among Japanese university athletes. Thus, we propose that athletes and their coaches should pay attention to an athlete's resilience to effectively cope with psychological stressors and stress responses.

Keywords: resilience; stressor; stress response; university athlete

1 INTRODUCTION

For competitive university athletes, success in their specialized sports is a great achievement and a highlight of their life in sports. However, athletes often have difficulties in performing at their best due to psychological stressors and stress responses. Over training [1], dropout from sports [2], burnout syndrome [3], depression [4] and eating disorders [5] are frequently reported as stress-related problems.

From the perspective of university athletes, they have to play multiple roles in university; at least as a competitive athlete and as a student. Thus, stressors arise in these roles. Oka et al., [6] reported that Japanese university athletes are frequently faced with at least following eight stressors: "human relationships in daily and competitive life," "competitive record," "expectation and pressure from others," "internal and social change in each individual," "content of club activity," and "economic condition and academic record." Moreover, according to Oka et al., [6] "human relationships in daily and competitive life," and "internal and social change in each individual" were associated with psychological responses to stress such as depression, anxiety, anger, confusion, staying at home all day, physical fatigue, and autonomic change. Thus, athletes and coaches need to learn to cope with stressors appropriately.

The process of arousing psychological stress may be understood using cognitive-evaluation theory [7]. This theory suggests that stressors cause stress responses via mediating factors consisting of the cognitive evaluation of stressors and subsequent coping behaviors, as shown Fig. 1. In this theory, individuals first perceive a stressor and then evaluate whether the stressor is a threat or not. Then, if the stressor is evaluated as a threat, attempting to cope with the stressor occurs, using problem-focused coping and emotion-focused coping. When the stressors are coped with appropriately, a stress response is not aroused, whereas if the coping behavior is ineffective, a stress response is aroused. Thus, if we could control stress-related cognition and stress coping, we may be able to regulate the level of the stress response. According to a previous study, personality is associated with coping strategies [8]. Extraversion and conscientiousness are related to problem-focused coping and cognitive restructuring, and neuroticism is related to emotion-focused coping. Thus, psychological traits affect the process by which psychological stress is aroused.

Resilience is defined as the successful adaptation and swift recovery after experiencing severe adversity, and has an important role in maintaining homeostasis in stress conditions [9]. For instance, stressful events (e.g., injury, illness, and disruption in interpersonal relations) occur suddenly and mental health is disturbed after the event. However subsequently, mental health recovers following the exposure, because of the individual's resilience. Oshio [10] highlighted the importance of understanding resilience as a psychological trait which is conducive to recovery from irreversible adversity, in order to understand individual adaptive development. Thus, in this study, we also adopted this view of resilience. Recently, since the importance of resilience in specific domains has been suggested, resilience in sports has been examined in many countries [11]. These previous studies have reported positive effects of resilience. Resilience may have a suppressive effect on negative stress-related cognition and stress responses.

Therefore, the aim of this study was to examine the relationship between resilience, stress-related cognition, and stress responses among Japanese university athletes. If resilience has a suppressive effect on stress-related cognition and stress responses, athletes and their coaches should give considerable attention to fostering resilience.

2 METHOD

2.1 Participants

We collected data from 511 Japanese university athletes (358 male, 153 female, $M = 20.0$ years of age, $SD = 1.5$ years). Their main sports events were individual sports (track and field, gymnastics, bicycle circuit, judo, Japanese art of fencing, and swimming) and team sports (soccer, volleyball, handball, squash, basketball, baseball, softball, and ice hockey). Average training time per day was 2.8 ± 0.9 hours, with practice 5.5 ± 1.6 days a week, on average.

2.2 Measurements

We collected the athletes' demographic information including sex, age, grade, main sports event, individual competitive level (regional level, prefectural level, district level, national level, and international level), and role on a team (regular player, non-regular player, and staff).

The Adolescent Resilience Scale (ARS) [10, 12] was used to measure resilience. This scale has 21 items and is composed of the following three factors: "novelty seeking," "emotional regulation," and "positive future orientation." Validity and reliability were confirmed. Participants responded to each item using a five-point Likert scale (1 = "disagree" to 5 = "agree"). The overall score was calculated as the sum of item scores. A high score indicates a high level of resilience.

The Daily and Competitive Stressor Scale (DCSS) [6] was used to measure the cause of psychological stressors in both daily life and sports-related activities. This scale has 35 items and assesses the following eight stressors: "human relationships in daily and competitive life," "competitive record," "expectation and pressure from others," "internal and social change in each individual," "content of club activity," and "economic condition and academic record." Participants responded to each item, using Likert scales, in terms of frequency (0 = "never" to 4 = "always") and feelings of repulsion (0 = "non-repulsive" to 4 = "repulsive"). The overall score was calculated by multiplying the frequency score by the repulsion score. A high score indicates a high level of stress-related cognition.

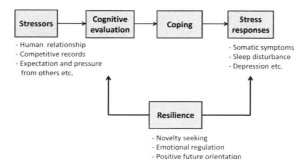

Figure 1.   Hypothetical model for this study.

233

The General Health Questionnaire-30 (GHQ-30) [13, 14] was used to measure psychological stress responses. This scale has 30 items consisting of the following six factors: "general illness," "somatic symptoms," "sleep disturbance," "social dysfunction," "anxiety and dysphoria," and "suicidal depression." Participants responded in terms of their agreement with each item using a five-point Likert scale (1 = "disagree" to 5 = "agree"). Overall score was calculated by summing item scores. High scores indicate worse psychological health.

The Self-Depression Scale (SDS) [15, 16] was also used to measure psychological stress responses. This scale has 20 items and assesses a single depression factor. Participants responded in terms of their agreement with each item using a four-point Likert-scale (1 = "disagree" to 4 = "agree"). The overall score was calculated by summing item scores. A high score indicates a high level of depression.

2.3 Ethical consideration

This study was approved by the Research Ethics Committee at the School of Health and Sports Science, Juntendo University. Prior to the study, we obtained written informed consent from all participants. Each participant was made aware of his or her right to decline cooperation at any time without repercussions, even after consenting to participate.

2.4 Analysis

First, to confirm the relationships between stressors and stress responses, we calculated Pearson's correlations between the total and subscale scores of the DCSS, GHQ-30, and SDS. Next, to examine the relationships between resilience and stressors, we calculated Pearson's correlations between the total and subscale scores of the ARS and DCSS. Finally, to examine the relationships between resilience and stressors, we calculated Pearson's correlations between the total and subscale scores of the ARS, GHQ-30, and SDS. We set the significance level at $p < 0.05$.

3 RESULTS & DISCUSSION

3.1 Relationship between stressors (DCSS) and stress responses (GHQ-30 and SDS)

Table 1 shows that there were significant inverse correlations between the total score of the DCSS and that of the GHQ-30 ($r = 0.46$, $p < 0.001$), and the subscale scores of the DCSS and those of the GHQ-30 ($r = 0.11$–0.49). The results showed that there were significant inverse

Table 1. Correlation coefficients between stressors (DCSS) and stress responses (GHQ-30 and SDS).

Stress responses	Stressors (DCSS)						
	HR	CR	EP	IS	CA	EC	TS
GHQ-30							
General illness	0.23**	0.21***	0.29***	0.22***	0.31***	0.21***	0.33***
Somatic symptoms	0.19**	0.18**	0.20***	0.14*	0.18**	0.18**	0.23***
Sleep disturbance	0.21***	0.19**	0.23***	0.11	0.26***	0.21***	0.27***
Social dysfunction	0.21***	0.16**	0.33***	0.32***	0.31***	0.14*	0.33***
Anxiety and dysphoria	0.39***	0.33***	0.49***	0.27***	0.39***	0.27***	0.48***
Suicidal depression	0.16**	0.15**	0.24***	0.14*	0.24***	0.14*	0.24***
Total score	0.35***	0.30***	0.44***	0.28***	0.41***	0.28***	0.46***
SDS, total score	0.20***	0.15**	0.28***	0.17**	0.22***	0.17**	0.27***

$*p < 0.05$, $**p < 0.01$, $***p < 0.001$.

Note: HR: human relationships in daily and competitive life, CR: competitive record, EP: expectation and pressure from others, IS: internal and social change in each individual, CA: content of club activity, EC: economic condition and academic record, and TS: total score.

correlations between the total score of the DCSS and that of the SDS ($r = 0.27$, $p < 0.001$), and the subscale scores of the DCSS and the total score of the SDS ($r = 0.15$–0.28). These results indicate that psychological stressors may cause stress responses. In particular, expectation and pressure from others (see column EP in Table 1) and content of club activity (see column CA in Table 1) are strongly linked to unhealthy conditions ($r = 0.44$ and 0.41, respectively) and depression ($r = 0.28$ and 0.22, respectively) as compared with other stressors.

3.2 *Relationship between resilience (ARS) and stressors (DCSS)*

Table 2 shows that there were significant inverse correlations between the total scores of the ARS and the DCSS ($r = -0.13$, $p < 0.01$), and the subscale scores of the ARS and DCSS ($r = 0.01$–0.19). These results indicate that resilience might suppress stress-related cognition slightly. In particular, emotional regulation may play a role in decreasing stress-related cognition (see column ER in Table 2). Positive future orientation was related to only expectation and pressure from others (see column PF in Table 2). Novelty seeking was not associated with stress-related cognitions (see column NS in Table 1).

3.3 *Relationship between resilience (ARS) and stress responses (GHQ-30 and SDS)*

Table 3 shows that there were also significant inverse correlations between the total score of the ARS and GHQ-30 ($r = -0.35$, $p < 0.001$), and the subscale scores of the ARS and GHQ-30. Moreover, there was a significant inverse correlation between the total score of

Table 2. Correlation coefficients between resilience (ARS) and stressors (DCSS).

Stressors (DCSS)	Resilience (ARS)			
	NS	ER	PF	TS
Human relationship in daily and competitive life	−0.02	−0.14**	−0.04	−0.08
Competitive record	−0.05	−0.12**	−0.05	−0.10
Expectation and pressure from others	−0.05	−0.19**	−0.19**	−0.17**
Internal and social change in each individual	−0.06	−0.11**	−0.09	−0.11
Content of club activity	−0.01	−0.11**	−0.05	−0.07
Economic condition and academic record	−0.03	−0.18**	−0.04	−0.11**
Total score	−0.04	−0.18***	−0.10	−0.13**

$*p < 0.05$, $**p < 0.01$, $***p < 0.001$.

Note: NS: novelty seeking, ER: emotional regulation, PF: positive future orientation, and TS: total score.

Table 3. Correlation coefficients between resilience (ARS) and stress responses (GHQ-30 and SDS).

Stress responses	Resilience (ARS)			
	NS	ER	PF	TS
GHQ-30				
General illness	−0.05	−0.10	−0.08	−0.10
Somatic symptoms	−0.03	−0.17**	−0.06	−0.12*
Sleep disturbance	−0.07	−0.21***	−0.20***	−0.19**
Social dysfunction	−0.17**	−0.20***	−0.33***	−0.27***
Anxiety and dysphoria	−0.09	−0.24***	−0.19**	−0.21***
Suicidal depression	−0.22***	−0.29***	−0.39***	−0.35***
Total score	−0.14**	−0.29***	−0.29***	−0.29***
SDS, total score	−0.45***	−0.50***	−0.57***	−0.59***

$*p < 0.05$, $**p < 0.01$, $***p < 0.001$.

Note: NS: novelty seeking, ER: emotional regulation, PF: positive future orientation, and TS: total score.

the ARS and the total score of the SDS ($r = -0.45$, $p < 0.001$). These results indicate that resilience might suppress stress responses. In particular, emotional regulation and positive future orientation may have a role in decreasing stress responses such as sleep disturbance, social dysfunction, anxiety and dysphoria, and suicidal depression (see columns ER, and PF in Table 3). Novelty seeking may prevent stress responses only related to social dysfunction and suicidal depression (see column NS in Table 3). All of subscales of resilience may have suppressive impact on depression as measured by the SDS.

Throughout the results, our hypothesis was supported. There is the indication that psychological stressors may lead to stress responses. In particular, expectation and pressure from others, and content of club activity predicted unhealthy psychological conditions better than other stressors. Thus, to control these two factors is very important in order to regulate stress responses among university athletes. The two factors may be especially impactful because high performance is always demanded of university athletes. Moreover, to cope with these stressors may be difficult because it is necessary for the athlete to consider his or her relationship with others.

Next, it was confirmed that resilience and stress-related cognition were slightly related. That is, resilience may suppress stress-related cognition slightly. In particular, emotional regulation and positive future orientation were related to expectations and pressure from others. Thus, to improve emotional regulation and positive future orientation may be effective in reducing the level of stress-related cognition, such as expectations and pressure from others. However, enhanced novelty seeking does not have any impact regarding stress-related cognition.

Finally, resilience was associated with stress responses. From the GHQ-30, positive future orientation and emotional regulation may suppress unhealthy conditions. Conversely, novelty seeking may not inhibit unhealthy conditions to as large an extent. However, from SDS, resilience may prevent the development of depression. Novelty seeking was not associated with unhealthy conditions as measured by the GHQ, but novelty seeking showed a relationship with depression as measured by the SDS. These results indicate that positive future orientation and emotional regulation have a role in preventing most stress responses, but novelty seeking is only effective for certain kinds of stress response.

4   CONCLUSIONS

We conclude that resilience might impact the cognition associated with stressors and stress responses among Japanese university athletes. Thus, we propose that athletes and their coaches should pay attention to an athlete's resilience in order to effectively cope with psychological stressors and stress responses.

5   LIMITATIONS AND FUTURE STUDY

This study showed relationships between resilience and psychological stressors and stress responses among university athletes. However, the following issues need to be considered. First, an examination of sex differences is needed. We could not investigate sex differences because we lacked female athletes. This point should be addressed in future research. Second, we need to address the relationship between resilience and stress coping. If this relationship is clarified, we might understand why someone who has high resilience does not show a marked stress response. Third, methods of improving resilience should be developed. For instance, providing success stories and roles models of successful athletes may be effective in enhancing positive future orientation. Further research including these issues will improve the possible applications of resilience research.

ACKNOWLEDGMENTS

We are grateful to all research assistants and graduate school students who contributed their discussions, help, and encouragement to this study. We wish to thank all volunteers for their donations of their time and subsequent data used in the study.

REFERENCES

[1] Fry RW, Morton AR, Keast D. Overtraining in athletes. *Sports Med* 1991;**12**:32–65.

[2] Fraser-Thomas J, Côté J, Deakin J. Understanding dropout and prolonged engagement in adolescent competitive sport. *Psychol Sport Exerc* 2008;**9**:645–62.

[3] Goodger K, Gorely T, Lavallee D, Harwood C. Burnout in sport: A systematic review. *Sport Psychol* 2007;**21**:127–51.

[4] Yang J, Peek-Asa C, Corlette JD, Cheng G, Foster DT, Albright J. Prevalence of and risk factors associated with symptoms of depression in competitive collegiate student athletes. *Clin J Sports Med* 2007;**17**:481–87.

[5] Byrne S, McLean N. Eating disorders in athletes: A review of the literature. *J Sci Med Sport* 2001;**4**:145–59.

[6] Oka K, Takenaka K, Matsuo N, Tsutsumi T. Development of daily and competitive stressors scale for university athletes and the relationship with mental health. *Japan J Phys Edu Heal Sport Sci* 1998;**43**:245–59. (in Japanese)

[7] Lazarus RS, Folkman S. *Stress, appraisal, and coping.* New York: Springer; 1984.

[8] Connor-Smith JK, Flachsbart C. Relations between personality and coping: A meta-analysis. *J Pers Soc Psychol* 2007;**93**:1080.

[9] Davydov DM, Stewart R, Ritchie K, Chaudieu I. Resilience and mental health. *Clin Psychol Rev* 2010;**30**:479–95.

[10] Oshio A, Nakaya M, Kaneko H. Development and validation of an Adolescent Resilience Scale. *Japan J Counsel* 2002;**35**:57–65. (in Japanese)

[11] Sarkar M, Fletcher D. How should we measure psychological resilience in sport performers? *Meas Phys Educ Exerc Sci* 2013;**17**:264–80.

[12] Oshio A, Kaneko H, Nagamine S, Nakaya M. Construct validity of the adolescent resilience scale. *Psychol Rep* 2003;**93**:1217–22.

[13] Goldberg DP, Blackwell B. Psychiatric illness in general practice: A detailed study using a new method of case identification. *Br Med J* 1970;**2**:439.

[14] Nakagawa Y, Daibo I. *The General Health Questionnaire.* Tokyo: Nihon Bunka Kagakusha; 1985. (in Japanese)

[15] Zung WW. A self-rating depression scale. *Arch Gen Psychiatry* 1965;**12**:63–70.

[16] Fukuda K, Kobayashi S. *Self-rating Depression Scale.* Kyoto: Sankyobo; 1983. (in Japanese)

New Ergonomics Perspective – Yamamoto (Ed.)
© 2015 Taylor & Francis Group, London, ISBN 978-1-138-02751-0

Analysis of sleep environment in Japanese young people

Hiroshi Yasuoka
Tokyo University of Information Science, Wakaba-ku, Chiba, Japan

Shigeka Shioji & Takeshi Sato
Jissen Women's University, Tokyo, Japan

Macky Kato
Waseda University, Mikajima Tokorozawa, Saitama, Japan

Shoji Igawa
Nippon Sport Science University, Yokohama, Kanagawa, Japan

ABSTRACT: It is one of the most important to create an optimal sleep environment. People who lived irregular hours, i.e. shift workers, global communication tasks, typically obtain 1–2 hour less sleep than the generally accepted target of 8 hours per day. Shift work has become increasingly common in the modern globalized economy. Many work settings such as factory work, service industries, power plants, internet communication management, health care and emergency services require shift work that includes work at night. Furthermore, not only shift workers but also young smart phone device users often experience an extended period of acute sleep deprivation at the start of a new day. Most sleep deprivation studies using complex cognitive tasks require the participants to complete the tasks alone. It was effected human health care and learning performance in young people. The purpose of the current study was to investigate how sleep environment changes between high school student and university student.

Jissen Womenn's Uiversity Ethics Committee approved this survey. There were 831 students (male: 387, female: 444) participate in this study. It was 16 sleep questionnaire items on our web server, i.e.: depth of sleep, difficulties in waking up, quality and latency of sleep, negative affect in dreams, and sleep irregularity, etc. It was collected data during our course activity.

There was no significant difference sleeping pattern in between high school and university student. Additionally it was on roughly the same level sleeping conditions of them. However it was a tendency toward to sleep deprivation onto nocturnal life.

In the current report we present results of a prospective study that used objective measures to evaluate sleep, it was consequently sluggishness due to the chronic sleep deprivation, to become a late-night students.

Keywords: Sleep environment; Daily life; Young people

1 INTRODUCTION

Recently, it was revised Sleep Guideline for Health Promotion 2014[1] by Ministry of Health, Labour and Welfare, Health Service Bureau. It was not only partly reported as related occupational ergonomics: patterns of performance degradation and restoration during sleep restriction and subsequent recovery[2]; effects of recovery sleep after one work week of mild sleep and performance[3], but also association between mobile device and sleep disturbances among Japanese adolescents[4]. Moreover it was well known that the risk factor of human error was to have relation sleep environments[5,6]. Thus, it was important the sleep for work irregular hours, i.e. shift workers, particularly in safety-critical workplaces, given that chronic sleep loss of this magnitude can substantially impair neuro-behavioural function during wake periods[7]. However it was not enough data from adolescents, previous study[4] was to not include over 20 years old as investigation object. The aim of this study was to investigate the comparative survey in sleep environment among high school age and university student.

2 METHODS

Subjects: There were 831 subjects (man: 387 female: 444, mean age ± sd: 17.7 ± 1.9 years old) participated in this survey. Informed consent was obtained from the subjects prior to the web access data collection. We had recruited the subjects on requested to chief information faculty staff in Japanese high school and university. The study was approved by the ethics committee in Jissen Women's University(No. H26-2) and conducted in accordance with the Declaration of Helsinki.

Data collection procedure: It was applied the Morningness-Eveningness questionnaire test[8] for exclusive use web server in Tokyo University of Information Science.

Data analysis: The SPSS computer packages was used. All ANOVAs were performed with all independent variables and a single dependent variable. All significant values are reported at $p < 0.05$.

3 RESULT

Figure 1 was illustrated distribution of questionnaire the moringness-eveneningness sleep environment test. There was no people in morningness sleep environment, 70% intermediate type and 30% nearly eveningness type, respectively. There was significant difference among ratio of nearly evening and intermediate sleep type. It was showed almost young people does not morning person. Also it was a few subjects under 18 years old evaluated eveningness

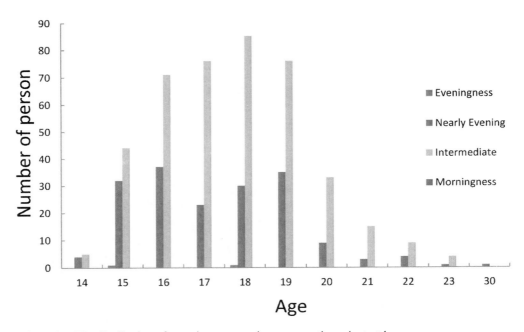

Figure 1. The distribution of morningness-eveningness questionnaire test by age.

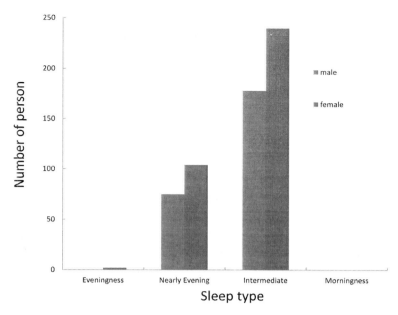

Figure 2.　The number of morningness-eveningness questionnaire test by gender.

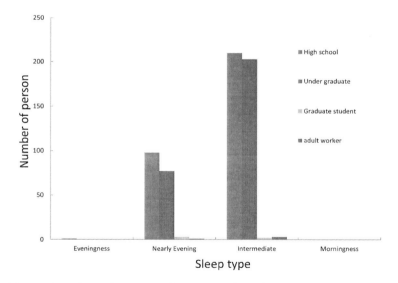

Figure 3.　The number of morningness-eveningness questionnaire test by academic year.

type. Figure 2 was illustrated the number of sleep type by gender. There was no significant difference sleep type in gender. Also there was no significant different by academic year in all age (Fig. 3).

4　DISCUSSION

In this study was supported Sleep Guideline for Health Promotion 2014[1] by Ministry of Health, unrelated to the age of sleep type among Japanese adolescents. It was considered that need for change a lifestyle to moringness sleep type in school age. However this study has limitation for analysis living environment, in particular investigate their activity in night. Furthermore it was essential survey together about information environment both qualitatively and quantitatively, related smart phone and game devices.

ACKNOWLEDGEMENTS

We would like to offer our special thanks to Miss. Mizuki Nakajima for helping this experiment.

REFERENCES

[1] Ministry of Health, Labour and Welfare, Health Service Bureau, ed. Sleep Guideline for Health Promotion 2014. Tokyo: Ministry of Health, Labour and Welfare, Health Service Bureau, 2014 (Japanese).
[2] Belenky G, Wesensten NJ, Thorne DR, Thomas ML, Sing HC, Redmond DP, Russo MB, Balkin TJ. Patterns of performance degradation and restoration during sleep restriction and subsequent recovery: a sleep dose-response study. J Sleep Res 2003;12, 1–12.
[3] Pejovic S, Basta M, Vgontzas AN, Kritikou I, Shaffer ML, Tsaoussoglou M, Stiffler D, Stefanakis Z, Bixler EO, Chrousos GP. Effects of recovery sleep after one work week of mild sleep restriction on interleukin-6 and cortisol secretion and daytime sleepiness and performance. Am J Physiol Endocrinol Metab 2013;305, 890–896.
[4] Munezawa T, Kaneita Y, Osaki Y, Kanda H, Ohtsu T, Minowa M, Suzuki K, Higuchi S, Mori J, Yamamoto R, Ohida T. The Association Between Use of Mobile Phones After Lights Out and Sleep Disturbances Among Japanese Adolescents: A Nationwide Cross-Sectional Survey. Sleep 2011;34, 1013–1020.
[5] Maia Q, Grandner MA, Findley J, Gurubhagavatula I. Short and long sleep duration and risk of drowsy driving and the role of subjective sleep insufficiency. Accid Anal Prev 2013;59, 618–622.
[6] Abe T, Komada Y, Nishida Y, Hayashida K, Inoue Y. Short sleep duration and long spells of driving are associated with the occurrence of Japanese drivers' rear-end collisions and single-car accidents. J Sleep Res 2010;19, 310–316.
[7] Belenky G, Wesensten J, Thorne R, Thomas L, Sing C, Redmond P, Russo B, Balkin J., Patterns of performance degradation and restoration during sleep restriction and subsequent recovery: a sleep dose–respose study. Journal of sleep Research 2003;12, 1–12.
[8] Kaneyoshi Ishihara, Akio Miyashita, Maki Inugami, Kazuhiko Fukuda, Katsuo Yamasaki, Yo Miyata, The results of inivestigateion by the Japanese version of Morningness-Eveningness Questionnaire, The Japanese Journal of Psycology 1986;57, 87–91.

New Ergonomics Perspective – Yamamoto (Ed.)
© 2015 Taylor & Francis Group, London, ISBN 978-1-138-02751-0

Quantitative confirmation of residents' everyday life behavior by acceleration measurement of facilities in a house

Macky Kato
Waseda University, Tokorozawa, Saitama, Japan

Yoshie Shimodaira
Nagano Prefectural College, Nagano, Nagano, Japan

Takeshi Sato
Jissen Women's University, Hino, Tokyo, Japan

ABSTRACT: In aged society, decreasing of worker population is one of the most important problems. Especially the farming population has been decreasing in recent years. On the other hand, the elderly people's families live away from them, because they cannot expect a reasonable profit from farming. In addition, it is hard to find other jobs in a rural area. Thus, the isolated farmers need somebody's help to take care of themselves. For example, the families can confirm their everyday lives by a remote system with monitor cameras. However, it can violate the residents' privacy. The appropriate system should have a confirmation function without violating the privacy. In this study, facilities in a house can be thought as sensors of everyday life in place of cameras. The acceleration measurement was held for quantification of the residents' behavior in a house. The residents in the subject house are 62-year aged husband, 58-year aged wife and 91-year aged father. The accelerometers have been fixed on the refrigerator and the door in the house. Approximately one-year lengths, acceleration of the facilities were recorded through 2013 and 2014. The fluctuation of acceleration of the refrigerator can indicate the resident's using in a day. It showed that the residents lived well-regulated lives. The refrigerator will be able to indicate the feature of their everyday life. On the other hand, the fluctuation of acceleration of the door can indicate the resident's movements inside out. The door movements depend on the seasons. The number of the door movements in each season is different. They kept the door open through the daylight in summer, but they close the door after going in winter. These results reveal that the acceleration measurement of facilities in a house can become the quantitative confirmation method of residents' everyday life stability.

Keywords: Everyday life behavior; Acceleration measurement; Quantitative confirmation; Aged society, Isolation of family

1 INTRODUCTION

One of the most important problems in recent years is the decreasing population in aged society of Japan. Some types of industry are suffering from a shortage of workers. Especially, decreasing of the farming population is quite severe. Many farms in the rural area of Japan consist of only elderly farmers. Most of their family lives away from the elderly parents, because they cannot expect a reasonable profit from small size farming. In addition, it is not easy to get other occupations in the rural area. The isolation problem is important not only in the rural area, but also in the city areas, for some other reasons. As results, the elderly people are forced to live by themselves. The most important features for them are the safety and health in everyday life. They need somebody's assists to take care of themselves. However, some of them do not have any neighborhood around them. In such a case, their family can confirm their everyday lives by a remote device such as monitor cameras. The studies in the past have used camera to confirm residents' safety remotely [1,2]. However, it is said that the confirmation with cameras has the possibility to violate residents' privacy. On the other hand, the remote wearable cardiovascular sensor was used to confirm the subject's health in the other studies [3,4]. It has no possibility to violate privacy, but the resident has to wear the device. One of the methods, which do not violate privacy without any restriction, is monitoring facility around the subjects. The studies in the past proved that the monitoring facility by sensor could estimate the residents' behavior [5]. Most of these studies in the past have function to monitor the residents' activity to extract accidents. In contrast, movement of the facility was confirmed by the sensors from a viewpoint of failsafe in the other study [6,7]. In brief, the evaluation of usual condition is required in order to extract unusual condition from the dataset. In this study, the purpose is the quantification of residents' regular everyday life for confirmation by long-term investigation. The facility movement would be measured by accelerometer for just under a year.

2 METHOD

2.1 *Acceleration measurement of facility*

The subjects of this study live in the west area of Nagano city, Japan. It is in the typical Japanese rural area. The family consists of 62-year aged husband, 58-year aged wife and 89-year aged father, who need someone's care. The door, which is frequently used to go through, was the first object to be measured the movement. The refrigerator is the second object. The objects were chosen based on the study in the past. These facilities had contributed to estimation of residents' behavior [7]. Attached acceleration transducers were AS-10GA (Kyowa Electric Instruments, Figure 1a), which was connected to the universal recorder EDX-100A (Kyowa Electric Instruments, Figure 1b). The directions of the transducers were set as same as the movement direction of the objects. Computer to record the measured data controlled the recorder for 24 hours in a day. Sampling frequency was 32 Hz. The length of measurement was 278 days through 2013 and 2014. Table 1 shows all the periods.

2.2 *Extracting unusual condition signs from the dataset*

Analysis procedure was as follows. The first step is the extracting the fluctuation of facility movement. The dataset was separated to the series of acceleration records on each day. Extra large oscillations in the series on each day were regarded as the facility movement signs of the objects. The number of the movement signs was counted on each hour in a day. The fluctuations of the movements in a day were driven as the evidence to estimate the residents' behavior.

The second step is the instituting the threshold of usual movement fluctuation. The influence of day of the week and fluctuation of temperature was ascertained as the external factors, which have possibility to obstruct the continuous criteria through a year. The correlation coefficient between the fluctuations of each day and the average fluctuation of the

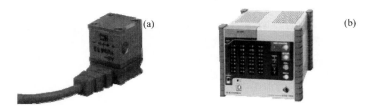

Figure 1.   (a) Acceleration transducer AS-10GA (Kyowa EI); (b) Data logger EDX-100A (Kyowa EI).

Table 1. Investigation periods.

Periods	Days	From	To
1	19	Sep.15.2013	Oct.3
2	20	Oct.5	Oct.24
3	19	Oct.27	Nov.14
4	34	Nov.17	Dec.20
5	30	Dec.23	Jan.21.2014
6	35	Jan.24	Feb.26
7	35	Mar.22	Apr.25
8	35	May 4	Jun.7
9	26	Jun.15	Jul.11
10	12	Jul.13	Jul.25
11	13	Jul.27	Aug.8
Total	278		

previous seven days were plotted on the management chart. It was regarded as the everyday life stability indicator in a day. The lower 2.5% of the indicators were regarded as the unusual condition signs, which means that some happening has been occurred.

The third step was comparing the extraction of unusual condition signs and the residents' behavior by interview with the residents.

The study ethics committee of Waseda University has already approved this study.

3 RESULT

3.1 Fluctuation of the movements of the objects in a day

Figure 2a shows the average of the door movements in a day. The x-axis means periods of time and the y-axis means the number of movement signs counted on each hour.

In the midnight, 0:00 h to 4:00 h, the door movement could not be observed very well. The door started to be moved around 5:00 h or 6:00 h. The peak of movements was observed around 8:00 h. In the daytime, the movements were fewer than the morning and the evening. In the evening around 16:00 h or 17:00 h, the second peak of movements could be observed. After 18:00 h, the movements decreased until midnight as the time went by. On the other hand, Figure 2b shows the average of the refrigerator movements in a day. The movement of the refrigerator made three peaks. The first peak could be observed around 6:00 h or 7:00 h. The second one could be observed around 12:00 h. And the third one could be observed around 17:00 h or 18:00 h. The movements in the midnight and daytime except the noon were quite fewer than the three peaks.

3.2 Effect of the external factors

3.2.1 Effect of the days of the week to the movement of the objects

The effect of the days of the week and fluctuation of temperature was ascertained as the external factors in this study. Figure 3 shows the relationship between the external factors and the average number of movements of the objects separated by days of the week. The x- and y-axis have the same meanings as those of Figure 2. The first peak of the door movement could be observed around morning 8:00 h in Figure 3a. Some smaller movements could be observed in the daytime. Every movement has different features. The second peak was observed around 16:00 h or 17:00 h. After the evening, the movements decreased on each day as time went by. In the midnight, few movements were observed except a little bit of movement around 0:00 h on Saturday. The peak of the movement of the refrigerator could be observed in Figure 3b as same as the Figure 2b. The first peak was observed around 6:00 h. The second was observed around 12:00 h. The third was observed around 17:00 h or 18:00 h. Each peak in the morning and those at 12:00 h are almost same. The peaks in the evening have two cased, 17:00 and 18:00.

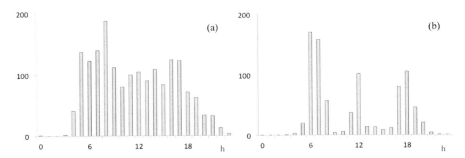

Figure 2. (a) Average fluctuation of the door movements in 24 h; (b) Average fluctuation of the refrigerator movements in 24 h.

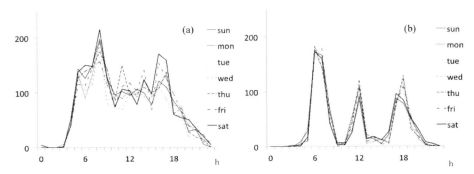

Figure 3. Relationship between the movements of the objects and the days of the week; (a) Average fluctuation of the door movements among days of the week; (b) Average fluctuation of the refrigerator movements among days of the week.

3.2.2 *Effect of the fluctuation of temperature to the movement of the objects*

The effects of fluctuation of temperature at Nagano city [8] were ascertained. The correlation coefficient between temperature and the door movement was 0.65. The correlation coefficient between temperature and the refrigerator was 0.23. Focusing on the door movement, Figure 4 shows the relationship between temperature and the door movement with the result of regression analysis. The negative regression line was calculated from the plots. It shows the movements increased as the temperature decreased.

Figure 5 shows the average movements classified under the temperature. All the classes have the same features in the morning, but the peaks are different between the classes. In addition, the differences in daytime were observed. The movements in warm days (20–) were few in daytime. On the other hand, the movements in cold days (−0, 0–10) were observed frequently in daytime. The moderate days (10–20) took middle position.

3.3 *Instituting the threshold of usual movement fluctuation*

Comparing everyday movement with the average in Figure 1 has possibility to decrease the appropriateness as the usual condition, because the fluctuation of temperature influenced the movement of the objects. Thus, the average movement of the previous seven days was used as the usual condition to be compared. The correlation coefficient between the series of movements and that of the previous seven-day average were calculated to extract the sign of something unusual.

3.3.1 *Correlation coefficient between the door movement and the previous seven days*

Figure 6 shows the fluctuation of the correlation coefficient between the door movement and the average of the previous seven days through all the investigation period. The average

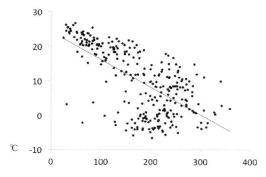

Figure 4. Regression analysis between temperature and average number of door movement signs.

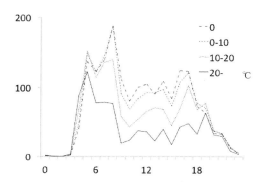

Figure 5. Average door movement signs classified under the temperature.

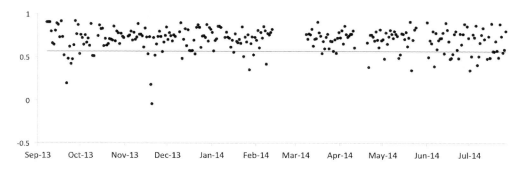

Figure 6. Extracting unusual conditions of the door movement with everyday life stability indicator.

correlation coefficient was 0.70 and the standard deviation was 0.14. Assuming a Gaussian distribution, the threshold was instituted in 0.57, which cut off approximately 2.5% of the distribution. The straight line in the Figure 6 is the threshold. As a result, 43 days were regarded as unusual conditions.

3.3.2 *Correlation coefficient between the refrigerator movement and the previous seven days*
Figure 7 shows the fluctuation of the correlation coefficient between the refrigerator movement and the average of the previous seven days through all the investigation period. The average was 0.82 and the standard deviation was 0.14. Assuming a Gaussian distribution, the threshold was instituted in 0.68. The straight line in the Figure 7 is the threshold. As a result, 30 days were regarded as unusual condition.

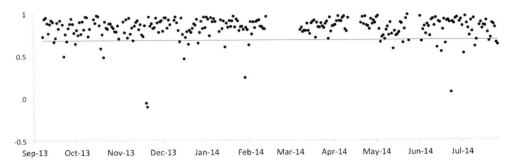

Figure 7. Extracting unusual condition of the refrigerator movement with everyday life stability indicator.

4 DISCUSSION

The results revealed that the quantification of everyday life could instituted thresholds which have discriminability of unusual conditions. Especially, when the residents have regular patterns in their life, the quantification would become easier as shown in the result of the refrigerator. The quantification of refrigerator movement will become an indicator of something unusual. However, a problem of the pattern of the refrigerator is that the using time is limited around the meal times. The feature of refrigerator cannot evaluate all the residents' condition just by itself. Even though the door movement can be influenced by the fluctuation of temperature, it will be used as the secondary indicator with some coordinate. Considering of effect of temperature, comparison of the daily fluctuation and the previous seven days can be regarded as a method of coordinate the influence for the evaluation. Absolutely, the combination of plural devices will improve the reliability. Naturally, the progress has a possibility of the error extract. However, some of the extracted sign would have the possibility to become the important alerts.

It was revealed from the interview to the residents that they were not home at the dates when some of the unusual conditions were found. Some of the signs were wrong. In brief, even when the observed value was under the threshold, it may not mean a trouble. However, the methodology in this study does not require the strict accuracy. It requires the function to extract unusual condition and evaluate the change in everyday life. Incidentally, a severe accident occurred in the 8th period. Their father fell and had his leg broken when he was going to the bathroom at midnight on May 26. The family was forced to take care of him after the accident. The door and the refrigerator movements on the next day showed different fluctuations from the usual one. The both indicators of Figure 5 and Figure 6 marked lower value than the thresholds. This unexpected accident proved that the methodology in this study has the function to extract unusual condition in everyday life.

Ultimately, the series of indicators, which is based on the correlation coefficient between the measured fluctuation and the average of the previous seven days, seems to have the enough function to extract unusual conditions. In addition, it also seems to have the function to evaluate the stability of everyday life. Thus, this series could be regarded as the Everyday Life Stability Indicator. The unit of evaluation in this study was "day". However, the dataset will be able to be divided into the "hour" unit to analyze finely. Especially, the fine dividing will be applied to the regular patterns. The unusual condition will be extracted under the regular series more precisely.

The periods of the related study in the past [6,7] are shorter than this study, however, they suggested the remote warning system by the Internet. The Integration of quantitative methodology of Everyday Life Stability Indicator into the network technology will contribute to complete the appropriate confirming system for isolated family.

ACKNOWLEDGEMENTS

We are pleased to acknowledge the long time cooperation by Mr. and Mrs. Yoshizawa and their family in Nagano city.

REFERENCES

[1] Taro S, et al., Effects of Cameras and Monitors on Caregivers' Work Stress in the Group Home, Institute of Electronics Information and Communication Engineering Technical Report, Japan, 2008, 107(555), 57–62.

[2] Roggen D, Calatroni A, et al, Walk-through the OPPORTUNITY dataset for activity recognition in sensor rich environments, *Adjunct Proceedings of the Eighth International Conference on Pervasive Computing*, 2010, Helsinki, Finland.

[3] Maki H, et al, A daily living activity remote monitoring system for solitary elderly people, Annual International Conference Proceedings of the IEEE Engineering in Medicine and Biology Society, Boston, 2011, USA, 5608–11.

[4] Dishongh TJ, McGrath M, Ben K, Wireless Sensor Networks for Healthcare Applications, Artech House, 2009, USA.

[5] Kasteren TV, Noulas A, et al, Accurate Activity Recognition in a Home Setting. Proceedings of the 10th international conference on Ubiquitous computing, Seoul, South Korea, ACM Press, 2008, 1–9.

[6] Macky K, Yoshie S, Takeshi S, Development of residential monitoring system by measuring fixtures acceleration, Proceedings of the IADIS International Conference Interfaces and Human Computer Interaction, 2011, 492–4.

[7] Macky K, Yoshie S, Takeshi S, Elderly people's health confirming by measurement of fixtures acceleration, Proceedings of the IADIS International Conference Interfaces and Human Computer Interaction, 2012, 335–8.

[8] Japan Metrological Agency, Wether data query, http://www.data.jma.go.jp/obd/stats/etrn/index.php, Japan, Accessed August 24, 2014.

New Ergonomics Perspective – Yamamoto (Ed.)
© 2015 Taylor & Francis Group, London, ISBN 978-1-138-02751-0

Considerations on diversity management for sexual minorities in the workplace

Nozomi Sato

Kinki University, Higashi-Osaka, Osaka, Japan

ABSTRACT: Diversity management in the workplace has attracted increasing interest in many organizations to overcome intense competition stemming from globalization of the market and to meet a wider range of customer needs. In addition, diversity management is conducted as part of Corporate Social Responsibility (CSR) from a human rights viewpoint. So far, major targets for diversity management have been gender (female workers), ethnicity (e.g., foreign workers), and age (older workers). However, diversity management for sexual minorities (i.e., Lesbian, Gay, Bisexual, and Transgender [LGBT]) has been avoided, partly because it is difficult to manage invisible diversity, and/or it is expected that critical issues related to sexual harassment of LGBT could occur in the workplace. However, over the past decade, practitioners and researchers have focused on the benefits of dealing with this topic. Consequently, an increasing number of organizations have implemented positive LGBT employment policies. Despite this, much remains to be considered to protect LGBT from disadvantages such as workplace discrimination and abuse that worsen their Quality of Life (QOL). A small number of studies focus on this issue. Thus, the aim of this paper is to provide a framework to broaden understanding of diversity management for LGBT in the workplace based primarily on a literature review. First, factors that possibly worsen mental health in LGBT are described. Second, challenges faced by LGBT in the workplace are delineated. Third, the roles of the organization, co-workers, and health care staff in improving LGBT's QOL is explained. Fourth, recent trends in diversity management for LGBT are highlighted. Finally, a framework for diversity management for LGBT in the workplace is provided.

Keywords: diversity management; sexual minorities; mental health

1 INTRODUCTION

With rapid globalization of the market and a wider range of customer needs, organizations face intense competition in the market. To overcome this critical situation, many organizations increasingly focus on diversity management in the workplace. Diversity management originated in a series of U.S. legislations in the 1960s and 1970s. Alongside the growth of the human rights movement, these legislations aimed to eliminate discrimination in the workplace. Furthermore, they led to the development of affirmative action and equal employment opportunity [1]. In addition to conducting traditional diversity management (i.e., eliminating discrimination in the workplace), organizations now recognize the necessity and benefits of creating a diverse workforce to resolve global and complex problems that are difficult to optimally solve through a homogeneous workforce. Successful diversity management presents potential opportunities to organizations, enabling them to gain competitive advantage, enter new markets, become more creative and innovative, and increase employee satisfaction [2].

In the workplace, core dimensions of diversity include age, ethnicity, gender, mental/physical abilities and characteristics, race, and sexual orientation [2]. Many organizations have made efforts to treat employees or applicants fairly without focusing on such dimensions. Thus, people who were excluded from the workforce have been gradually included under the protection of the law and organizations' strategies to be more productive. However, diversity management for sexual minorities (i.e., Lesbian, Gay, Bisexual, and Transgender [LGBT]) lags behind other dimensions, partly because it is difficult to manage invisible diversity, and/or it is expected that critical issues related to sexual harassment of LGBT could occur in the workplace.

For the past decade, practitioners focused on the many benefits of dealing with diversity management for LGBT. Consequently, an increasing number of organizations implemented positive LGBT employment policies. However, it was pointed out that sexual orientation remains the last to be accepted and a remaining prejudice [3]. Therefore, to develop comfortable working environments for LGBT, it is important to recognize the difficulties they face in the workplace. Until now, a small number of studies on this issue have been conducted. Thus, the aim of this paper is to provide a framework to broaden understanding of diversity management of LGBT in the workplace. To achieve this aim, the following topics are discussed in this paper: (1) Factors that could worsen mental health in LGBT, (2) challenges faced by LGBT in the workplace, (3) the role of organizations, co-workers, and health care staff in improving LGBT's QOL, and (4) recent trends in diversity management for LGBT.

2 MENTAL HEALTH STATUS OF SEXUAL MINORITIES

It was suggested that LGBT are at a higher risk for poor mental health than heterosexual individuals. Lewis conducted a meta-analysis of state- or region-based studies on sexual minorities and mental health, noting that compared to heterosexual men, gay men more frequently experienced depression, anxiety, thoughts of suicide, and other disorders [4]. Marshal et al. also conducted a meta-analysis to examine the risk of suicide and symptoms of depression in sexual minority youth. They reported higher rates of suicide and depression for sexual minority youth than heterosexual youth [5]. By analyzing pooled data from 2005 and 2007, Bostwick et al. indicated that sexual minority youth experienced more feelings of sadness, and were at a higher risk for suicide-related behavior than heterosexual youth [6].

Previous research suggests that harassment related to sexuality is associated with poor mental health in LGBT. Furthermore, victimization based on sexuality, such as verbal insults and threats regarding disclosure of sexual orientation positively correlated with psychological distress [7]. In addition, perceived discrimination toward sexual orientation was associated with emotional distress [8]. Although self-disclosure of sexuality to others plays an important role in LGBT identity development [9], it is a difficult process, as they may be targeted for harassment, discrimination, and rejection from others [10], which may lead to poor mental health. On the other hand, if they do not disclose their sexuality, which sometimes requires pretending to be heterosexual, their sense of self may weaken. This situation may also worsen mental health.

3 CHALLENGES FACED BY LGBT IN THE WORKPLACE

Many organizations have made efforts to prevent discrimination or disadvantages related to sexuality. Nevertheless, practically, LGBT still faces many challenges in the workplace, such as discrimination, unintentional practice, and negative attitudes. Previous research associated workplace discrimination with negative health conditions [11] and linked minority stressors (e.g., discrimination, expectation of stigma) to psychological distress [12]. Ragins and Cornwell associated perceived discrimination based on sexual orientation with negative attitudes toward job and career [13]. Ozeren conducted a systematic literature review on sexual orientation discrimination in the workplace, identifying "coming out" as a major theme in this area [3]. As mentioned, "coming out," which means the same as self-disclosure [14], is a serious event for LGBT. However, a positive aspect is that disclosure in the workplace is associated with higher job satisfaction and affective commitment [10, 15] and lower job anxiety [10]. Summarizing previous studies, Ozeren suggests that employees carefully evaluate the risks and benefits of coming out [3].

4 ROLES OF ORGANIZATIONS, CO-WORKERS, AND HEALTH CARE STAFF IN IMPROVING QOL IN LGBT

Research on individual differences in disclosure of sexuality suggests that it is impacted by supportiveness of the organization and co-workers. Previous research confirms that lesbian, gay and bisexual workers are more likely to disclose sexual orientation when they perceive the organization or their work groups as more supportive [10, 16] and when organizations demonstrate supportive policies and practices [13]. Social support from co-workers after disclosure also positively affects job satisfaction and commitment to organizations [17]. Organizational and co-worker support acts as a mediator in improving LGBT's QOL in the workplace. Therefore, organizations should foster a culture and climate of fair treatment by providing opportunities to learn about diversity issues and creating policies to prevent discrimination and harassment related to sexuality. Such efforts may increase the supportiveness that facilitates LGBT's perceived security and comfort in the workplace. Even if these situations realize, it is expected that LGBTs will still be at risk in terms of physical or mental health. Therefore, it is important that organizations have support systems in place, including professional/clinical staff such as counselors and occupational health physicians knowledgeable on health issues pertaining to LGBT.

5 RECENT TRENDS IN DIVERSITY MANAGEMENT FOR LGBT

Traditional diversity management as part of Corporate Social Responsibility (CSR) from a human rights viewpoint has been focused on to protect LGBT from harassment and discrimination related to sexuality in the workplace through legislation and organizational policies. Besides this, recent diversity management for LGBT emphasizes recruitment of LGBT who offer potential benefits to organizations (e.g., higher levels of critical analysis of assumptions and implications of decisions, improved creativity and innovation, etc., [2]). This trend could benefit LGBT in terms of widening the labor market from which they were excluded based on their sexuality [18]. However, excessive publicity emphasizing the acceptance of LGBT employees is evident in some organizations. This could unintentionally induce feelings of discomfort regarding using sexuality for organizational public relations, and that sexuality is put forward more than necessary against their will, even if it is not disclosed. From an LGBT perspective, organizations must consider what constitutes appropriate treatment for LGBT employees and candidates.

6 DIVERSITY MANAGEMENT FRAMEWORK FOR LGBT

Fig. 1 illustrates a framework for diversity management of sexual minorities based on the issues discussed in this paper. In this framework, non-disclosure is associated with negative outcomes regarding poor physical or mental health. However, decisions for disclosure/non-disclosure of sexual orientation should be respected in all situations.

To facilitate disclosure, as mentioned, organizations must establish a solid supportive system for LGBT and provide continuous training opportunities for employers and employees to learn about stereotyping, prejudice, and discrimination.

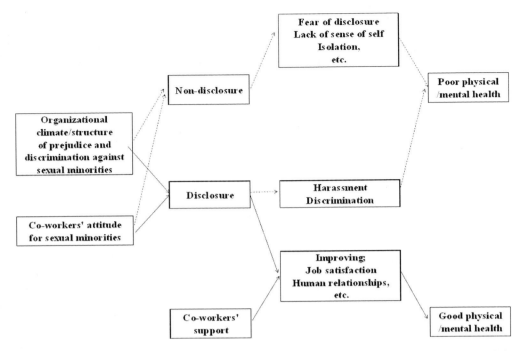

Figure 1. Brief framework of diversity management for sexual minorities. Solid line means possible positive effects and dot line means possible negative effects.

7 CONCLUSIONS

Diversity management of sexual minorities is difficult, as individual sexuality is invisible and prejudice and discrimination against LGBT remains inside and outside organizations. However, it is inevitable that organizations deal with diversity management in light of the recent increased interest in CSR and organizational needs to survive in global and complex market situations. To successfully manage diversity, organizations should prioritize establishing a secure, comfortable work environment free from prejudice and discrimination.

REFERENCES

[1] Gardenswartz L, Rowe A. Managing diversity. 3rd ed. Alexandria: Society for Human Resource Management; 2010.
[2] Hubbard EH. The manager's pocket guide to diversity management, Amherst: HRD Press; 2004.
[3] Ozeren E. Sexual orientation discrimination in the workplace: A systematic review of literature. *Procedia Soc Behav Sci* 2014;**109**:1203–15.
[4] Lewis NM. Mental health in sexual minorities: Recent indicators, trends, and their relationships to place in North America and Europe. *Health Place* 2009;**15**:1029–45.
[5] Marshal MP, Dietz LJ, Friedman MS, Stall R, Smith HA, McGinley J, et al. Suicidality and depression disparities between sexual minority and heterosexual youth: A meta-analytic review. *J Adolesc Health* 2011;**49**:115–23.
[6] Bostwick WB, Meyer I, Aranda F, Russell S, Hughes T, Birkett M, et al. Mental health and suicidality among racially/ethnically diverse sexual minority youths. *Am J Public Health* 2014;**104**:1129–36.
[7] Mustanski B, Newcomb M, Garafalo R. Mental health of lesbian, gay, and bisexual youth: A developmental resiliency perspective. *J Gay Lesbian Soc Serv* 2011;**23**:204–25.
[8] Almeida J, Johnson RM, Corliss HL, Molnar BE, Azrael D. Emotional distress among LGBT youth: The influence of perceived discrimination based on sexual orientation. *J Youth Adolesc* 2009;**38**:1001–14.

[9] Baiocco R, Laghi F, Di Pomponio I, Nigito CS. Self-disclosure to the best friend: Friendship quality and internalized sexual stigma in Italian lesbian and gay adolescents. *J Adolesc* 2012;**35**:381–7.

[10] Griffith KH, Hebl MR. The disclousure dilemma for gay men and lesbians: "Coming out" at work. *J Appl Psychol* 2002;**87**:1191–99.

[11] Bauermeister JA, Meanley S, Hickok A, Pingel E, VanHemert W, Loveluck J. Sexuality-related work discrimination and its association with the health of sexual minority emerging and young adult men in the Detroit Metro Area. *Sex Res Social Policy*. 2014;**11**:1–10.

[12] Velez BL, Moradi B, Brewster ME. Testing the tenets of minority stress theory in workplace contexts. *J Couns Psychol* 2013;**60**:532–42.

[13] Ragins BR, Cornwell JM. Pink triangles: Antecedents and consequences of perceived workplace discrimination against gay and lesbian employees. *J Appl Psychol* 2001;**86**:1244–61.

[14] Kaplan DM. Career anchors and paths: The case of Gay, Lesbian & Bisexual workers. *Hum Resour Manag R* 2014;**24**:119–30.

[15] Day NE, Schoenrade P. Staying in the closet versus coming out: Relationships between communication about sexual orientaion and work atitudes. *Pers Psychol* 1997;**50**:147–63.

[16] Ragins BR, Singh R. Cornwell JM. Making the invisible visible: Fear and disclosure of sexual orientation at work. *J Appl Psychol* 2007;**92**:1103–18.

[17] Law CL, Martinez LR, Ruggs EN, Hebl MR, Akers E. Trans-parency in the workplace: How the experiences of transsexual emplyees can be improved. *J Voc Beh* 2011;**79**:710–23.

[18] Drydakis N. Sexual orientation discrimination in the labour market. *Labour Econ* 2009; **16**:364–72.

5 *Human computer interaction*

New Ergonomics Perspective – Yamamoto (Ed.)
© *2015 Taylor & Francis Group, London, ISBN 978-1-138-02751-0*

Designing understandable vibration patterns for tactile interface

Daiji Kobayashi & Kana Takahashi
Chitose Institute of Science and Technology, Chitose, Japan

ABSTRACT: In our previous study, the way of designing the perceptual vibration patterns was researched through experiments using the mouse-type tactile interface from the vibration perception and cognition. As the results, the threshold of vibration duration and gaps between the duration were estimated statistically. Further, the requirements for designing memorable vibration pattern such as the rhythmical vibration pattern called "vibration rhythm" for elderly people were proposed from perceptibility; however, the validity of the vibration rhythms in the context of assumed use have not been cleared and therefore it is required to reveal the validity of vibration rhythm for the actual use. Then, we assumed a context of the actual use which the messages from home appliances are presented to the user by a vibrating interface. In the case of using the vibration rhythms for informing messages, the user has to identify the vibration rhythm presented and able to recall the corresponded message of the vibration rhythm they have learned before. Therefore, we conducted experiments using a custom vibrating interface presenting four vibration rhythms and research the characteristics of understandable vibration rhythm according to the following procedure. The participants were ten young individuals and ten elderly persons. First, we evaluated and selected the distinguishable vibration rhythms from seven vibration rhythms including some memorable vibration rhythms we assumed based on our previous knowledge. Then the selected vibration rhythms were related to messages from assumed home appliances of each and the participants learned the message using our custom software. After the learning, the participants tried to identify and answer the message of presented vibration rhythm. The experimental results showed that all of the participants were able to understand the messages of two vibration rhythms regardless of their age. These results suggest that the vibration rhythms are less retrieval cue of recalling the message for the participants. Therefore, it is required to the way of designing vibration patterns regardless of vibration rhythms having more retrieval cue of recalling the message in order to use the vibrating interface for the actual use we assumed.

Keywords: Tactile perception; Vibration rhythm; Tactile interface

1 INTRODUCTION

1.1 *Vibrating interface's accessibility*

ISO 9241-910 [1], framework for tactile and haptic interaction, mentions that it is important to consider the age of potential users of tactile/haptic devices, since there is a considerable decline in haptic sensitivity with age. From this viewpoint, we assumed that to explore the tactile device's usability for elderly persons should be required. As far as the vibration patterns are concerned, ISO 9241–910 suggests that the perception of an event can be enhanced by a careful choice of patterns of oscillatory bursts. Thus, the vibration patterns should be perceptibility by a wide range of ages in order to improve the vibrating device's accessibility. Therefore, designing the vibration patterns for elderly people should be considered.

For presenting more complex information by tactile interfaces, some ideas have been proposed and evaluated using linear tactile actuators [2–3]. Almost of the tactile interfaces proposed were used by wearing around the waist or attaching on the forearms or the wrists; however, our previous study found that there are Japanese senior citizens who are aversion to high-tech gizmos [4]. Thus it is preferable to touch the tactile interface for the older people rather than to wear or to attach the tactile interface on their body part. Accordingly, we made a mouse-type tactile interface with a little familiar computer mouse for the Japanese senior citizens in our previous study described later [5].

1.2 Vibration perception

Vibration perception have been considered from physiological and medical viewpoints. For instance, the relation between vibration perception threshold and age, height, and etc. has been investigated using a biothesiometer [6]. The biothesiometer which is used for measuring large nerve fiber function of patients produce the varied amplitude of vibrations. As the results, the significant factor for vibration perception threshold was age rather than sex and etc.; therefore, it is assumed that the higher amplitude of vibration or the higher vibration velocity is perceptible in other parts of the older person's body such as palms. Hence the vibration velocity as described variable v in Fig. 1 should be as high as possible for the aged to perceive the vibration patterns.

On the other hand, a minimum perceptible duration of vibration (d) as well as a minimum perceptible gap (r) between vibration durations as shown in Fig.1 may change with not only the perceptive aspect of users but also the characteristics of vibration from the vibrating device. Therefore our previous study estimated statistically such the two thresholds (the minimums of d and r) for designing perceivable vibration patterns using our custom tactile interface as described below in detail [5]. As the results, the minimum vibration duration included in the vibration pattern was determined as 50 milliseconds in round figures and the threshold of the gap was estimated at 20 milliseconds in round figures.

1.3 Processing vibration information

Although it is assumed that the memory ability among older adults varies with their individual, the sensed information could process at working memory. The model of working memory system is proposed by Baddeley and the system includes "visuospatial scratch-pad" and "articulatory loop" or "phonological loop" [7]. Although these systems are not for just tactile information, the tactual information could be processed based on higher-order non-tactual information [8]. In other words, the tactual information processing could be related to the characteristics of vibration patterns such as images from the tactile rhythms rather than the characteristics of the oscillation within the vibration patterns. Further, the memorability could be prompted by the skill of catching vibration patterns such as musical skill. In this case, the memorability means ease of recalling the vibration patterns correctly. Therefore, to introduce the musical rhythm patterns into vibration patterns could be an idea and worth a try because the most people are familiar with music and songs regardless of age and then our previous study research the rhythmical vibration pattern from the view of memorability for older people [5]. In addition, we called the rhythmical vibration pattern "vibration rhythm".

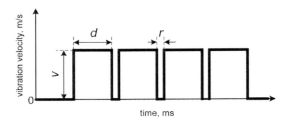

Figure 1.   Variables specifying vibration patterns.

1.4   Memorable vibration rhythm

The characteristics of memorable vibration rhythms for older persons were revealed through our previous experiment [5]. In addition, we found that the characteristics was as well as the case of young individuals. From these results, the guidelines of designing accessible vibration rhythms assumed as the followings:

- Many music notes are not included;
- Varied music notes are not included;
- The same music notes are not used repeatedly.

It is consider that the above guidelines are effective as far as when the tactile interface user memory a vibration rhythm; however, the memorability of vibration rhythms is left with following problem in the context of actual use. When the user memory multiple vibration rhythms, the user have to memory the multiple vibration rhythms and the multiple messages corresponding to the each vibration rhythms such as "You got a mail", "Your got a call", and etc. Therefore it is assumed that the user has to catch the vibration rhythms and has to identify the message represented by the vibration rhythm practically; however, the limits of the older person or young individual's capacities for identifying the multiple vibration rhythms has not been clear.

1.5   Aim of this study

In this study, the limits of the capacities for tactile interface users identifying multiple vibration rhythms were measured through experiments in order to reveal the requirements for using vibration rhythms in the context of actual use.

2   METHOD

To measure the limits of the memorability of the multiple vibration rhythms with messages, we conducted an experiment using the custom tactile interface as well as our previous study [5].

2.1   Designing Vibration mouse

The system of the tactile interface we used was simple and composed by the computer mouse (DELL USB mouse) including a vibration motor on the substrate and a high-precision analog I/O terminal (CONTEC AIO-160802 AY-USB) controlled by a personal computer using our custom software. We called such mouse-type tactile interface "vibration mouse" as shown in Fig. 2.

The vibration motor in the vibration mouse rotated within a range of 0.3–0.7V and oscillate the vibration mouse. The power voltage for activating the vibration motor was controlled using a high-precision analog I/O terminal (CONTEC AIO-160802 AY-USB) and a personal computer (DELL Vostro 1500) running Windows 7 Professional Japanese edition. The voltage applied to the vibration motor was controlled by the I/O terminal with our custom software. The wave form of the amplitude of vibration on the top of the vibration mouse was not a sine curve but very rough as shown in Fig. 3. The maximum resonant frequency of the amplitude of vibration was ranging from 74 to 116 Hz in accordance with the voltage applied to the vibration motor in the vibration mouse. In addition, the vibration mouse functioned also as the computer mouse with two buttons and a scroll wheel.

2.2   Designing perceptual vibration rhythm

Although the vibration patterns has been expressed using musical note in previous tactile studies, the musical rhythm or musical sound have used in mobile devices such as phone

cover   vibrating motor
         on the substrate

Figure 2.   Vibration mouse presenting vibration rhythm.

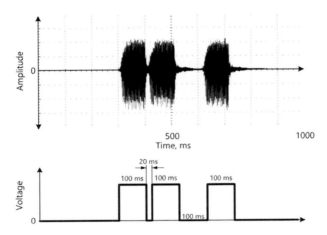

Figure 3.   Example of wave form of the amplitude of vibration on the top of the vibration mouse (above) by the voltage applied to the vibration motor (below).

ringing. However, temporal gaps among the notes is necessary in order to express the rhythm by the vibration as shown in Fig. 4 and the duration of the gaps should be as short as possible for taking no account of the gap, and then it is possible to consider that the vibration rhythm is in order. Thus the duration of the gap should be determined according to the minimum perceptible gap so that musical rhythms are made into vibration patterns. The minimum perceptible gap we determined was 20 ms as described above.

In our previous research, we found that the vibration rhythm for older persons should be slower-paced and determined that the minimum vibration duration included in the vibration rhythm is 50 milliseconds in round figures; therefore, the shortest musical note in the vibration rhythm such as sixteenth note in the vibration rhythm should be represented by vibration duration for 50 milliseconds and then the duration of eighth note is 100 milliseconds. In addition, the duration of gap was determined as 20 milliseconds as described above.

2.3   *Choosing distinguishable vibration rhythms*

To indicate some vibration rhythm having message by a tactile interface, the vibration rhythms should be distinguishable for the users; therefore, we assumed the seven vibration rhythms including two memorable vibration rhythms such as pattern-C and pattern-F from our previous study as shown in Table 1 and verified through experiment. The participants were five male and five female young individuals as well as older persons. The young ranged from 20 to 21 years of age (mean = 20.7, SD = 0.5) and the elderly persons ranged from 68 to 82 years of

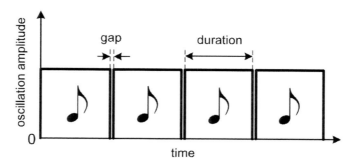

Figure 4.   Correspondence relation between a vibration rhythm represented by the vibration mouse and musical notes.

Table 1. Percentages of the participants who evaluated the vibration rhythm was easy to distinguish.

Pattern	Vibration rhythms	The young (n=10)	Elderly (n=10)
A		40	20
B		20	10
C		90	80
D		20	0
E		60	60
F		100	80
G		60	60

age (mean = 74.7, SD = 4.2). All of the participants were able to perceive the vibration stimuli from the vibration mouse and were able to catch the vibration rhythms.

In the experiment, the vibration rhythms was presented in order from the pattern-A to H as shown in Table 1 and the participant perceived the vibration rhythms using the vibration mouse. The vibration velocity was 2.3 m/s as well as the experimental condition of our previous study. The respective vibration rhythms were presented repeatedly on demand until the participant recalled and represented the vibration rhythm or gave up the trial. The way of answering the perceived vibration rhythms was describing using a simple code. However, a few elderly participants struggled to write using the code and their performances made it more difficult to recall and to represent the vibration rhythms. Thus, the way of answering for the elderly group was to sing the recalled vibration rhythm and the same researcher judged whether the sung rhythm by elderly was right or not. As the results of trials, the percentages of the participants who answered the vibration rhythm was easy to distinguish from the others as indicated in Table 1.

Although the percentages of participants recalling and answering the vibration rhythm accurately differed between the young and elderly participants, the percentages by pattern-C, E, F, and G were higher than by pattern-A, B, and D regardless of their age. Thus the memorable vibration rhythms such as pattern-C and F assumed as distinguishable among the seven vibration rhythms. Further, the vibration rhythms such as pattern-E and G composed of the same musical note were given as positive by the participants regardless of their age. This feature of the vibration rhythm corresponded the requirement of designing memorable vibration rhythm. Therefore, we found that the distinguishable vibration rhythms were easy to memory and had features such as a sense of the rhythm's speed.

From the above results, we assume the pattern-C, E, F, and G were distinguishable vibration rhythm and used in the next experiment for evaluating understandable vibration rhythms.

2.4 *Evaluating understandable vibration rhythms*

Considering an intended use of vibration rhythm, we made an application software for vibration mouse which inform two to four messages from home electric appliances such as "The plate in the microwave warm up.", "The laundry is done.", "The rice is cooked.", and "The laundry dry in the dryer.". Although these messages are informed by sounds or alarms in daily life, the application software informs the messages through the vibration mouse to the user. The vibration rhythms for informing messages were the distinguishable vibration rhythms such as pattern-C, E, F, and G as indicated in Table 1. The participants in this experiment were all participated in former experiment and the procedure of this experiment was as described below.

First, the participants intend to learn four messages relating to respective vibration rhythms using learning software as below. The software indicated a window to the participants on screen including four buttons with the messages of each and the correspondent vibration rhythm was indicated by the vibration mouse when the participant selected the button. The participants were learned the messages of four vibration rhythms of each using the learning software repeatedly until they gained the relations. After the learning, the four vibration rhythms were randomly-presented to the participants one by one and the participants tried to identify and orally answered the vibration rhythm's message presented accordingly. These four trials were repeated more two times continuously. Thus the participants tried to identify the vibration rhythm totally twelve times in first step of the experiment. The each answers by the participants were checked by an observer.

Second, after relaxing a while, the participants intend to learn the messages of three vibration rhythms including pattern-C, E, and F as well as before step. These vibration rhythm's messages were as well as the first step. After learning three vibration rhythms, the subject tried to identify the randomly-presented vibration rhythms and orally answered three times of each; therefore, the participants tried to identifying the messages of vibration rhythm nine times in the second step and their answer were checked by the observer.

Last, the participants intend to learn the messages of two vibration rhythms including pattern-C and E, and then they tried to identify the learned vibration rhythm's message following the procedure of first or the second step. Thus they answered for the trials six times.

After the experiment, the participants' comment about the trials were heard and noted by the observer.

3 RESULT

As the results of the experiment in three conditions such as identifying from four, three, and two vibration rhythms, the percentages of the participants' correct answers were calculated and compared. The percentages in those conditions are indicated in Tables 2, 3 and 4.

Table 2. Percentages of the participant's correct answers in case of identifying from four vibration rhythms.

Pattern	Correponded meaning	Vibration rhythms	The young	Elderly
C	The plate in the microwave warm up.		57	20
E	The laundry is done.		78	33
F	The rice is cooked.		87	67
G	The laundry dry in the dryer.		63	23

Table 3. Percentages of the participant's correct answers in case of identifying from three vibration rhythms.

Pattern	Correponded meaning	Vibration rhythms	The young	Elderly
C	The plate in the microwave warm up.		100	73
E	The laundry is done.		93	50
F	The rice is cooked.		90	40

Table 4. Percentages of the participant's correct answers in case of identifying from two vibration rhythms.

Pattern	Correponded meaning	Vibration rhythms	The young	Elderly
E	The laundry is done.		100	100
F	The rice is cooked.		100	100

Although we used distinguishable vibration rhythms for the evaluation experiment, Table 2 shows lowest percentages in the case of using four vibration rhythms regardless of the participant's age. In this regard, using multiple vibration rhythms for informing messages could lack of utility for actual use. As well as Table 2, Table 3 shows that it was difficult for elderly participants to understand the message from three vibration rhythms. On the other hand, Table 4 shows that using two vibration rhythms for informing messages is acceptable for the young and older persons. In these regard, some older participants complained that we had difficulty with learning the messages corresponding to the each vibration rhythms, though, we could identify the vibration rhythms.

4 DISCUSSION

From the above results, we may say that we found the requirements for designing memorable and distinguishable vibration rhythms; however, the experimental results also indicated that the messages of two vibration rhythms beating very slow and high-paced were understandable. Further, the older participants' comment described above suggest the retrieval cues for learned messages were just the rhythm's tempo. Therefore, we have to say that the way of using the vibration rhythms informing messages could be of limited use. Thus it is required to design alternative vibration patterns for informing various messages in a characteristic manner instead of vibration rhythms. In other words, we proposed a hypothesis that the vibration pattern which was a kind of musical rhythm was a characteristic manner of understandable vibration pattern; however, we have understood that the other constituent factors of retrieval cues which made the vibration patterns became more understandable should be researched from the viewpoint of cognitive process for vibration stimuli. As described above, the vibration stimuli could be processed based on higher-order non-tactual information as well as the other tactual information. Thus the vibration patterns having phonological character would be expected as the alternative vibration rhythm.

5 CONCLUSION

The experimental results indicated that only two messages could be informed using the vibration rhythms while we selected and used the distinguishable vibration rhythms. As this results suggest, we have to conclude that the present vibration rhythms having musical rhythm characteristics could be not useful for informing messages to physically unimpaired people. However, the vibration rhythms presented by vibrating interfaces could be alternative way of informing a few messages according to the user's physical characteristics such as auditory and/or visual disturbance. Therefore, we have to extend the roll of vibration patterns. In this regard, the experimental results suggest that to enhance the retrieval cues for recalling the message corresponded to the vibration rhythm could be a key to address the challenge.

REFERENCES

[1] ISO 9241–910. Framework for Tactile and Haptic Interaction; 2011.
[2] Brown LM, Brewster SA, Purchase HC. Multidimensional tactons for non-visual information presentation in mobile devices. ACM Int. Conf. Proc. Series 2006; **159**:231–38.
[3] Hoggan E, Brewster S. Designing Audio and Tactile Crossmodal Icons for Mobile Devices. In: ICMI 2007; 12–5.
[4] Kobayashi D, Yamamoto S. Usability Research on the Older Person's Ability for Web Browsing. In: Kumashiro M, editor. *Prom. of W. Ability Towards Productive Aging*, London: CRC; 2009, p. 227–35.
[5] Kobayashi D. Study on Perception of Vibration Rhythms. In: Yamamoto S, editor. HIMI 2014, Part I, LNCS 8521, Springer Int. Pub. Switzerland; 208–16.
[6] Wiles PG, Pearce SM, Rice PJS, Mitchell JMO. Vibration Perception Threshold: Influence of Age, Height, Sex, and Smoking, and Calculation of Accurate Centile Values. Diabetic Medicine 1991; **8**, 157–161
[7] Baddeley AD, Logie RH. Working Memory–The Multiple-Component Model. In: Miyake A, Shah P, editor. *Models of Working Memory: Mechanisms of Active Maintenance and Executive Control*, New York: Cambridge University Press; 1999, p. 28–61.
[8] Kaas AL, Stoeckel MC, Gorbrl R. The neural bases of haptic working memory. In: Grunwald M, editor. *Human Haptic Perception–Basics and Applications*, Basel: Birkhäuser; 2008, p. 113–30.

New Ergonomics Perspective – Yamamoto (Ed.)
© 2015 Taylor & Francis Group, London, ISBN 978-1-138-02751-0

Study on information display in driving support devices

Kimihiro Yamanaka
Tokyo Metropolitan University, Hino, Japan

Yutaro Nakamura & Mitsuyuki Kawakami
Kanagawa University, Yokohama, Japan

ABSTRACT: Recent years have brought about the development of practical operational driving support systems such as antilock braking systems and pre-crash braking systems, and informational driving support systems such as car navigation systems and driving safety support systems. There have also been advances in various safe driving support technologies, and stricter police enforcement of traffic laws. The result is a decrease in the number of traffic accidents and fatalities.

However, car navigation system displays place a load on the information processing resources of the driver, and several studies have shown the possibility of this leading to dangerous behavior. For example, the information display of a hazard prediction system can focus the driver's attention on a specific point, leading to attention bias that can lead to an accident. Issues such as this make methods of information presentation for hazard prediction systems a topic of utmost importance. Yet, to provide indices for accurately evaluating attention bias, there is little previous research in this field regarding the target and level of driver attention.

The present study investigates the relation between information display methods and attention bias, by simultaneously measuring P300 and Contingent Negative Variation (CNV) event-related potential as obtained by electroencephalography. It is known that CNV is related to prediction and P300 to task difficulty and attention level, and here we aim to develop a new evaluation method that combines these indices as a way to investigate the relation between information display method and attention bias.

Keywords: Attention bias, Car navigation system, Contingent Negative Variation (CNV), Event-related potential (P300)

1 INTRODUCTION

Recent years have brought about the development of practical operational driving support systems such as antilock braking systems and pre-crash braking systems, and informational driving support systems such as car navigation systems and driving safety support systems. There have also been advances in various safe driving support technologies, and stricter police enforcement of traffic laws. The result is a decrease in the number of traffic accidents and fatalities[1].

However, car navigation system displays place a load on the information processing resources of the driver, and several studies have shown the possibility of this leading to dangerous behavior. For example, the information display of a hazard prediction system can focus the driver's attention on a specific point, leading to attention bias that can lead to an accident. Issues such as this make methods of information presentation for hazard prediction systems a topic of utmost importance. Yet, to provide indices for accurately evaluating attention bias, there is little previous research in this field regarding the target and level of driver attention[2].

For driving safety, support systems such pre-crash braking have recently come into greater use. These systems, which are typically referred to as Intelligent Transportation Systems (ITS) and Driving Safety Support Systems (DSSS), are likely to become more widespread in the future[3-8]. Table 1 presents a list of DSSS. These systems acquire information about signaling or the direction of travel of other nearby vehicles by means of communication between vehicles or between vehicle and road, and help to prevent accidents by presenting the driver with this information as images or speech via the onboard automatic navigation system. DSSS of this type are useful for accident prevention, but researchers have pointed out that information support in automatic navigation systems could itself create risks. In particular, research conducted by the Japan Safe Driving Center has indicated that the use of automatic navigation systems could create attentional biases in drivers. Drivers must direct their attention to many targets such as oncoming vehicles, pedestrians, traffic lights, and road signs while driving. When using information-based support systems such as ITS and DSSS, the driver's attention tends to become biased toward the information support system itself and items displayed as hazards, leading the driver to give insufficient attentional resources to risk anticipation. It is therefore necessary to clarify the biasing effect of automatic navigation system usage on driver attention. However, methods for qualitatively evaluating this attentional bias have not been evaluated before now. In this study, we developed a method for simultaneously evaluating driver attention and anticipation by using Event-Related Potential (ERP), which is temporally related to attention and anticipation in humans. Our aim was to experimentally investigate whether attentional bias in drivers could be detected.

Table 1.   List of Driving Safety Support Systems (DSSS).

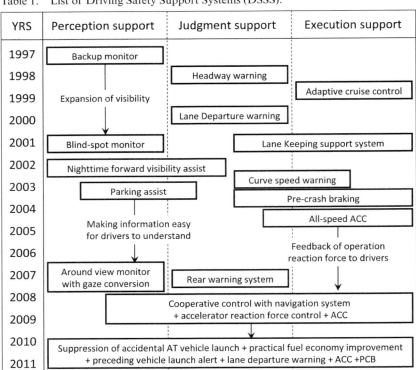

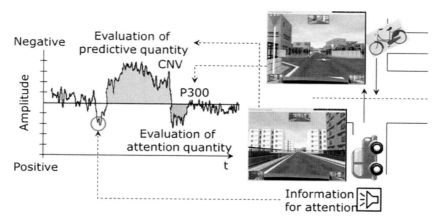

Figure 1. Driving situation and quantitative evaluation by physiological signal.

2 EXPERIMENTAL SETUP AND CONDITIONS

A participant watched a driver's-view video while seated in front of a rear projector placed 150 cm in front of the eyes. The video was created using a driving simulator scenario editor. The scenarios, which were randomly presented, showed an oncoming vehicle going straight ahead or turning right or left when the driver was entering an intersection. Participant had to anticipate the direction of travel of the other vehicle as they watched the video.

The two experimental conditions were the presence and absence of simulated support from an automatic navigation system. In the condition with support, the direction of travel of the oncoming vehicle was communicated to the participant about 5 s before the participant's vehicle entered the intersection, by displaying an image combined with a vocal warning over the driving video. these details are shown in figure 1.

The following vocal warning was created by voice-reproduction software: "A vehicle is _____. Be careful," where the blank was replaced with "going straight ahead," "turning right," or "turning left."

The participants were 10 male university students (mean age 22.55 ± 1.4 years). Each participant completed 30 trials in each condition. The number of signal-averaged measurements for ERP detection in each condition was therefore also 30. Electroencephalography (EEG) was performed using a multi-telemetry system in a shielded room, with Fz, Cz, and Pz points used for the recording electrodes in accordance with the international 10–20 system[3]. The reference electrode was placed at the right earlobe, and the ground electrode was placed at the supraclavicular region. The EEG sampling frequency was 1 kHz, with a low-frequency filter of 0.1 Hz and high-frequency filter of 50 Hz. EEG data were recorded using a data recorder.

3 EVALUATION INDICES

As shown in figure 1, the ERP components used as evaluation indices in this study were the area of Contingent Negative Variation (CNV), which is thought to be evoked by anticipation, and the amplitude of P300, which is thought be evoked by attention[9,10]. CNV area and P300 amplitude were also calculated from the baseline derived at an interval of 100 ms before initiation of information support.

In this study, CNV was evaluated as the negative potential change in the EEG in the period from initiation of information support to presentation of stimulus. If we consider the initiation of information support as the warning stimulus and the appearance of a vehicle as the imperative stimulus, then risk anticipation should be induced in the condition with information support acting as a warning stimulus, but not induced, or only minimally induced, in the condition without information support as a warning stimulus. The detected CNV area would therefore be expected to be smaller without information support with information support.

In humans, P300 is a positive potential change observed for about 250 to 500 ms when an event occurs toward which attention is directed. The amplitude increases with the amount of attention given to the event. In the experiment, we expected the condition with information support to evoke a large amount of attention toward the vehicle's path, and therefore a large P300 amplitude.

4   RESULTS AND DISCUSSION

Figures 2 and 3 show the results for CNV area and P300 amplitude at Fz. The figures show means and standard deviations relative to the baseline (information support condition) for 9 participants, with 1 participant excluded because ERP could not be properly measured due to the influence of noise from body movement.

Figure 2 shows that, contrary to our expectations, there was no difference in CNV area between the two conditions. We attribute this result to the participants predicting the appearance of a vehicle as a hazard irrespective of the presence or absence of information support, since vehicles were the only hazards appearing in the driving simulation in this experiment.

In contrast, figure 3 shows that the P300 amplitude was much larger when information support was present. A test of difference in mean values between the two conditions showed that the difference was significant ($p < 0.01$, t-test). We attributed this finding to strong attentional bias in the driver (participant) toward the direction of travel of the other vehicle when using the information support. Since the results for CNV reflect anticipation of appearance vehicle appearing, there were no differences in results with or without information support, whereas using P300 amplitude we were able to evaluate attentional bias toward the vehicle's path. These results suggest that by evaluating P300 in drivers, it is possible to detect attentional bias induced by information support systems.

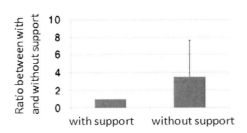

Figure 2.   Results for CNV area.

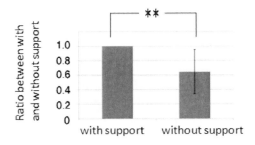

Figure 3.   Results for P300 amplitudes (**: $p < 0.01$).

5 CONCLUSION

We conducted an experiment with the aim of using ERP to detect in drivers attentional bias induced by use of ISS. The results suggest that attentional bias in drivers can be detected using P300 amplitude as an evaluation index. The effectiveness of the proposed method now needs to be experimentally verified in situations where multiple hazards are present in driving footage. We also plan to investigate information support methods in automatic navigation systems that do no induce attentional bias in drivers.

REFERENCES

[1] Policy on Cohesive Society, White paper on road safety 2012, http://www8.cao.go.jp/koutu/taisaku/h24kou_haku/pdf/zenbun/gen1_1_1_01.pdf, (4/Feb. 2013 access).

[2] Metropolitan Police Department, Statistics of metropolitan Police Department 2011, http://www.keishicho.metro.tokyo.jp/toukei/bunsyo/toukei23/pdf/toukei_H23_1.pdf, (4/Mar. 2013 access).

[3] H. Takahashi, "A Survey on Key Technologies Underlying Evolutions of Driver Support Systems in Japan", Proc. of The 10th Asia Pacific Conference on Computer Human Interaction, 2012, pp.343–351.

[4] K. Sakai, "Around View Monitor: the First Technology in the World", Journal of Automotive Engineers, Vol.62, No.3, 2008, pp.100–101.

[5] K. Shimizu and Y. Hanada, "Rear Side Obstacle Warning System (Rear Vehicle Monitoring System)", Journal of Automotive Engineers, Vol.63, No.12, 2009, pp.34–37.

[6] Y. Uemura, M. Nakamura, et al.," Development of Adaptive Cruise Control (Navigation-enabled Function)", Journal of Automotive Engineers, Vol.63, No.12, 2009, pp.52–55.

[7] M. Kobayashi and H. Uno, "The Development of Safety Technology", Journal of Automotive Engineers, Vol.63, No.12, 2009, pp.17–22.

[8] S. Usui, "Driving Assist Technology by Forward Recognition Sensor", Journal of Automotive Engineers, Vol.63, No.12, 2009, pp.30–33.

[9] Donchin. E, "Event-related brain potentials: A tool in the study of human information processing. In H. Begleiter (Ed.)", Evoked potentials and behavior, 1979, pp.13–75.

[10] H. Gray, N. Ambady, W. Lowenthal and P. Deldin, "P300 as an index of attention to self-relevant stimuliq", Journal of Experimental Social Psychology, 40, 2004, pp.216–224.

New Ergonomics Perspective – Yamamoto (Ed.)
© *2015 Taylor & Francis Group, London, ISBN 978-1-138-02751-0*

Ergonomics design requirements of human interaction with virtual display through smart glasses

Chiuhsiang Joe Lin
Department of Industrial Management, National Taiwan University of Science and Technology, Taipei, Taiwan, R.O.C.

ABSTRACT: Several computerized visual glasses with 3D capability have been seen on the Internet as commercially soon available new kinds of interfaces. These glasses claim to allow human users to interact with the menu (2D) or objects (3D) displayed virtually through the visual glasses. Since precise interaction is needed in this scenario, one fundamental question is whether the object seen by the user through the glasses will have the distance as expected. If the distance is shorter or longer than expectation, the user is not able to make an accurate click on the menu or object of selection. A literature review that looks into reports concerning inaccuracy of distance estimation in virtual reality is undertaken and current results in virtual reality distance estimation are compiled and reported in this paper. Design guidelines considering inaccuracy of interaction in virtual reality scenarios are discussed in this paper.

Keywords: Virtual reality; Computerized visual glasses; Human computer interaction

1 INTRODUCTION

Wearable computers are being advocated and promoted as one of the most popular "hitech" gadgets that are almost commercially available now due to the fast advancement of display technology. One such computerized visual glass is the Google Glass [1]. The Google Glass is a glass device that can be worn over the face just like a pair of usual eye glasses except that it is a computer and connected with the Internet. Further, it has an augmented small display that can be used to display information that is downloaded immediately from the Internet. The augmented display can be mounted over an usual eye glasses in a spot that would not block the most used visual field of the eye glasses. With this newly developed device, one is able to wear the computer display anywhere and the displayed information can help the user search information needed anytime and anywhere, navigate on the road, and even use social network.

Some other glasses take this idea further by installing two small computerized displays and the merge of the two images displayed by the glasses will become one stereo image, that is, it turns two displayed images into a virtual stereoscopic image that may be floating in front of the eyes. One such device can be exemplified by the metaPro glasses [2]. The device is claimed to be able to display information with depth. For example, a computer menu or a navigation map may be displayed in front of the eyes. With further instrumented cameras and other devices, the user is able to actually interact with the displayed menu by selecting any item on the menu. Similarly, the user can thus use the fingers and hand gestures to zoom in or zoom out on the displayed map in front of the eyes. This is truly amazing compared to the existing 3D virtual reality technology that has been available for quite some time in the form of Head Mounted Display (HMD) where similar visual and optical technology is utilized to merely display stereoscopic objects. Two additional such products can be found at the Atheer Labs [3] and the Technical Illusions [4]. These smart glasses offer interaction capabilities that are not seen with the conventional computer applications. One characteristic is the freehand interaction, that is, the user does not need to hold and hand held device or to touch any physical button during the interaction. The user simply wave the hand or move the finger (gesture control) or point to and touch the target item on the virtual menu or object (virtual touch). The former topic on gesture control has been researched in the literature for quite some time and is not within the scope of the current paper. The latter is termed as virtual touch and will be subject to discussion as to its human factors design considerations with respective to the current literature review over the potential problems that may be encountered during such virtual touch interactions with the virtual object.

2   STEREOSCOPIC DISPLAY AND VIRTUAL REALITY

One of the most important characteristics of virtual reality is its capability to create stereoscopic images of scenes using the advanced software and hardware computing technology. Most applications of virtual reality today refer to this display capability, in which users can see optical objects and scenes as if the real solid objects are being perceived, hence the name virtual reality. Therefore, creating and displaying images as real as possible to what is being simulated becomes the major goal of virtual reality developers. There are in general two types of displaying technology available in terms of hardware display devices, the Head Mounted Display (HMD) and the Projection Based System (PBS). The HMD is equipped with two small displays housed in an eye goggle with each eye sees its own image independently with the other. The HMD is usually heavy and bulky and the eyes are fully immersed in the HMD, hence it can be referred to as immersive stereoscopic display. Recently, the HMD is developed into lightweight eye glasses as in those smart glasses in which one is able to view the stereoscopic display and the real world simultaneously. This is referred to as the see-through smart glasses. The second type is developed with the projector in which the projected images are split into two images apart. The user wears a pair of lightweight glasses that has two lenses. The lenses are synced with the projected images in a way that when one image is to be seen by the left eye, that left lens is open and the right lens is closed. At the next time instant, the left lens is closed and the right lens opens to allow the other image to be seen. The switching of the left eye image and right eye image is quickly enough so that one cannot notice the switch and the two images are merged into one stereoscopic image. With this type of virtual reality display one is able to see other real world images together with the virtual image, hence, it is also capable of seeing through and maybe viewed as non-immersive. With the see-through capability, augmented reality is possible, that is, the user sees both the real images and virtual images mixed and simultaneously.

3   INTERACTION CATEGORIES IN STEREOSCOPIC DISPLAY

One major goal in the applications of these stereoscopic virtual reality display devices or smart glasses is to allow free hand interaction with the object or images rendered in the display. The interaction with the virtual images can be classified into several categories. The first category is the viewing application where the user can only see the displayed virtual images, no matter it is a static image or a dynamic image as in a movie. The user, through additional physical buttons or gesture of the fingers, may be able to control the change of the displayed content, but the user cannot actually touch or click on the images. The second category is the single touch application where the user can reach the virtual image and use the finger to touch the image. One typical application is the menu selection. The smart glasses display a virtual menu at a distance reachable from the finger and the user can touch and therefore select the items on the menu. The third category involves the planar multiple touch on the virtual image. There are more than one contact points on the displayed scene but they are all on the same plane. The fourth category deals with the planar sliding and rotation of the virtual image when the fingers are engaged in touching the images. The fifth category deals with the three dimensional manipulation of the virtual object, that is, to rotate the object and to translate the object. Most direct interaction of the fingers with the virtual display belongs to the above five movements. The techniques used to realize these interaction manipulations mostly include gesture and movement capturing, classification, and analysis in real time. The current study discusses the basic problems encountered and would need further ergonomics design considerations in coming up with high usability solutions.

4   ERGONOMICS REQUIREMENTS FOR HIGH USABILITY INTERACTION

Some general ergonomic factors that may affect users' willingness to use the smart glasses including weight, fit, aesthetics, ease of control, ... and many others, of the device. However, these general requirements related to usability of the device can and will become better under the competitive market environment. This paper focuses on some additional factors that are discussed here because the authors think they are critical factors and would not be easy to come to a good solution which the user may be satisfied, as discussed in the following.

One important fact in the above analysis of interaction with virtual environment is that the image or object is in fact "virtual", that is, it is an optical illusion that is seen by the eye. One can reach toward the virtual target but cannot actually feel the touch as the finger reaches it. There is no physical contact point possible. The lack of tactile feedback in this type of interaction makes it difficult to provide a closed loop during the human information processing. The "reach and touch" action gives no tactile feedback and one has to rely solely on visual feedback. Whether the target is reached and touched is dependent on the visual judgement by continuously comparing the position of the fingers involved with the appeared position of the virtual target. This behaviour is unnatural and can be very different from the behaviour when interacting with the real object where tactile feedback is immediately available upon touch. The movement control of the hand during the reach action toward a target has been studied extensively in ergonomics. In general, when the position of a target is determined, the movement has two stages. The first stage is a ballistic movement characterized by a predetermined trajectory formed immediately after the target position is perceived. The second stage is the homing action when the target is closely approached. The movement speed is slowed down to increase the accuracy of landing of the finger on the correct position. Finally, the touch feedback determines the end of the movement loop. With the interaction in virtual environment, there is no touch feedback. Everything depends on visual perception. Therefore, the most important question is whether the perception of position of a target is accurate in the virtual display. The second question to ask is how to end the movement loop when there is no touch feedback. What are the perceptual criteria since the decision as to a successful landing on the target is dependent on visually comparing the finger position and the perceived target position. In short, the question seems to boil down to whether the visual estimation of the position of the virtual target and the control of the movement of reaching to that position is accurate.

5   CURRENT RESULTS IN THE PERCEPTION OF TARGET POSITION
IN VIRTUAL ENVIRONMENT

The research on the usability and perceived quality of the virtual reality system is numerous. However, it is only recently that people began to address the issue of ill judgment on the perceived position or location of the object appeared in the virtual reality display. A recent review about this topic can be found in [5], [6], and [7]. In [5], through a comprehensive literature review, it was found that at least 60% of the studies in the literature showed benefits in using the stereoscopic display over the traditional non-stereo display in their tasks. The major advantage is that the stereoscopic display provides depth perception that is not available or poorer in the traditional 2D display. On the other hand, in [6], problems have also been indentified regarding the perception, cognition, usability, and visual discomfort and syndromes.

Table 1. Distance estimation in three environments.

Environment	Perceived Distance/Actual
Real	94%
Virtual reality (Stereoscopic)	76%
Augmented reality	89%

One major category of problems that has been reported is the distance underestimation in virtual environments. The review found that this effect is consistently present regardless of the measuring method used by those studies reviewed. Therefore, distance underestimation is almost certain. This phenomenon has important implications as to the questions proposed in the last section—if one is to successfully reach and touch the virtual target as displayed in the smart glasses, the distance estimation has to be accurate. However, the literature consistently shows underestimation. In another mini literature review work [8] we attempted to compile and analyze the available data reported in the literature. We collected a total of 39 studies from the last decade and perform an analysis with an aim to come up with an average value of underestimation percentage. The mean experimental data or results from each study were obtained from either the tabulated data and the text provided in the study or estimated from the figure as accurately as possible. Since each study is done with different number of subjects, the mean was weighted by the number of subjects. The overall average discrepancy is somewhere around 76% as shown in Table 1, that is, one sees an object in the virtual environment at a distance 75 cm away from the eye is in fact 100 cm away. A relatively small number of studies in the literature were conducted in the augmented reality environment and the data were analyzed similarly. It was found that the underestimation (89%) was not as bad as in the virtual reality environment. The smart glass environment is in fact more close to the augmented reality because the user can see through, that is, what is seen is a mixture of virtual images and real images.

6   CONCLUSION

This paper discusses the current issues regarding the ergonomics design considerations in the use of computerized stereoscopic displays or smart glasses. While smart glasses and stereoscopic displays are gaining market acceptance with their advocated potential applications, there are also perceptual and safety issues. This paper focuses in particular on the perceptual factor that has been found in many reported studies that there is a distance underestimation phenomenon. It has been shown that the underestimation is as much as 25% in virtual reality environment. This can have important implications in the usability of the smart glasses when there is physical interaction with the target shown in the stereoscopic display such as touching an item of a menu or reaching an object trying to manipulating it. Its effects on the interaction performance and design consideration are discussed in this paper.

ACKNOWLEDGEMENTS

The presentation of this paper is partially funded by a project from the Ministry of Science and Technology (MOST 103-2221-E-011-100-MY3).

REFERENCES

[1]   https://www.google.com/glass/start/what-it-does/.
[2]   https://www.spaceglasses.com/.

[3] https://www.atheerlabs.com/.

[4] http://technicalillusions.com/.

[5] McIntire JP, Havig PR, Geiselman EE. Stereoscopic 3D displays and human performance: A comprehensive review. Displays 35, 2014, 18–26.

[6] Renner RS, Velichkovsky BM, Helmert JR. The perception of egocentric distances in Virtual Environments—a Review. ACM Computing Surveys, published online 2013, DOI: http://dx.doi.org/10.1145/0000000.0000000.

[7] Urvoy M, Barkowsky M. How visual fatigue and discomfort impact 3D-TV quality of experience: a comprehensive review of technological, psychophysical, and psychological factors, Annals of Telecommunications, 68, 2013, 641–655.

[8] Lin CJ, Widyaningrum R, Haile Bereket, Partiwi, SG. How Accurate is Distance Estimation in Virtual Reality Displays? The 1st Asias Conference in Ergonomics and Design, May 21–24, Jeju, South Korea.

New Ergonomics Perspective – Yamamoto (Ed.)
© 2015 Taylor & Francis Group, London, ISBN 978-1-138-02751-0

Fatigue sensation of Eye Gaze Tracking System users

Yasuhiro Suzuki, Sou Yamamoto & Daiji Kobayashi
Chitose Institute of Science and Technology, Chitose, Hokkaido, Japan

ABSTRACT: Although the methods and visual user interfaces for Eye Gaze Tracking-System (EGTS) have been studied, almost of the studies discussed about the design of button or pull-down menu for the Eve Gaze Tracking-System. However, little is known about the influence of the EGTS on eye fatigue. Therefore, we designed two-types of selection screen indicating different button layout and the subjective eye fatigue using the selection screen of each types was compared. The task for participants was to select buttons on the selection screen by gazing over 400 milliseconds and the participants tried the task one thousand times repeatedly. The target button for the task was indicated on the upper center of selection screen. In the experiment, the time for selecting buttons of each indicated, number of errors, and movement distance of the viewpoint were measured. Further, the participants' fatigue sensation of eye were investigated using a questionnaire every one hundred trials. From the experimental results, it was showed that the almost of participants became conscious of that their eyes were dried among the tasks and it was assumed that the visual fatigue was caused regardless of whether of selection screen. Concretely, almost of the participants realized dryness of the eye while the participants selected buttons 400 times requiring around 13 minutes as well as they realized visual fatigue while they selected buttons 700 times requiring around 22 minutes. In this regard, we were able to estimate that less than around 10 minutes could be recommended for the users executing concentrated task as far as using our two types of custom selection screens for EGTS.

Keywords: Eye Gaze Tracking System; Interface Design; Haptic and Tactile Interaction

1 INTRODUCTION

Although the methods and visual user interfaces for Eye Gaze Tracking-System (EGTS) have been studied, almost of those studies discussed the design of button or mechanisms of pull-down menu for the EGTS [1–4]. However, little is known about the influence of using EGTS on the eyes. In order to reveal the eye fatigue, we compared the fatigue sensation of the eyes by using two different types of selection screen of EGTS.

2 APPARATUS

The EGTS, we made up was composed of three main components such as an eye-mark recorder (NAC System Technology, EMR-8), a desktop personal computer (DELL XPS 8500) running Microsoft Windows 8 professional Japanese edition and a 65-inch display (SONY BRAVIA KDL-65H920) as shown in Fig. 1(b). The display which indicated the selection screen of EGTS had 1920×1080 pixels resolution. The eye-mark recorder detected the gaze points of left eye in the coordinate space of 640×480 units as well as the participant's visual field and the coordinate data were transferred to the personal computer in real time. The coordinate data of the gaze points were constantly estimated and indicated the

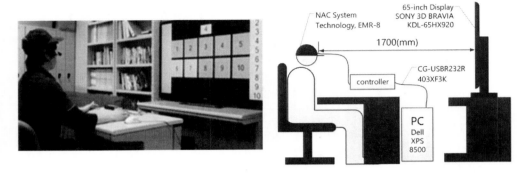

Figure 1. (a) Experimental environment (the left photograph); (b) Components of the EGTS (the right figure).

gaze point on the 65-inch display as well as the selection screen by our custom software which was made using Microsoft Visual Studio 2010 Professional edition. Hence, the appropriate visual distance between the screen and the participants was considered in order to estimate the gaze point more accurately and we decided that the visual distance was 1700 millimetres for our experimental environment as shown in Fig. 1(b). However, the calibration of the gaze points of respective participants were executed before experiment.

3 METHOD

Although the task for the participant was to select button on the selection screen, it was required to determine the appropriate minimum duration for gazing on the target button in order to accept the selection. Hence, we had tried to accommodate the minimum duration for gazing based on the some participants' satisfaction and we assumed the minimum duration was 400 milliseconds. As the result, the gazed button was accepted after the gaze point continued to be inside the same target button 400 milliseconds.

In the experiment, the target button of selection tasks was randomly indicated on the upper center or center of selection screen as shown in Fig. 2 (a) and (b). The selection screen the participants tried was Type-B (Fig. 3) following Type-A (Fig. 2). The Type-A indicated two-by-four-arranged buttons and Type-B's buttons were arranged in a circle.

The target button indicated were changed after the participants completed the selection task before and the task was repeated one thousand times. However, we gave the participant's eyes a break every 100th trial and heard about his fatigue sensation of eye in order to reveals the influence of the task using EGTS.

To investigate the eye fatigue is difficult because the eye fatigue was from various factors relating to work environment, physical characteristics, and/or psychological matters; therefore, the eye fatigue could be vary from individual to individual. Thus, we investigated the fatigue sensation that was subjective feeling about fatigue relating the selection tasks. In this regard, Working Group for Occupational Fatigue in Japan published the subjective symptoms investigation questionnaire. Hence, we arrange the questionnaire by extracting the evaluation items relating to eye fatigue and rated using nine Likert items as follows: Feel heaviness of the head, Feel dryness of the eyes, Feel headache, Feel stiff shoulder, Feel daze, Feel blurred vision, Feel visual fatigue, Feel backache, Feel bleary eyes.

The experiment was a within-subjects design and the performance data such as the time for a task and the number of errors as well as the movement distance of the gaze point for the each selection screen were measured. After the experiment, the participant's opinion about the task was heard.

Participants were seven male students ranging from 20 to 25 years of age and they had adequate vision.

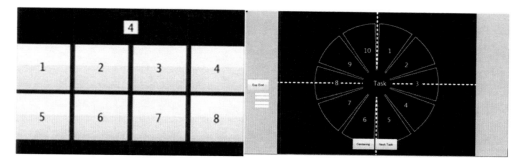

Figure 2.   Tested layout of two selection screens: (a) Type-A (left) and (b) Type-B (right).

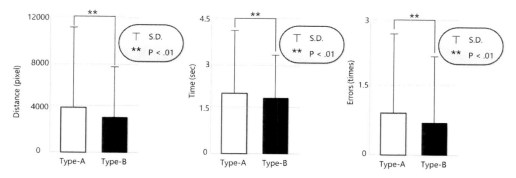

Figure 3.   (a) Movement distance of the viewpoint ($n = 7$); (b) Times per select ($n = 7$); (c) Numbers of errors per task ($n = 7$).

4   RESULT

4.1   *Performance*

The experimental results showed that the total of movement distance of the gaze point in a selection screen, the time for a task, and number of errors per task were significantly reduced in case of using Type-B as shown in Fig. 3. Thus, we found that the efficiency of the participants using the Type-B was heigher. However, some participants pointed out the fatigue sensation of eye in case of using Type-B was as well as the case of using Type-A as descibed below.

4.2   *Fatigue sensation of eye*

As the results of rated subjective eye fatigue sensation by the participant at every 100th trial, the averaged Likert scales of respective evaluation items were shown in Fig. 4. Comparing the cases between using Type-A and Type-B, it is seemed that the subjective eye fatigue sensation is not different significantly. In this regard, the participants' opinions showed that the participants became conscious of that their eyes were dried and visual fatigue was caused regardless of whether of the selection screen type as show in Fig. 4 (a) and (b). After the participants repeated the task 400 times in around 13 minutes, almost of the participants complained of feeling dryness of the eyes. Further, almost of the participants also complained of feeling visual fatigue after 700 tasks in 22 minutes. Therefore, we were able to estimate that less than around 10 minutes could be recommended for the users executing concentrated task as far as using both types of selection screens for EGTS.

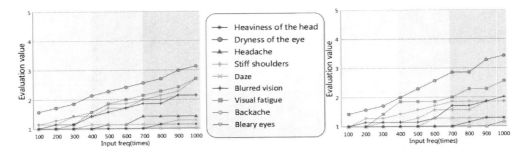

Figure 4.   (a) Subjective fatigue using Type-A; (b) Subjective fatigue using Type-B.

5 CONCLUSION

Although the EGTS has advantages of selecting menu items without using hands, the user of EGTS often feels eye-fatigue sensation. In this study, the eye-fatigue sensation of the user were tried to measure based on nine Likert items of our custom questionnaire sheet. As the result, the subjective eye fatigue sensation occurred in 13 minutes using both types of selection screen though the selection performance using both types differed significantly. However, it was not cleared that the relation between the subjective eye-fatigue sensation and the user's selection performance such as erroneous selection. Therefore, the design of less-stressful selection screen and selection method should be researched.

REFERENCES

[1] Ohno T. Quick Menu Selection Task with Eye Mark. *Information Processing Society of Japan* 1999; **40**: 602–12.
[2] Murata A, Miyake T, Moriwaka M. Effectiveness of Eye-Gaze Input System. *The Japanese Journal of Ergonomics* 2009; **145**: 226–35.
[3] Murata A, Miyake T, Moriwaka M. Effectiveness of the Menu Selection Method for Eye-Gaze Input System: Comparison between Young and Older Adults. *The Japanese Journal of Ergonomics* 2011; **147**: 20–30.
[4] Murata A, Hayashi K, Moriwaka M. Optimal Scroll Method to Browse Web Pages Using an Eye-Gaze Input System. *The Japanese Journal of Ergonomics* 2011; **147**: 127–38.

New Ergonomics Perspective – Yamamoto (Ed.)
© 2015 Taylor & Francis Group, London, ISBN 978-1-138-02751-0

Development of real-time acquisition system of UX curve

Takuro Hanawa & Nobuyuki Nishiuchi
Graduate School of System Design, Tokyo Metropolitan University, Hino, Japan

ABSTRACT: In the rapid technical development, it has been very difficult for designers and developers to make a distinctive difference of a product and a service by functionality. As a consequence, a valuable experience which is named UX (User Experience) has been needed to develop new product and service. Generally, Ethnography, Persona, UX curve, ESM (Experience Sampling Method) and DRM (Day Remember Method) have been used for UX evaluation methods. By these methods, it is able to extract the needs of the user for the development of a new product and service. However, it is necessary to extract potential needs from the users who already used it to be able to add a value for releasing product and service.

In this study, we developed the real time acquisition system of detail and long-term UX curve without simplification and bias. In the experiment, comparing the characteristics between the traditional UX curve and proposed system, the change of the subject's feelings that were a part of UX was obtained while watching a short movie. When the subject feels the change of emotion, the degree of emotion and event was inputted to the proposed system by the subject as a kind of UX curve. And after watching the movie, the subject wrote a general UX curve by handwriting. From experimental results, although the shape of curve with both methods was similar, much simplification and bias were observed in the UX curve by the traditional method.

Keywords: User experience; UX curve; Interface; UX evaluation method

1 INTRODUCTION

1.1 *Background*

In the rapid technical development, it has been very difficult for designers and developers to make a distinctive difference of a product and a service by functionality. As a consequence, a valuable experience which is named UX (User Experience) has been needed to develop new product and service. A usability evaluation method has been used for the general evaluation of products and services, and helped their improvement of the usability. However, for current diversification of products and services, the usability is still the level of the adaptation standard when we consider a charm of the product and service. Therefore, a concept of UX was also proposed to raise the charm of them. The UX can be defined as a creative usability (Fig. 1) or a good experience itself to be comfortable.

In addition, the UX is classified by the period of experience [1] shown in Fig. 2. The experience before the use of products or services is the anticipatory UX. The experience at using them is the temporary UX. The experience after the use of them is the episodic UX. Finally, the whole experience is the cumulative UX.

1.2 *Evaluation methods of UX*

Generally, Ethnography, Persona, UX curve [2], ESM [3], [4] (Experience Sampling Method) and DRM (Day Remember Method) have been used for UX evaluation methods.

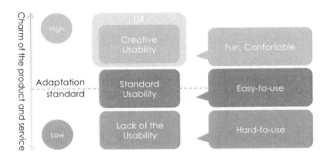

Figure 1.   What is the user experience?

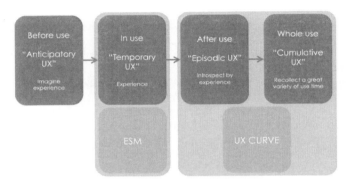

Figure 2.   Period of the user experience.

On Ethnography and Persona, a developer and a designer have to make a conjecture for the potential needs which user oneself doesn't notice. Therefore, it is necessary to use the UX (temporary UX) for the investigation of the potential needs. On UX curve and DRM, because those methods rely on the memory (episodic UX) of the user, the user's experience during a long term might be including a simplification and bias. On ESM which can acquire detailed data, the method will disturb user's experience because users are asked their experience many times during the method. Therefore, a smart and portable system that is able to input UX curve sequentially will be expected to develop. By these traditional methods, it is able to extract the needs of the user for the development of a new product and service. However, it is necessary to extract potential needs from the users who already used it to be able to add a value for releasing product and service.

1.3 Objective

In this study, we developed the real time acquisition system of detail and long-term UX curve without simplification and bias. The study was organized as follows. In Section 2, the application of UX measurement is explained in detail. Section 3 presents the verification experiment which compares the characteristics of UX curve between by the traditional method and by the proposed system. Finally, conclusions are discussed in Section 4.

2   APPLICATION OF UX MEASUREMENT (UXPLOT)

We developed the application [5], [6] of UX measurement (UXPLOT) using a smart device (iPhone5 s; Apple, Inc.). As shown in Fig. 3 (a), the interface was constituted by two buttons to input the UX curve and a graph which shows UX level (from -100% to 100%) on the

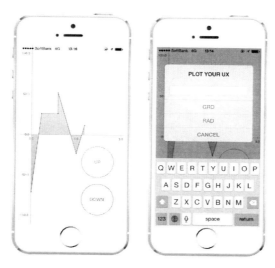

Figure 3.   (a) Interface to input UX curve; (b) Popup window to input change reason.

vertical axis and the time on the horizontal axis. After inputting the UX curve, popup window appears to input the reason of UX change and to select the gradient of UX change from two types (RAD or GRD) shown in Fig. 3 (b). GRD means gradually change of UX curve, and RAD means radically change of UX curve. Through a sequence of these operations, detail UX curve can be obtained. Data including time, UX value and the reason of UX change can be outputted as a log file.

3   VERIFICATION EXPERIMENT

3.1   *Summary and objective of the experiment*

In the experiment, comparing the UX curve between using the traditional method and the proposed method using UXPLOT, user's emotion which is a part of UX was focused. Short movie was used in the experiment to obtain the user's emotion, because Hassenzahl [7] said that user's instantaneous emotion during the interaction with a product or a service is an important term of UX.

The subjects were 10 university students who were in their 20's and mid-20's men and women. The specifications of the experimental apparatus were shown in Table 1.

The experimental procedure is as follows. The subject firstly uses the UXPLOT until getting familiar with the operation, and answers the questionnaire about general 6 emotions getting from the package and the title of the short movie. In the experiment, the subject inputs the emotion to the UXPLOT while watching the short movie whenever the subject feels any emotion change from the short movie. After watching the short movie, the subject writes the UX curve by handwriting based on the subject's memory as the traditional method, and answers the same questionnaire before watching the short movie. The 6 questions of the questionnaire are shown in Table 2.

3.2   *Experimental results*

Subjects were divided into groups based on the correlation coefficient r between the UX curve using the traditional method and the proposed method, such as high correlation ($r > 0.7$), middle correlation ($0.4 < r < 0.7$), low correlation ($0.2 < r < 0.4$), and no correlation ($r < 0.2$). In this experiment, three subjects were involved in the high correlation group, four subjects were involved in the middle correlation group, two subjects were involved in the low

Table 1. Experimental apparatus.

	Specification
Smart device to input UX curve	Apple iPhone5S (iOS 7.0.4)
Development environment	Apple Xcode 5.0.2, (Open Source Library: CorePlot)
Short movie	AsmikAce "The Red Balloon" Albert Lamorisse Remaster Edition (36 minutes)
Reproduction apparatus for short movie	Apple MacBook Pro 15-inch, Late 2011, 2.4GHz Intel Core i7, 16GB 1333 DDR3

Table 2. Questions of the questionnaire.

Question number	Question
Q1	Did you feel surprise with this movie?
Q2	Did you feel sorrow with this movie?
Q3	Did you feel nausea with this movie?
Q4	Did you feel pleasure with this movie?
Q5	Did you feel anger with this movie?
Q6	Did you feel fear with this movie?

correlation group, and one subject was involved in the no correlation group. The average of the correlation coefficient of each group is shown in Fig. 4. In this study, we focused on the high correlation group and low correlation group, and compared their results.

Fig. 5 (a) shows the UX curve using the traditional method (square dots) and UXPLOT (circle dots) of user A in high correlation group, and Fig. 5 (b) shows them of user B in low correlation group. Fig. 6 (a) shows the difference between the traditional method and UXPLOT of user A, and Fig. 6 (b) shows them of user B. Fig. 7 (a) and Fig. 7 (b) show the gradient of Fig. 6 (a) and Fig. 6 (b) respectively. And Fig 8 (a) and Fig. 8 (b) show the results of subjective evaluation about the degree of expectation before watching the short movie and the impression after watching it.

3.3 Considerations

As shown in Fig. 5 (a), although there is a small phase shifting and difference of scale of amplitude, high correlation coefficient was gotten because of the similarity of the changing points of user's emotion. On the other hand as show in Fig. 5 (b), rough curves of the graph were similar, but there was a simplification due to the changing points of the user's emotion (from 8 to 23 minutes) in the traditional method, so the correlation coefficient was low.

In the results of Fig. 6 (a) and Fig. 6 (b), there was a tendency that the difference was small in the beginning part but gradually became bigger. Moreover, as shown in Fig. 6 (b), the simplification appeared as the sharp changes in amplitude. The values of integral difference between each UX curves of Fig. 6 (a) and Fig. 6 (b) were $S_{high} = 1084.40$ and $S_{low} = 1131.93$. It is considered that these values can be used for the evaluation for the simplification.

As shown in Fig. 7 (a) and Fig. 7 (b), pulse waves which coincident with the sudden change of user's emotion appeared. Especially, in Fig. 7 (b), many pulse waves were observed in the range from 8 to 23 minutes.

Finally, as shown in the Fig. 8 (a), the degree of expectation before watching the short movie had bigger score than the impression after watching the short movie, or was the same score with it. On the other hand, as shown in the Fig. 8 (b), the impression after watching the short movie had bigger score than the degree of expectation. These results mean that the subjects of the low correlation group had more meaningful experience with the short movie than the degree of the expectation.

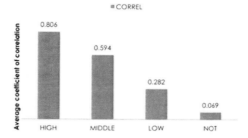

Figure 4. Correlation Coefficient of each groups.

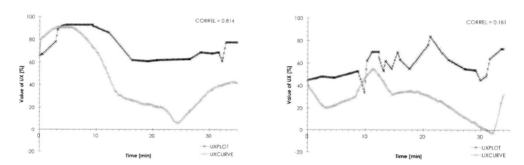

Figure 5. (a) UX Curve (High correlation group); (b) UX Curve (Low correlation group).

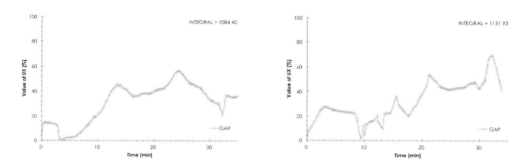

Figure 6. (a) Difference between UX curves (High correlation group); (b) Difference between UX curves (Low correlation group).

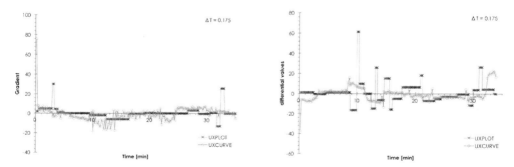

Figure 7. (a) Gradient of UX curve (High correlation group); (b) Gradient of UX curve (Low correlation group).

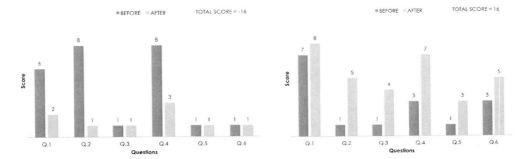

Figure 8. (a) Subjective evaluation (High correlation group); (b) Subjective evaluation (Low correlation group).

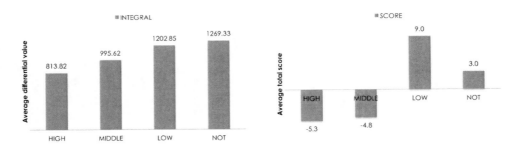

Figure 9. (a) Value of integral difference between UX curves of each groups; (b) Average of total score of subjective evaluation of each groups.

On the middle and no correlation groups, the value of integral difference between UX curves is shown in Fig. 9 (a) and the average of total score of subjective evaluation is shown in Fig. 9 (b). From these results, it is considered that our considerations can be applied to the middle and no correlation groups. In this experiment, the short term UX was discussed. In the actual use of the UX evaluation, the longer term UX will be expected, then the more number of subjects of low correlation group will be observed. Especially in the low correlation group, many simplifications were observed in the traditional method in this experiment, therefore, the UXPLOT allowed users to input the various changes of user's emotion.

4 CONCLUSIONS

We proposed the real-time acquisition system of UX curve. The experiment using short movie was conducted, and the UX curve using the traditional method and the proposed method using the application UXPLOT were compared. In the experimental results, when the users had meaningful experience, the shape of UX curve became complicate. By the traditional method, it was difficult to express the detail change of user's emotion, and the simplification and bias were observed. Therefore, the effectiveness of our proposed method using the application UXPLOT was confirmed.

REFERENCES

[1] ROTO, V., et al. User experience white paper-bringing clarity to the concept of user experience. Retrieved November, 2011, 7: 2011.
[2] KUJALA, Sari, et al. UX Curve: A method for evaluating long-term user experience. Interacting with Computers, 23.5: 473–483, 2011.

[3] LARSON, Reed; CSIKSZENTMIHALYI, Mihaly. The experience sampling method. New Directions for Methodology of Social & Behavioral Science, 1983.

[4] CONSOLVO, Sunny; WALKER, Miriam. Using the experience sampling method to evaluate ubicomp applications. Pervasive Computing, IEEE, 2.2: 24–31, 2003.

[5] SHNEIDERMAN, Ben. Shneiderman's eight golden rules of interface design. Retrieved July, 2005, 25: 2009.

[6] Apple Inc., iOS Human Interface Guidelines, http://developer.apple.com/library/ios/#documentation/UserExperience/Conceptual/MobileHIG/, May, 2013. (2013.7.20 acces).

[7] HASSENZAHL, Marc. User Experience (UX): towards an experiential perspective on product quality. In: Proceedings of the 20th International Conference of the Association Francophone d'Interaction Homme-Machine. ACM. 11–15, 2008.

New Ergonomics Perspective – Yamamoto (Ed.)
© 2015 Taylor & Francis Group, London, ISBN 978-1-138-02751-0

Study on the effect of expanding advertisement for internet users

Yoshifumi Ikemoto & Nobuyuki Nishiuchi
Graduate School of System Design, Tokyo Metropolitan University, Hino, Japan

ABSTRACT: Studies regarding web advertisement have just been undertaken. However, most of them are from the viewpoint of the advertising clients who are only considering how much attention these web advertisements get from internet users. In this study, we evaluated the web advertisement from the viewpoint of users and focused on so-called expanding advertisement which is one of the rich media advertisement. Thus, the expanding speed of the expanding advertisement was validated. We conducted two experiments. In the first experiment, we made an experimental portal site and two types of advertisements; banner advertisement, and expanding advertisement. During the experimental tasks, the gazing points of the user were obtained by the eye tracker system. After that, a subjective evaluation was also done. In the second experiment, based on the results of the first experiment, we proposed a new expanding advertisement in which we controlled the expanding speed. From the results of the second experiment, it was clarified that the user felt less stress from the proposed expanding advertisement than the normal expanding advertisement, and the gaze duration of the proposed advertisement was increased.

Keywords: Web advertisement; Banner advertisement; Expanding advertisement; Eye tracking; Gaze duration

1   INTRODUCTION

In recent years, the web advertisement market has been rapidly expanding. Studies regarding web advertisement have just been undertaken [1], [2]. Fujita et al. [3] discussed the effects of web advertisement to users and analyzed how to make better web pages. As an evaluation of the web page, a model of the user's interest in the web page was also formulated. However, these approaches are from the viewpoint of the advertising clients who are only considering how much attention these web advertisements get from internet users. On the other hand, Kishi et al. [4] researched the internet advertising from the viewpoint of user behavioral. However, in this study, the banner advertisement was only used. Rich media advertisement which has various gimmicks has been used in these days.

In this study, we evaluated the web advertisement also from the viewpoint of users and focused on so-called expanding advertisement which is one of the rich media advertisements. Then, the expanding speed of the expanding advertisement was validated. We conducted two experiments. In the first experiment, user reaction was observed in the experimental task using the banner advertisement and the expanding advertisement in a portal site. After the experiment, we conducted a subjective evaluation about the advertisement, the task and the site. Then, each result of the advertisement was analyzed and compared. In the second experiment, the expanding speed of the expanding advertisement was focused. Normal speed and proposed speed of the expanding advertisement were compared.

The present study is organized as follows. In Section 2, the expanding advertisement used in the study is explained in detail. Section 3 presents the first evaluative experiment which compares the banner advertisement and expanding advertisement. Section4 presents the second experiment which is focused on the expanding speed of the expanding advertisement. Finally, conclusions are discussed in Section 5.

Figure 1. (a) Banner advertisement and expanding advertisement before expanding; (b) Expanding advertisement.

2 BANNER ADVERTISEMENT AND EXPANDING ADVERTISEMENT

As shown in Fig. 1 (a), the banner advertisement is generally located at the upper-right of a web page and at the top page of a portal site, and is in square shape. When a user clicks the advertisement, the web page is moved to the advertising client's site.

The expanding advertisement is generally located at the upper-right of the page and at the top page of a portal site, and usually looks like a banner advertisement. The expanding advertisement expands to a larger size when the user clicks on it or places the cursor within the advertisement range. When it expands, the advertisement becomes almost the same size as the portal site, then user won't be able to see pictures and text of the portal site by the expanded advertisement.

The portal site which had similar structure of "Yahoo! JAPAN" was made and used in our experiment (Fig. 1), because the portal site of Yahoo! JAPAN has banner advertisement and the expanding advertisement, and has highest share percentage (63%; Nov. 16, 2013) in Japan [5]. Also, we made two types of advertisements: banner advertisement (350×240 pixels), and expanding advertisement (350×240 pixels, expanding to 950×600 pixels). They had quite simple contents which consist of white text and gray background.

3 EVALUATIVE EXPERIMENT OF EXPANDING ADVERTISEMENT AND BANNER ADVERTISEMENT

3.1 *Objective of the evaluative experiment*

When the expanding advertisement is expanded at an experimental task, user reaction is observed. Also, the experiment using the banner advertisement is conducted for a comparative data. From the results of the experiment, the effect of the expanding advertisement for users is validated.

3.2 *Experimental methodology*

The subjects were 10 university students who were healthy men and women from 21 to 25 years old. All subjects were familiar with the fundamental operation of the web browser and were using it almost every day. The monitor size used in the experiment was 17.3 inches (DELL; Precision M6700), and the distance between the monitor and subject's eye was about 50 cm.

The experimental task was a set of operations that were required to discover and click the designated icon or text from the top page. If the subject performed the operations in sequence shown in Fig. 2, it is associated with a high probability that the mouse cursor overlaps with the expanding advertisement range. In the actual experiment, the expanding advertisement was expanded in the task by all users. The contents of the task were different between the

Figure 2.　Path of task in our experiment.

tasks using the banner advertisement and the expanding advertisement respectively. However, the difficulty levels of the task were almost the same. Moreover, the order of the task using the banner advertisement and the expanding advertisement was changed on the subjects.

During the experimental task, the gazing points [6] of the user were obtained using the eye tracker system (Tobii X1 Light Eye Tracker). After the task, a subjective evaluation about the advertisement, the task and the portal site was done. Questions in the questionnaire were shown below:

(About the advertisement):

1. Did you see the advertisement? (Ad. Vison)
2. Do you remember the advertisement? (Ad. Memory)
3. How is your impression about the advertisement? (Ad. Impression)
4. How bothersome did you feel about the advertisement? (Ad. Stress)

(About the task):

5. How smooth could you do the task? (Task Efficiency)
6. How difficult did you feel about the task? (Task Difficulty)
7. How bothersome did you feel about the task? (Task Stress)

(About the portal site):
8. How bothersome did you feel about the site? (Site Stress)

The words enclosed in parentheses express the terms of questions, but in the actual experiment, they weren't shown in the questionnaire. Answer for the questions of "1) Ad. Vision" and "2) Ad. Memory" is Yes/No, and the other answers were five-grade evaluation.

3.3　Results and considerations

3.3.1　Subjective evaluation
Table 1 shows the results of the subjective evaluation. The score of questionnaire means that higher value is equivalent to better evaluation. Except for the results of "1) Ad. Vision" and "2) Ad. Memory", the upper value shows the average and the lower value enclosed in parentheses shows the standard deviation.

As shown in Table 1, even though more subjects recognized the expanding advertisement than the banner advertisement, all scores of the expanding advertisement were worse than the banner advertisement. As previously mentioned, in the actual experiment, the expanding advertisement was expanded in the task by all users. So, it is considered that the expanded advertisement gave botheration to subjects and made worse impression.

Table 1. Results of subjective evaluation.

	1) Ad. vison	2) Ad. memory	3) Ad. impression	4) Ad. stress	5) Task efficiency	6) Task difficulty	7) Task stress	8) Site stress
Banner Advertisement	Yes: 50%	Yes: 40%	3.4 (0.8)	3.7 (0.5)	3.8 (0.7)	3.1 (1.0)	3.3 (0.7)	3.0 (0.5)
Expanding Advertisement	Yes: 90%	Yes: 50%	1.9 (0.7)	1.9 (1.3)	3.1 (0.8)	2.8 (1.3)	3 (1.2)	2.8 (1.1)

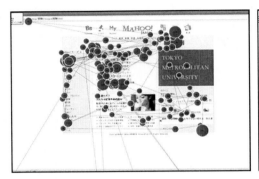 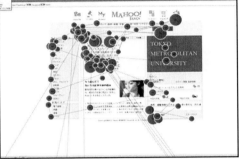

Figure 3. (a) Gazing points of banner advertisement; (b) Gazing points of expanding advertisement.

As shown in the results of "6) Task Difficulty", the difference of difficulty between the banner and expanding advertisement was very small. So the content of the tasks using the banner and expanding advertisement was the same. But as shown in the results of "4) Ad. Stress", the stress caused by the expanding advertisement was clearly more than that by the banner advertisement.

On the other hand, if the expanding time was short, some subjects didn't feel stress from the expanding advertisement. It was considered that the expanding time is related with the user's stress.

3.3.2 Locus of cursor's movement

In the task of the expanding advertisement, some users' reaction in which the cursor avoids the advertisement range were observed. And the some subjects who initially did the task using the expanding advertisement avoided the advertisement range in the task using the banner advertisement. The expanding of the advertisement was working not for catching the user's interest but for making the user avoid the advertisement.

3.3.3 Eye movement

The results of gazing points of the banner advertisement and the expanding advertisement are shown in Fig. 3 (a) and Fig. 3 (b) respectively. The size of the circle shows the gaze duration. From the results, there was no big difference on the eye movement between the tasks using the banner advertisement and the expanding advertisement, so the difficulty of the tasks was the same. The results of gaze duration of the banner advertisement and the expanding advertisement is shown in Table 2. The expanding advertisement has longer gaze duration than the banner advertisement. However, 50% of the subjects didn't remember the contents of the advertisement (Table 1). During expanding the advertisement, the gazing points were moving to the target of the next task to avoid the expanded advertisement shown in Fig. 4. The same tendency were observed between the movements of cursor and gazing points. From this result, it is considered that user's eye movement is also related with stress.

Table 2. Results of gaze duration.

	Average of gaze duration [sec.]	Standard deviation [sec.]
Banner Advertisement	1.0	1.1
Expanding Advertisement (Before expanding)	1.8	1.5
Expanding Advertisement (In expanding)	1.4	1.1
Expanding Advertisement (Total)	3.2	2.3

Figure 4. Gazing points when the expanding advertisement was expanded.

4 EXPERIMENT OF EXPANDING SPEED OF EXPANDING ADVERTISEMENT

4.1 Objective of the experiment

From the results of the first experiment, although the expanding advertisement has higher visual attraction than the banner advertisement, the expanding advertisement caused stress which was related with the expanding time. Then, we proposed the new expanding advertisement, which controlled the expanding speed, that has high visual attraction and less stress for user, and then we validated the effectiveness of it.

4.2 Experimental methodology

The new expanding advertisement was designed with the concepts below:

a. Shorten the time interrupting the user's line of sight: Don't interrupt the user's line of sight by the slow expanding speed in the beginning.
b. Keep high visual attraction: Perceive the existence of the advertisement to user by the motion of expanding of the advertisement.

Under these concepts, the expanding speed was designed shown in Table 3. The initial size of the advertisement was 360 × 240 pixels.

The experimental site, contents of the task and experimental procedure in this experiment were the same with the first experiment. But the subjects were different from the first experiment and were also 10 university students who were healthy men and women from 21 to 25 years old. All subjects were also familiar with the fundamental operation of the web browser.

Table 3. Expanding speed of proposed expanding advertisement.

	Time [sec.]	Final size [pixel]	Expanding speed [degree/sec.]
Slow expanding part	0.0–0.9	425 × 285	2
Quick expanding part	0.9–1.0	900 × 600	160

Table 4. Results of subjective evaluation.

	1) Ad. vison	2) Ad. memory	3) Ad. impression	4) Ad. stress	5) Task efficiency	6) Task difficulty	7) Task stress	8) Site stress
Proposed expanding advertisement	Yes: 80%	Yes: 10%	3.0 (0.7)	3.3 (1.1)	3.7 (1.2)	3.7 (0.5)	3.3 (1.2)	3.2 (1.1)

Table 5. Results of gaze duration.

	Average of gaze duration [sec.]	Standard deviation [sec.]
Proposed Expanding Ad. (Before expanding)	2.1	1.7
Proposed Expanding Ad. (In expanding)	0.2	0.5
Proposed Expanding Ad. (Total)	2.3	1.8

4.3 *Results and considerations*

Table 4 shows the results of the subjective evaluation of the proposed expanding advertisement. The score of questionnaire means that higher value is equivalent to better evaluation. Except for the results of "1) Ad. Vision" and "2) Ad. Memory", the upper value shows the average and the lower value shows the standard deviation. All scores of the proposed expanding advertisement were better than the normal expanding advertisement. Especially, the score of "4) Ad. Stress" was 3.3, it was almost the same with the score of the banner advertisement (3.7), and much better than the normal expanding advertisement (1.9).

The results of gaze duration of the proposed expanding advertisement are shown in Table 5. The gaze duration of the proposed expanding advertisement was longer than the normal expanding advertisement. When the cursor was moved in the range of the proposed advertisement and the advertisement gradually expanded, users directed their attention to the proposed advertisement and then gazed on it. It was clear that the proposed expanding advertisement still got the user's high attention. Additionally, eight out of ten users stopped expanding the advertisement by moving the cursor out of the range of the advertisement and didn't expand the proposed advertisement until full sized (900 × 600 pixels). From this result, it was considered that the advertisement didn't interrupt the user's line of sight.

5 CONCLUSIONS

In this study, we evaluated the expanding advertisement from the viewpoint of users and compared it with the banner advertisement. From the results of first experiment, when the expanding advertisement was expanded, the user felt stress and avoided the advertisement. And from the results of the second experiment, it was clarified that the user felt less stress from the proposed expanding advertisement which controlled the expanding speed, and had high attention to it.

REFERENCES

[1] Atsushi Toda. Internet Advertising in Japan-Historical Changes and its Essence. Japan Information-Culturology Society 17(1), 49–54, 2010.

[2] Akihiro Fujita. A new model of advertising business on the Web: Possibility of new advertising media. Japan Society of Information and Knowledge 16(4), 23–32, 2006.

[3] Hironori Fujita, Sennosuke Kuriyama, Toyokazu Nose, Sadaya Kubo. A Study on the Effects of Advertisement through the Internet. Japan Industrial Management Association 51(6), 587–593, 2001.

[4] Kouji Kishi, Akira Sakamoto, Yasuhisa Sakamoto. Internet Advertising Effectiveness Study from a User Behavioral Viewpoint. Information Processing Society of Japan 98(22), 1–6, 1998.

[5] Japanese share of search engine and portal site, http://www.webcreate.ga-pro.com/search.html, accessed 10 August, 2014.

[6] Ryoko Fukuda. An experimental consideration on the definition of a fixation point. Ergonomics Research 32(4), 197–204, 1996.

New Ergonomics Perspective – Yamamoto (Ed.)
© 2015 Taylor & Francis Group, London, ISBN 978-1-138-02751-0

Usability evaluation of interfaces: 2D & 3D input interfaces both with 2D display

Kenshiro Fujii & Nobuyuki Nishiuchi

Graduated School of System Design, Tokyo Metropolitan University, Hino, Japan

ABSTRACT: In recent years, various input interfaces such as the mouse, touch panel and the Three-Dimensional (3D) motion capture have been developed by the rapid progress of information and communication technology. However, in general operation of PC such as the click and scroll, it is thought that the operability of these interfaces is not sufficient in many situations and it is hard to use the 3D input interface combined with the 2D display. It is considered that the different dimensions between input and output interfaces have affected the operability.

In this research, the usability of Two-Dimensional (2D) and Three Dimensional (3D) input interfaces both combined with 2D display was evaluated. In our experiment, using these interfaces, the operation time was measured while the click task and the scroll task as basic operations of PC, and the subjective evaluation about the satisfaction and fatigue was also conducted. In the scroll task, there is a dynamic graphical reaction against the user's operation. So the scroll task includes a new element compared with the traditional click task. On the click task, the data of the experimental results was modeled using the conventional Fitts' Low. From the results of the click task, the efficiency of the touch panel was the best and the 3D motion capture was the worst. In the scroll task using the touch panel was also the best. In conclusion, it was clarified that the touch panel had high usability although there still was the need for improvement on the fatigue. On the other hand, the 3D motion capture needed to be significantly improved on the usability and fatigue.

Keywords: Input interface; Usability evaluation; Fitts' low; Mouse; Touch panel; 3D motion capter

1   INTRODUCTION

In recent years, various input interfaces such as the mouse, touch panel and the Three-Dimensional (3D) motion capture have been developed by the rapid progress of information and communication technology [1]. The mouse which used to be the most popular as a Two-Dimensional (2D) input interface has been replaced by the touch panel because it is easier and more convenient to use. On the other hand, the 3D motion capture as a new 3D input interface of PC has also been expanded in recent years. The 3D input interface allowed us to operate PC by the 3D motions including abscissa, ordinate, and applicate motion. However, in general operation of PC such as the click and scroll, it is thought that the operability of these interfaces is not sufficient in many situations and it is hard to use the 3D input interface combined with the 2D display. It is considered that the different dimensions between input and output interfaces have affected the operability. There were the researches [2], [3] which evaluated the satisfaction and effectiveness of different dimensional input interfaces. However, the efficiency of the interfaces was not evaluated sufficiently in these researches.

In this research, the usability of the 2D and 3D input interfaces both combined with the 2D display was evaluated. In our experiment using these interfaces, the operation time was measured with the click task and the scroll task as representative operations of PC, and the subjective evaluation about the satisfaction and fatigue was conducted. The present research is organized as follow. In Section 2, the conducted experiment is explained in detail. Section 3 presents the results and considerations of the experiment. Finally, conclusions are discussed in Section 4.

2 EXPERIMENT FOR USABILITY EVALUATION

2.1 *Objective of the experiment*

The usability of the 2D and 3D input interfaces both combined with 2D display is evaluated. The mouse and touch panel is used as the 2D input interface, and the 3D motion capture is used as the 3D input interface. Through the experiment, the operation time of the click task and the scroll task is compared between these interfaces from the view point of the efficiency, while satisfaction and fatigue are also evaluated by the questionnaire.

2.2 *Experimental setup*

In our experiment, the mouse (SANWA SUPPLY; MA-117HR) and the 21.5-inch touch panel (DELL; S2240Tb 1920 × 1080 pixels) were used as the 2D input interface, and 3D motion capture (LEAP Motion; Leap motion Controller [4]) was used as the 3D input interface. And the touch panel was also used for 2D display as shown in Fig. 1.

Using the 3D motion capture, the position of cursor can be controlled by the position of user's finger at the virtual panel which is not visible, placed vertically above the device of 3D motion capture and parallel to the subject. The click and drag operation can be done by moving user's finger toward the monitor. The device of 3D motion capture was placed at 20 cm distance from the monitor (Fig. 1(c)). Before the experiment, each input interface was used by the subjects until the subjects became familiar with its operation of them. At that time, the position and angle of the display were adjusted by each subject.

2.3 *Experimental procedure*

The subjects were five university students (21–25 years old) who were familiar with the mouse and touch panel operation, and were beginner users of the 3D motion capture. All subjects were right-handed and practiced the operation of the 3D motion capture. We conducted two kinds of task. As a basic operation of PC, the click task and scroll task were performed by subjects with their right hand. After the experiment, the questionnaire as a subjective evaluation was conducted. The details for each task are as follow:

Fig. 2 shows the image of the click task. When the subject clicks the start button at the center of the monitor (Fig. 2(a)), one target icon appears at random direction and distance from the start button, and the icon size is also changed at random (Fig. 2(b)). The subject clicks the target icon promptly, and then the start button in the center of the monitor appears again like Fig. 2(a). The sequence of these operations is defined as one set, the operation time

Figure 1.   Evaluated input interfaces (a) mouse; (b) touch panel; (c) 3D motion capture.

of this set is measured. One subject performed 40 sets and used three kinds of interface, so the total is 120 sets.

The directions of the icon from the start button are eight directions such as upper-lower, left-right and oblique directions. There are three distances of the icon from the start button such as 22.1 mm, 57.6 mm, 90 mm, and there are three sizes of the target icon such as 12.2 mm (48 pixels), 21.6 mm (85 pixels), and 30.5 mm (120 pixels) squared. The distance and the size of the target icon are set up based on the index of difficulty of the Fitts' low [5], [6] such as 1.0, 1.5, 2.0, 2.5, and 3.0.

Fig. 3 shows the image of the scroll task. In the scroll task, there is a dynamic graphical reaction against the user's operation. So this task includes a new element compared with the traditional click task. After the subject clicks the start button at the center of the monitor (Fig. 3(a)), the subject scrolls the screen to the indicated direction until the target icon appears, and then the subject clicks the icon promptly (Fig. 3(b)). The direction of the scroll has been indicated beforehand to the subject. In the scroll task using the mouse, the mouse wheel was not used for the scroll operation but a scroll bar. The sequence of these operations is defined as one set, the operation time of this set is measured. One subject performed 32 sets and used three kinds of interface, so the total is 96 sets.

The directions of the icon from the start button are eight directions such as upper-lower, left-right and oblique directions, and the distance of the icon from the start button are two kinds such as short distance: 254 mm (1000 pixels) and long distance: 635 mm (2500 pixels). The size of the all target icons is 25.4 mm (100 pixels) squared.

After the each task, the subjective evaluation about the general usability and fatigue was conducted.

Figure 2.   (a) Start button of click task; (b) target icon which appears at random direction and distance from the start button.

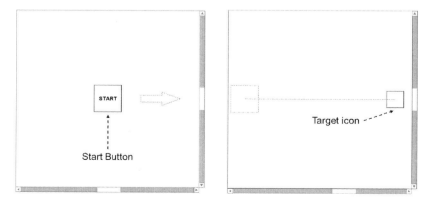

Figure 3.   (a) Start button of scroll task; (b) target icon for scroll task.

3 EXPERIMENTAL RESULTS AND CONSIDERATIONS

3.1 *Click task*

On the click task, the data of the experimental results was modeled using the conventional Fitts' Low [6].

$$mt = a + b \log_2(d/s + 1.0)$$

where mt is the time to move from the start button to the target icon, and d is the distance from the start button to the target icon, and s is the size of the target. The parameters a and b represent empirical constants determined by the linear regression. The term of $\log_2(d/s + 1.0)$ is the index of the difficulty ID. The determined parameters a, b, and the r^2 of the linear regression between mean mt and index of difficulty ID are shown in Table 1. They are calculated by using all data derived from directions. These results show that the efficiency of the touch panel was the best and the 3D motion capture was the worst. It is considered that 3D motion capture requires more time for the click operation which is the motion towards the monitor while keeping the vertical and horizontal position.

The operation times for each direction are shown in Fig. 4 wherein each line shows the ID. In the results of touch panel and 3D motion capture, the operation times for lower side and lower right directions are longer than the other directions as shown in Fig. 4(b), (c) because it was observed that the target icon was sometimes hidden by the user's right hand.

The subjectivity evaluation of the click task is shown in Fig. 5. On the terms of "Operability" and "Satisfaction", higher score means better evaluation. While on "Fatigue", higher score means feeling more fatigue. The subjects felt that 2D input interfaces were easy-to-use, but 3D motion capture was not. In addition, subjects felt more fatigue in using the touch panel and 3D motion capture than the mouse.

Table 1.   Obtained parameters of Fitts' Low.

	Mouse	Touch panel	3D motion capture
a	0.39	0.46	1.07
b	0.18	0.09	0.46
r^2	0.83	0.59	0.62

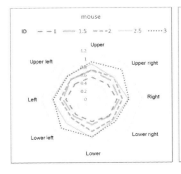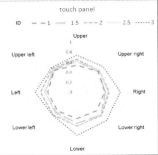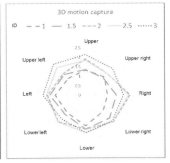

Figure 4.   Operation time of each direction of the click task [sec.] (a) mouse; (b) touch panel; (c) 3D motion capture.

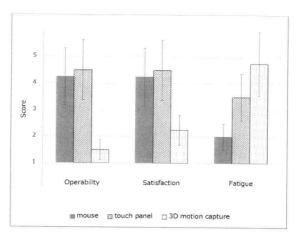

Figure 5. Subjective evaluation of the click task.

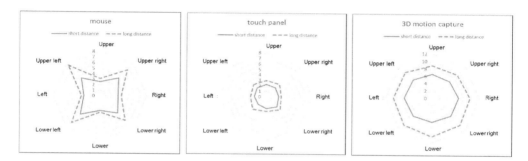

Figure 6. Operation time of each direction of the scroll task [sec.] (a) mouse; (b) touch panel; (c) 3D motion capture.

3.2 Scroll task

The operation times for each directions are shown in Fig. 6 in which the each line shows the distance from the start button to the target icon. As for the case of the click task, the operation time of the touch panel was the best and the 3D motion capture was the worst.

In the scroll task using the mouse, the operation time for the oblique directions was about two times longer than the other directions as shown in Fig. 6(a) because the subjects had to move both the horizontal and vertical scroll bars for the oblique directions. If the mouse wheel is used in the scroll task for the upper-lower direction, a part of result will be improved.

In the scroll task using the touch panel and 3D motion capture, as for the case of the click task, the target icon was sometimes hidden by the user's hand. However, the influence of it was smaller than the click task as shown in Fig. 6(b), (c) because the subjects recognized beforehand which direction the target icon is.

The operation time of the 3D motion capture was longer than the mouse and touch panel. Because, it was often observed that the subject mistakenly scrolled to the former position caused by the perception error of applicate motion.

The subjectivity evaluation of the scroll task is shown in Fig. 7. The subjects felt fatigue as for the case of the click task, however, the difference between the touch panel and mouse for "Operability" and "Satisfaction" was clearer.

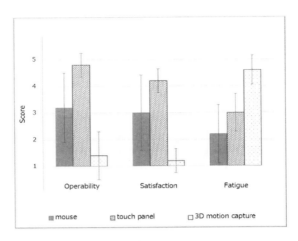

Figure 7.   Subjective evaluation of the scroll task.

4   CONCLUSIONS

We conducted the usability evaluation for the 2D and 3D input interfaces both combined with 2D display. In our experiment, the click task and the scroll task were conducted using the mouse, touch panel and 3D motion capture. From the experimental results, the touch panel has high usability although there still is the need for improvement on the fatigue. On the other hand, the 3D motion capture needs to be significantly improved on the usability and fatigue.

In recent years, the 3D display has been also widely used [7]. The 3D display offers us the more realistic experience than the conventional 2D display. As a next step, the combination between 2D & 3D input interfaces both with the 3D display will be evaluated.

REFERENCES

[1] Ministry of Internal Affairs and Communications "Communications Usage Trend Survey", http://www.soumu.go.jp/johotsusintokei/statistics/pdf/HR201000_001.pdf (August 1, 2014 access).
[2] Chih-Hung Ting, Teng-Yao Tsai. Human Factor Research of User Interface for 3D Display. Human-Computer Interaction, Part IV, LNCS8007, 506–512, 2013.
[3] Shunichi Tano, et al. Is 3D Space Useless for Design Work? -3D Sketch System Based on New "Life-sized and Operable" Concept-. IEICE Technical Report, IE2011–154, MVE2011–116(2012–3), 121–126, 2012.
[4] Leap motion controller, https://www.leapmotion.com/ (August 1, 2014 access).
[5] Hirokazu Iwase, Atsuo Murata. Extending the Model of Fitts' Law to a Three Dimensional Pointing Movement Caused by Human Extremities. IEICE TRANSACTIONS on Information and Systems 85(11), 1336–1346, 2002.
[6] MacKenzie, I. S. A note on the information-theoretic basics for Fitts' law. Journal of Motor Behavior, 21, 323–330, 1989.
[7] NPD Display Search, 3D Display Technology and Market Forecast Report, 2012.

New Ergonomics Perspective – Yamamoto (Ed.)
© 2015 Taylor & Francis Group, London, ISBN 978-1-138-02751-0

Selecting a function by how characteristic shapes afford users

Makoto Oka
Tokyo City University, Setagaya, Tokyo, Japan

Masafumi Tsubamoto
YAHOO! Japan, Minato, Tokyo, Japan

Hirohiko Mori
Tokyo City University, Setagaya, Tokyo, Japan

ABSTRACT: In this study, we focused on shapes in accordance with the affordance theory. We would like to propose as a substitute for a mouse a new interface that enables humans to instinctively select functions. Instead of making the optimally shaped devices for each function of computers, the aim is to instinctively select functions with a minimum device. To this end, we verify what shape of devices would enable humans to imagine and select all the functions. The present study takes a close look at "how devices are held"; and we conduct experiments, focusing on musical instruments that are held in different manners.

Keywords: affordance theory; select function; menu

1 INTRODUCTION

Humans select shapes that are suitable for specific functions. For example, how do humans use sticks? They sometimes write characters and pick up nuts with sticks, but do not use the latter as chairs. As such, constrained by the shape and size of things, humans subconsciously narrow down their functions. As time goes by, originally simple-shaped tools become fragmented depending on their respective functions, and keep evolving until they become the most appropriate shape.

On the other hand, taking a close look at computers, one may notice that they have evolved in the opposite direction of the reality. For example, one computer has multiple functions, such as "drawing pictures", "writing characters", and "calculating". These functions are expressed as icons and hierarchically structured menus, and manipulated with a mouse. Computers are distinct from the history of humans that have essentially conceived functions based on shapes. Humans cannot imagine the usage of mice that they directly touch.

The limitations that humans subconsciously have with respect to shapes are called the affordance theory [1]. Some alternative shapes to a mouse prepared with the use of affordances must enable humans to instinctively select and manipulate functions. Following the current hierarchical menu structure of computers, the phase of selecting the shape of such an alternative mouse would correspond to the first layer, and the phase of holding it in different ways would determine the second layer. Research on such "holding" interactions is underway.

We call the "possibility of behavior constrained by the limitations that humans subconsciously have with respect to shapes and sizes" an affordance. To maximize the use of affordances, we would like to propose as a substitute for a mouse a new interface that enables humans to instinctively select functions. Instead of making the optimally shaped devices for each function of computers, the aim is to instinctively select functions with a minimum device. To this end, we verify what shape of devices would enable humans to imagine and select all the functions. The present study takes a close look at "how devices are held"; and we conduct experiments, focusing on musical instruments that are held in different manners. The study verifies whether users can identity a wider range of musical devices when they are given the opportunity to select and combine appropriately shaped devices for many musical instruments, compared to a case in which they are allowed to use a device of one single shape.

2 RELATED WORK

Taylor proposed a function selecting method called "Grasp-Recognition" [2]. He made a device equipped with 72 touch sensors on its surface, displays on the front and back, and an acceleration sensor (Fig. 1). Those 72 touch sensors detected the points where fingers were touching the device. He examined how the subjects held the device in cases of a camera, a cell phone and a music player, as examples. He extracted data from 13 subjects, and analyzed the recognition rates by machine learning. From the way they held the device, 70% of the recognition rate was obtained for each of a camera, a cellphone, and a music player. Taylor revealed that grasping could be one of the guidelines in selecting functions.

There have been studies of selecting the best functions by analyzing the grasping of a single shape of a device with multiple functions added. However, one single shape of a device limits the number of functions available to select from. We think that, thinking in the hierarchical menu, there would be too many functions in one hierarchy to select the functions by ways of grasping.

If we think about things that have evolved into the best shapes, it is difficult to think a knife to have evolved from a ball, or a glass to have evolved from a stick. If we trace things back to their origins, it is unthinkable that they all go back to one shape. Based in the affordance theory, it can be thought, by using a hierarchy of 'shape' before the hierarchy of 'grasping', we can increase number of functions. We think that combining objects of multiple shapes expands the range of functions and enables intuitive selections of functions.

3 PROPOSED SYSTEM

We made the system with which we can presume the function that a user selected from the way for the user to hold a device. We prepared a total of five kinds of devices: cylindroid, cube, cone, thick and long stick, and short and thin stick (Fig. 2). We defined the choices of functions as 11 kinds of musical instruments. From the shape of a device the user selected, the way an user hold it, and the way the user take a posture with it, the system can presume the function (of a musical instrument) that the user intended to play.

Figure 1.   Graspable device (The Bar of Soap [2]).

Figure 2. The devices for experimental system (a) cube; (b) cone.

The system can detect where ten fingers touch a device to let us know how a user hold the device. Furthermore, the device can detect whether the palm touches a device or not in order to distinguish when a user hold the device by the finger tips and when the user does in the whole palm. This system can measure the way a user holds a devise by detecting how ten fingers and palm touch the device. We call this digital contact information. This system measures the angle of three axes of a device for the ground in order to know how a user holds the device for his/her body. We developed five kinds of devices equipped with various sensors: cylindroid, cube, cone, thick and long stick, and short and thin stick. We also developed a bimanual glove-type device. Each device can be connected to a PC through Arduino.

In this study, we replaced the selection of a musical instrument which causes a large change in the way to hold it with the matter of the functional selection in order to facilitate the detection of the ways a user hold the device and take a posture with it. The functions to be set as choices are 11 kinds of musical instruments: violin, guitar, violoncello, trumpet, saxophone, flute, piccolo, clarinet, recorder, ocarina, and harmonica.

We estimated the musical instrument which a user intended to play using k-Nearest Neighbor Algorithm from the device (shape) a user selected, the ways the user hold it and take a posture with it. We applied the KStar of Weka3.6 as k-Nearest Neighbor Algorithm. Study subjects were 30 university students. We did not require that they had played the 11 musical instruments; however, we required that they knew them as premises. When a subject intended to play each musical instrument, we acquired selected-device information and the ways the user hold it and take a posture with it as data. We utilized the acquired data for learning data and inspection data using Cross-validation.

4 EVALUATION EXPERIMENT

The subjects hold a device to intend to play the musical instrument which experimenter specified. The subjects could voluntarily select any devices among five kinds shaped devices. They were allowed to select multiple devices. At that time, we measured the ways the user held it and took a posture with it at the same time. Also, we carried out a similar experiment using only one device (the stick and long stick) as a control group in order to know the effectiveness of the shapes. In either case, we discarded those data when a subject did not know a musical instrument which was appointed by experimenter.

5 RESULT

5.1 Selected devices

When multiple shapes of devices are used, some instruments had divided tendencies of shape selections. In case of the violin, the selections of shapes were divided into two. One was cylindroid and the other was long stick.

307

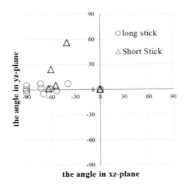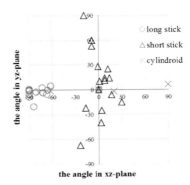

Figure 3. The result of device angle (a) the clarinet; (b) the violoncello.

In the case of the guitar, some subjects picked one device, the long stick; and others picked two devices, the long stick and the cylindroid, to express the instrument. The combination of the long stick and the cylindroid were thought to be more accurate for the shape of the guitar, but as it restricted the way of holding, more subjects picked the long stick only to express the instrument. For the cello, most of the subjects used two devices; the long stick and the short stick. Only one device was picked to express; the saxophone, the flute, the piccolo, the clarinet, the recorder, the ocarina and the harmonica. The long stick was used by the saxophone and the clarinet, the short stick was used by the flute, the piccolo and the recorder, the cylindroid was used by the ocarina, and the cuboid was used by the harmonica.

5.2 The way of taking a posture

We understood that players took a similar posture for many musical instruments. When human beings stand straight, we think about xyz space which consists of the right hand direction as x-axis, the head top direction as y-axis, and the forward direction as z-axis. In Figure 3, "the angle in the xz-plane" is defined as the angle between x-axis and the device in the xz-plane. We understood that clarinet turns to the minus direction of the y-axis (Fig. 3(a)). We understood that in the case of violoncello, the long stick turned to the minus direction of the y-axis and the short stick was present in an xz-plane (Fig. 3(b)).

5.3 Classification results

From the data obtained by the experiment, we classified musical instruments and calculated precision, recall, and F-measure. Table 1 show identification results when single device was selected. Table 2 shows identification results when multiple devices were selected.

The F-measure was low of 38.7% when single device was selected. As the reason why the F-measure was low, it can be explained by the misclassification between flute and piccolo, and that among saxophone, clarinet and the recorder. These musical instruments were misclassified each other, which resulted in lower F-measure. This was coincided that some subjects said, "I cannot distinguish musical instruments each other" when we carried out this study.

The F-measure was high of 64.1% when multiple devices were selected. By these facts, it was understood that the precision of the functional selection using multiple musical instruments was higher than that using single one.

6 DISCUSSION

As for the piccolo and clarinet in the table 2, F-measures of them resulted in low values of 25.0%. We found out that in these two musical instrument data, such the data that some subjects held them in obviously wrong way were included. Therefore, we removed outliers and then carried out classification again. Table 3 shows the results. The F-measures in the table 3 improved to 71.5%.

This shows that it is necessary for us to implement modeling after we increase the number of the data and clean learning data when we really use the data as input system.

Table 1. Identification result of single device classification.

	Precision	Recall	F-measure
violin	0.571	0.640	0.604
guitar	0.615	0.615	0.615
violoncello	0.583	0.824	0.683
trumpet	0.323	0.385	0.351
sax	0.300	0.261	0.279
flute	0.500	0.391	0.439
piccolo	0.067	0.100	0.080
clarinet	0.000	0.000	0.000
recorder	0.294	0.370	0.328
ocarina	0.571	0.462	0.511
harmonica	0.176	0.120	0.143
average	0.387	0.396	0.387

Table 2. Identification result of multiple device classification.

	Precision	Recall	F-measure
violin	0.824	0.519	0.636
guitar	0.828	0.857	0.842
violoncello	1.000	0.526	0.690
trumpet	0.654	0.630	0.642
sax	0.429	0.545	0.480
flute	0.500	0.667	0.571
piccolo	0.286	0.222	0.250
clarinet	0.217	0.294	0.250
recorder	0.633	0.679	0.655
ocarina	0.815	0.815	0.815
harmonica	0.800	0.769	0.784
average	0.670	0.634	0.641

Table 3. Identification result of multiple device classification (The removal of outliers).

	Precision	Recall	F-measure
violin	1.000	0.444	0.615
guitar	0.842	0.842	0.842
violoncello	0.750	0.429	0.545
trumpet	0.733	0.647	0.688
sax	0.500	0.444	0.471
flute	0.464	0.650	0.542
piccolo	0.250	0.286	0.267
clarinet	0.545	0.667	0.600
recorder	0.800	0.923	0.857
ocarina	0.933	1.000	0.966
harmonica	1.000	1.000	1.000
average	0.749	0.717	0.715

7 CONCLUSION

In this study, we paid attention to functional selection using the shape based on an Affordance Theory. Using the multiple devices with different shapes, we tried to carry out the functional selection by the way the subjects held them. We narrowed down the function to musical instrument selection and set the 11 kinds of representative musical instruments as the study targets. When we compared the recognition rates between the cases using single device and using multiple devices, the latter rate was 71.5% which was higher by about 33 points than that of the former rate. It can be said from these results that we could express the intention of the subjects intuitively by preparing for multiple devices.

8 FUTURE WORK

It may be planned for us to carry out a new study by changing the classification algorithm without changing the data of the shapes, of the ways of holding the devices and taking a posture with them. We will study on such a hybrid technique as carry out the classification of the way subjects hold a device after do the classification among shapes, because it is expected that impact of shapes is strong.

We carried out the study by replacing the functional selection with the selection of a musical instrument this time; however, it is necessary for us to try whether we can perform a study by replacing the functional selection of software with shape selection of it in the future.

REFERENCES

[1] James J. Gibson: The Theory of Affordances: "Perceiving, Acting, and Knowing", John Wiley & Sons Inc. (1977).
[2] Brandon Taylor: "Grasp Recognition as a User Interface", CHI '09 Proceedings of the 27th international conference on Human factors in computing systems, pp. 917–925, (2009).

6   *Current issue and ergonomics approach*

New Ergonomics Perspective – Yamamoto (Ed.)
© 2015 Taylor & Francis Group, London, ISBN 978-1-138-02751-0

The influence of cooling forearm/hand on maximal hand grip strength and the time needed to reach the maximal strength

Yuh-Chuan Shih & Yue-Jin Tsai

Department of Logistics Management, National Defense University, Taipei City, Taiwan

ABSTRACT: This paper intended to examine the influence of cooling forearm/hand on grip MVC and the associated time needed to reach the MVC (denoted as T_{MVC}). Twenty volunteers, including 10 males and 10 females, were recruited for this study. All were right-handed, healthy, and free of musculoskeletal disorders in the upper extremities. The apparatus and materials used included a water bath, a submersible cooler, a digital thermometer and hygrometer, a digital 4-channel thermometer, and a grip gauge with a load cell. The grip span of the grip gauge was set at 5 cm and the handles were wrapped in bandages to prevent slippage during exertion. Participants' grip MVCs were measured before and after a cooling immersion, in which participants were asked to immerse their dominant hands into the 14°C-water bath up to the elbow joint for 30 minutes. After 30-min cooling, the hand skin temperature was about 14.2°C for both genders. The ANOVA results indicated that lowering the skin temperature on forearm/hand could reduce hand grip MVC, but did not shift the T_{MVC}. In summary, the influence of cooling the HST is just on the magnitude of grip force, but not on the time consumed to generate the grip MVC.

Keywords: MVC; hand skin temperature; cold pressor; time needed to reach the MVC

1 INTRODUCTION

Handgrip force is one of the most essential forces for manual operation. Besides poor postures and repetitive motions, force demands have been consistently considered as main risk factors associated with work-related musculoskeletal disorders (Silverstein, Fine, & Armstrong, 1987). In addition, several epidemiologic studies have shown that cold may be a risk factor for the occurrence or aggravation of musculoskeletal disorders, such as in the fish-processing industry (Chiang et al., 1993; Nordander et al., 1999) and meat-processing factories (Kurppa, Viikari-Juntura, Kuosma, Huuskonen, & Kivi, 1991; Piedrahíta, Punnett, & Shahnavaz, 2004). Wiggen and colleagues indicated that even petroleum workers must often be exposed to harsh and extreme environments while performing not only heavy lifting tasks but also tasks demanding grip strength and dexterity, for which such workers have to remove their gloves (Wiggen, Heen, FaeREVIK, & Reinertsen, 2011). Therefore, it is unavoidable that bare hands will be exposed in a cold environment. A report by the European Agency for Safety and Health at Work also noted that the risk of musculoskeletal disorders increases with work in cold environments (Schneider et al., 2010).

The hands are possibly the most used part of the body because they offer the most effective means to accomplish complex tasks, given their ability to perform specialized tasks that require dexterity, manipulability, and tactile sensitivity. Therefore, the hands are often exposed to different environments. In many cases, such exposure is related to changes in Hand/Finger Skin Temperature (HST/FST). Unfortunately, lowering HST/FST has been considered the vital factor in the reduction of tactile sensitivity (A. Enander, 1984), hand dexterity (Cheung, Montie, White, & Behm, 2003; A. E. Enander & Hygge, 1990; Heus,

Daanen, & Havenith, 1995; Riley & Cochran, 1984; Schiefer, Kok, Lewis, & Meese, 1984), and tracking performance (Goonetilleke & Hoffmann, 2009).

Besides above, Petrofsky and Lind (1980) indicated that under the same submaximal exertion level, surface Electromyography (sEMG) amplitude decreased as forearm muscle was cooled by after 30-min immersion in 10 and 20°C water. Chen, Shih, and Chi (2010) also indicated the sEMG decreased with the reduction in ST on the extensor digitorum and flexor digitorum superficialis during execution of same-hand dexterous tasks. Specifically, local cooling of the hands and arms also affects muscle activity through a decrease in ATP utilization, which decreases contraction velocity, maximal strength, and time to exhaustion (Heus et al., 1995). A reduction in handgrip strength caused by lowering HST was slao revealed in the past (Augurelle, Smith, Lejeune, & Thonnard, 2003; Chi, Shih, & Chen, 2012). Moreover, a reduction in muscular co-ordination was found when the forearm was cooled (Bergh & Ekblom, 1979; Oksa, Rintamäki, & Rissanen, 1997). Therefore, we can hypothesis that lowering the HST will reduce the hand grip strength.

Furthermore, the data on how much time is needed to reach the MVC (denoted as T_{MVC} hereafter) may be a useful and interesting index for evaluating the rate of strength generation and/or be an index of resistance-response time, that is, how much time is needed to overcome a given resistance. Tsaousidis and Freivalds (1998) evaluated the rate of force development of grip, pinch, and torque. They indicated that the corresponding T_{MVC} values for the peak grip, pinch, and torque strength under bared hand were 2.04, 1.42, and 2.14 sec, respectively. Jung and Hallbeck (2004) used the T_{MVC} as one of a set of selected criteria to quantify the effects of instruction type, verbal encouragement, and visual feedback on static and peak handgrip strength. They indicated that the time to reach the peak strength among three instruction types—free instruction, fast contraction and immediate release, and fast contraction and maintain—were insignificant (around 1 sec), and they were faster than those of slow contraction and immediate release and of slow contraction and maintain (around 2 sec). Chen et al., (2004) used the change in T_{MVC} between pre- and post-exertion to assess the muscle fatigue. Later on, Shih, Lo, and Huang (2006) indicated that wrist posture affected T_{MVC}, in which the longest T_{MVC} (1.069 sec) occurred when flexing wrist at 30 degrees, but the postures of forearm and wearing splints or not did not change T_{MVC}. Unfortunately, related information on T_{MVC} seems to be still less well-documented, despite the facts that gripping tasks have been found to be highly correlated with musculoskeletal disorders.

Petrofsky and Lind (1980) indicated that cooling the forearm could cause reductions in muscular co-ordination, nerve conduction speed, and strength. Altered muscular co-ordination due to cooling was also revealed by Bergh and Ekblom (1979) and Oksa, Rintamäki, and Rissanen (1997). Changes in muscle performance can affect manual manipulation through changes in power, contraction speed, or muscle endurance. Therefore, lower HST could be hypothesized to have longer T_{MVC}.

2 METHODS

2.1 *Participants*

A convenience sample of 20 volunteers, including 10 males and 10 females, was recruited for this study. All were right-handed, healthy, and free of musculoskeletal disorders in the upper extremities. The means (Standard Deviation, SD) for age, weight, and height for males and females were 28.7 (5.5) and 24.2 yr. (3.6); 68.8 (6.9) and 53.9 kg (4.7); and 172.8 (2.1) and 162.7 cm (4.6), respectively. Significant differences were found between genders in age, weight, and height ($p < 0.05$). During the experiment, each participant was dressed in a short-sleeved T-shirt, short pants, and sports shoes.

2.2 *Apparatus and materials*

The apparatus and materials used in previous studies (Chen, Shih, & Chi, 2010; Chi et al., 2012) were employed in the current study. They included a water bath, a submersible cooler, a digital thermometer and hygrometer, a digital 4-channel thermometer, and a grip gauge with a load cell. The grip span of the grip gauge was set at 5 cm and the handles were wrapped in bandages to prevent slippage during exertion.

2.3 Experimental procedures and data acquisition

First, the experimental procedure was explained, and all participants signed an informed consent. The experiment contained three successive stages: initial grip MVC (Maximal Volitional Contraction) measurement, 30-min cooling, and cooled grip MVC measurement.

Prior to formal measure, the probes of the thermometer were attached by sponge tape on the dorsal side of the middle phalanx of the middle finger (namely FST), on the middle of the third metacarpal of the dorsal side of the hand (namely HST), and on the muscles of the Extensor Digitorum (ED) and Flexor Digitorum Superficialis (FDS) of the forearm (namely FAST-E and FAST-F, respectively).

When thermometer probes were attached properly, participants were first asked to exert the maximal volitional contraction (MVC) of grip three times. The measurement protocol followed that employed by (Shih & Ou, 2005). Next, participants were asked to immerse their dominant hands into the 14°C-water bath up to the elbow joint for 30 minutes, and then the grip MVC was measured again. After 30-min cooling stage, the skin temperatures of forearm/hand was listed in Table 1.

For handgrip MVC measurement, each participant sat erect in a chair with the elbow at approximate 90° flexion and the upper arm parallel to the trunk. Each measurement of handgrip 5-sec MVC was replicated three times, and a 2-min rest was given between successive trials to avoid muscular fatigue. The maximal value of each 5-sec contraction was recorded, as well as the corresponding time reaching MVC (denoted as T_{MVC}).

2.4 Experimental design and data acquisition

The force output was acquired by a computer program with a sampling rate of 1000 Hz, and the load cell was zeroed prior to each measurement. Subjects were asked not to exert until a tone produced by the computer, and data were recorded immediately after that. In order to determine when exertion started, a 95% confidence interval for the first ten data points was calculated. When five consecutive points after these ten data points exceeded the upper bound of this confidence interval, the first of them was considered to be the start of the force development (Shih & Ou, 2005). The first maximum strength over the duration was considered as the MVC due to force generation is not monotonic. The consumed time from exertion start to reach MVC is, therefore, the T_{MVC}.

In this study the factorial ANONA was used with independent variables of sex and HST; the dependent variables were MVC and T_{MVC}. The level of significance (α) was set 0.05.

Table 1. The skin temperatures (°C) of different forearm/hand locations at two stages.

Locations	Sex	Intial ST		Cooled ST	
		Mean	SD	Mean	SD
FST	Male	31.3	1.1	13.7	0.3
	Female	31.3	0.8	13.8	0.2
HST	Male	31.5	1.2	14.4	0.2
	Female	31.9	1.3	14.4	0.2
FAST-E	Male	31.8	1.3	19.3	0.4
	Female	31.9	1.2	19.0	0.9
FAST-F	Male	32.0	1.1	19.6	0.6
	Female	32.7	1.0	19.4	0.6

3 RESULTS AND DISCUSSIONS

The ANOVA results indicated that, for grip MVC, the main effects of gender ($p < 0.001$) and HST ($p < 0.001$) were significant, but their interaction was not. Table 2 shows the female MVC was about 64% of male MVC (32.29 vs. 47.11 kgw). In addition, MVC decreased as HST lowered, with 11% reduction (36.45 vs. 40.94 kgw).

As to T_{MVC}, ANOVA results revealed that none of main effects were significant, nor was their interaction. The average T_{MVC} was 1.756 sec (SD = 0.928) and the descriptive statistics about T_{MVC} was shown in Table 3. Shih and Ou (2005) indicated that the gender effect on T_{MVC} is significant under normal HST (uncooled). Average male T_{MVC} is 1.828 s, and it is 1.346 s for females. Present paper even did not find the significant gender effect, but under normal HST (initial HST) male T_{MVC} (1.779 sec) was slightly longer than that of female (1.530 sec). This fact was consistent with the finding of Shih and Ou (2005). Of interest, the difference in T_{MVC} between gender diminished after cooling (male: 1.883 vs. female: 1.831).

Additionally, HST effect was not significant, which did not support our hypothesis. But what noticeable was lowering HST seemed to delay T_{MVC}. For male, it shifted from 1.779 sec to 1.883 sec; for female, it shifted from 1.530 sec to 1.831 sec.

4 CONCLUSIONS

Cooling forearm/hand to lower HST could reduce the grip MVC apparently, but did not delay gripping T_{MVC} significantly. On the other hand, male MVC was larger than female MVC, but T_{MVC} was indifferent between genders. In summary, the influence of cooling the

Table 2. The descriptive statistics for grip MVC (kgw) (Standard Deviation, SD).

Factors	Levels		Mean	SD	Female/male (%)	14°C/Initial (%)
Sex	Male		47.11	6.47	64%	
	Female		30.29	4.52		
HST	Initial		40.94	10.14		89%
	14°C		36.45	9.65		
Sex × HST	Male	Initial	49.60	6.41	65%	90%
		14°C	44.62	5.59		
	Female	Initial	32.29	3.63	63%	88%
		14°C	28.29	4.48		

Table 3. The descriptive statistics for T_{MVC} (sec) (Standard Deviation, SD; Coefficient of Variation, CV).

Factors	Levels		Mean	SD	CV
Sex	Male		1.831	1.060	58%
	Female		1.681	0.777	46%
HST	Initial		1.655	0.918	55%
	14°C		1.857	0.935	50%
Sex × HST	Male	Initial	1.779	1.057	59%
		14°C	1.883	1.078	57%
	Female	Initial	1.530	0.752	49%
		14°C	1.831	0.784	43%

HST is just on the magnitude of grip force, but not on the time consumed to generate the grip MVC.

REFERENCES

Augurelle, A.-S., Smith, A.M., Lejeune, T., & Thonnard, J.-L. (2003). Importance of cutaneous feedback in maintaining a secure grip during manipulation of hand-held objects. Journal of neurophysiology, 89(2), 665–671.

Bergh, U., & Ekblom, B. (1979). Influence of muscle temperature on maximal muscle strength and power output in human skeletal muscles. Acta Physiologica Scandinavica, 107(1), 33–37.

Chen, W.-L., Shih, Y.-C., & Chi, C.-F. (2010). Hand and finger dexterity as a function of skin temperature, EMG, and ambient condition. Human Factors: The Journal of the Human Factors and Ergonomics Society, 52(3), 426–440.

Cheung, S.S., Montie, D.L., White, M.D., & Behm, D. (2003). Changes in Manual Dexterity Following Short-Term Hand and Forearm Immersion in 10C Water. Aviation, space, and environmental medicine, 74(9), 990–993.

Chi, C.-F., Shih, Y.-C., & Chen, W.-L. (2012). Effect of cold immersion on grip force, EMG, and thermal discomfort. International Journal of Industrial Ergonomics, 42(1), 113–121.

Chiang, H.C., Ko, Y.C., Chen, S.S., Yu, H.S., Wu, T.N., & Chang, P.Y. (1993). Prevalence of shoulder and upper-limb disorders among workers in the fish-processing industry. Scandinavian journal of work, environment & health, 126–131.

Enander, A. (1984). Performance and sensory aspects of work in cold environments: a review. Ergonomics, 27(4), 365–378.

Enander, A.E., & Hygge, S. (1990). Thermal stress and human performance. Scandinavian journal of work, environment & health, 44–50.

Goonetilleke, R.S., & Hoffmann, E.R. (2009). Hand-skin temperature and tracking performance. International Journal of Industrial Ergonomics, 39(4), 590–595.

Heus, R., Daanen, H.A.M., & Havenith, G. (1995). Physiological criteria for functioning of hands in the cold: a review. Applied ergonomics, 26(1), 5–13.

Kurppa, K., Viikari-Juntura, E., Kuosma, E., Huuskonen, M., & Kivi, P. (1991). Incidence of tenosynovitis or peritendinitis and epicondylitis in a meat-processing factory. Scandinavian journal of work, environment & health, 32–37.

Nordander, C., Ohlsson, K., Balogh, I., Rylander, L., Pålsson, B., & Skerfving, S. (1999). Fish processing work: the impact of two sex dependent exposure profiles on musculoskeletal health. Occupational and environmental medicine, 56(4), 256–264.

Oksa, J., Rintamäki, H., & Rissanen, S. (1997). Muscle performance and electromyogram activity of the lower leg muscles with different levels of cold exposure. European journal of applied physiology and occupational physiology, 75(6), 484–490.

Piedrahíta, H., Punnett, L., & Shahnavaz, H. (2004). Musculoskeletal symptoms in cold exposed and non-cold exposed workers. International journal of industrial ergonomics, 34(4), 271–278.

Riley, M.W., & Cochran, D.J. (1984). Dexterity performance and reduced ambient temperature. Human Factors: The Journal of the Human Factors and Ergonomics Society, 26(2), 207–214.

Schiefer, R., Kok, R., Lewis, M., & Meese, G. (1984). Finger skin temperature and manual dexterity—some inter-group differences. Applied ergonomics, 15(2), 135–141.

Schneider, E., Irastorza, X., Copsey, S., Verjans, M., Eeckelaert, L., & Broeck, V. (2010). OSH in figures: Work-related musculoskeletal disorders in the EU—Facts and figures. Luxembourg: European Agency for Safety and Health at Work.

Shih, Y.-C., Lo, S.-P., & Huang, W.-S. (2006). The Effects of Wrist Splints on When Females Reach the Peak Grip Strength under Different Wrist and Forearm Positions. Journal of the Chinese Institute of Industrial Engineers, 23(5), 435–442.

Shih, Y.-C., & Ou, Y.-C. (2005). Influences of span and wrist posture on peak chuck pinch strength and time needed to reach peak strength. International Journal of Industrial Ergonomics, 35(6), 527–536.

Silverstein, B.A., Fine, L.J., & Armstrong, T.J. (1987). Occupational factors and carpal tunnel syndrome. American Journal of Industrial Medicine, 11(3), 343–358.

Wiggen, O.N., Heen, S., FaeREVIK, H., & Reinertsen, R.E. (2011). Effect of cold conditions on manual performance while wearing petroleum industry protective clothing. Industrial health, 49(4), 443–451.

New Ergonomics Perspective – Yamamoto (Ed.)
© 2015 Taylor & Francis Group, London, ISBN 978-1-138-02751-0

Using cloud computing to support customer service in the automobile industry: An exploratory study

Chu-Chai Henry Chan, Yuju Lo, Cheng Young Chen & Ping Chen Tsai
Department of Industrial Engineering and Management, Chaoyang University of Technology, Taichung County, Taiwan, R.O.C.

ABSTRACT: This study proposes applying the concept of social network in customer service to create higher customer values. As most people know, Facebook creates a big interactive social network to link people together. Social network connects each customer with the other customers with a huge cloud. A social network is an important interactive process for a specific consumer group. From the interactive process, corporations would be able to find out the problems or demands for a group of consumers with similar preferences. From the social network, corporations can find the real hidden needs of their customers.

Two steps were implemented in this study to find the consumers' demands. First, we collected data from social-network-based websites of automobile retailers such as Toyota Taiwan and Nissan Taiwan. Second, data mining with statistics was applied to find the correlation between consumers' preferences and demands to segment social groups. Then, marketing strategies for managing social groups would be formulated. After that, the better strategies were formulated from consumers' word-of-mouth communication that responded to the real needs of customers. The process of value creation by a social network is the aim of this study. Finally a case study is implemented in well-known automobile retailers.

Keywords: Cloud Computing Services; Social Networks; Customer Relationship Management; Service Innovation

1 INTRODUCTION

After Taiwan joined World Trade Organization (WTO), many dealers imported a lot of cars from other countries and faced stronger competition than ever, so most dealers had to find a profitable model for surviving. In addition to selling cars, many dealers also provide maintenance service and it's a significant source of profit. Thus, increasing the value of maintenance service is an important task for the dealers.

Over the years, the automobile industry has already accumulated a lot of customers. Among these customers, it's important for the firms to be able to pinpoint high value customers and retain them through promotional strategies. For low value customers, value analysis is needed to determine whether to abandon or to promote the type of customers.

The purpose of customer relationship management is to increase customer loyalty and satisfaction, and create profit for the company. Thus, providing high quality service to create loyalty and satisfaction is essential to profitability. This study proposes using cloud computing with social network to analyze the demands of customers and to develop a high quality service process. In addition, company such as COKE and other major firms have already started to use social network in their customer relationship management to promote products. Integrating social network in customer relationship management to enhance company image and promote marketing will be an important issue in the future.

Table 1. Definition of cloud computing services.

	Year	Definition
Rajkumar Buyya [1]	2009	A Cloud is a type of parallel and distributed system consisting of a collection of inter-connected and virtualized computers that are dynamically provisioned and presented as one or more unified computing resource(s) based on service-level agreements established through negotiation between the service provider and consumers.
Peter Mell & Tim Grance [2]	2009	The capability provided to the consumer is to use the provider's applications running on a cloud infrastructure. The applications are accessible from various client devices through a thin client interface such as a web browser (e.g. web-based e-mail). The consumer does not manage or control the underlying cloud infrastructure including network, servers, operating systems, storage, or even individual application capabilities, with the possible exception of limited user-specific application configuration settings.
IDC Frank Gens [3]	2009	Global Overview program also analyzes how quickly cloud services will be adopted, and by which customer segments, how cloud will impact vendor business models and service offerings, and the customer benefits and challenges surrounding cloud services.
S. Subashini, V. Kavitha [4]	2010	Cloud computing is a way to increase the capacity or add capabilities dynamically without investing in new infrastructure, training new personnel, or licensing new software.
Sean Marston, Zhi Li, Subhajyoti Bandyopadhyay, Juheng Zhang, Anand Ghalsasi [5]	2011	The emergence of the phenomenon commonly known as cloud computing represents a fundamental change in the way Information Technology (IT) services are invented, developed, deployed, scaled, updated, maintained and paid for.

Customers in the auto industry usually make appointments through phones to obtain service. But with the increasing use of Internet, automobile retailers such as Toyota and Nissan began to provide online reservation service. In addition, a lot of companies now run a fans group on Facebook, closely interact with the customers and receive feedback from them on the Internet instead of from phones. The study focuses on analyzing the social group feedback to understand customers' demands.

2 RELATED WORK

To dig out real needs from targeted consumers, this study proposes merging cloud computing services with the concept of networking. This section briefly reviews related literature on both topics.

2.1 Cloud computing services

Recent literatures of cloud computing services included Rajkumar Buyya (2009) [1], Peter Mell & Tim Grance (2009) [2], S. Subashini (2010) [4] and Sean Marston (2011) [5]. Frank Gens (2009) [3] proposed applying cloud computing services to meet the demands of different customer segments.

2.2 Social network

Since the rise of social networking sites, many consumers have used social networking service to share data and interact with others. The definition of social network is summarized in Table 2.

Table 2. Definition of social network.

	Year	Definition
Isobel Claire Gormley, Thomas Brendan Murphy [6]	2010	Social network data represent the interactions between a group of social actors. Interactions between colleagues and friendship networks are typical examples of such data.
Saeedeh Shekarpour, S.D. Katebi1 [7]	2010	A social network is a set of people, organizations or other social entities connected by a set of social relationships, such as friendship, co-working or information exchange.
Hsiang-Lin Cheng [8]	2010	A social network member not only recognizes institutional pressures from its general macro-environment via network information channels (Selznick, 1957) but also conforms to "specific pressures" that develop in its embedded social network. Firms may adopt prevailing new practices for the sole purpose of gaining legitimacy in a "micro" institution, such as in a network (Kostova, 1999).
Yusoon Kim, Thomas Y. Choi, Tingting Yan, Kevin Dooley [9]	2011	A network is made up of nodes and ties that connect these nodes. In a social network, the nodes (i.e., persons or firms) have agency in that they have an ability to make choices. With its computational foundation in graph theory (Cook et al., 1998; Kircherr, 1992; Li and Vitanyi, 1991), SNA analyzes the patterns of ties in a network.

3 METHODOLOGY

The study proposes applying the concept of cloud computing data mining to social network in order to find better strategies to improve the quality of service.

The study approach is summarized as follows:

- Review the literature on cloud computing, social network and Customer Relationship Management (CRM).
- Search for the best cloud computing service model for the automobile industry.
- Build a good business model.
- Collect data from well-known social network websites in the automobile industry such as Toyota and Nissan.
- Use statistics to do data-mining to explore the real needs of consumers.
- After data analysis, provide feedback and strategy to meet customer demands.

4 RESULTS AND DISCUSSION

The study collected data from messages posted by Toyota Taiwan (176 messages) [10] and Nissan Taiwan (117 messages) [11] on their Fan Page during a five-month period. Based on the content of the messages, they were divided into five categories: promotional activities, product recommendation, knowledge sharing, scenery sharing and others. The distribution of the messages in these five categories is presented in Table 3.

For both sites, the highest percentage of messages fell into the category of promotional activities, followed by product recommendation. The Nissan fan page also posted messages regarding scenery while the Toyota site did not provide such information. From the data above, we see that both companies still manage their fan page in the traditional marketing fashion, using promotional activities to attract customers. However, messages posted by the

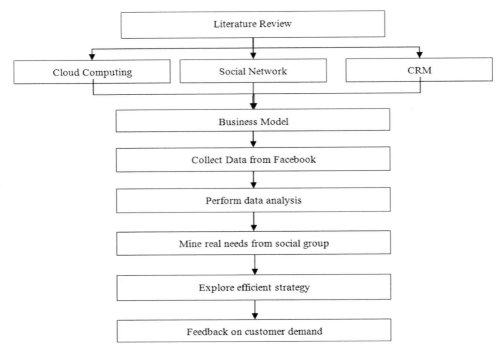

Figure 1. Study approach.

Table 3. Message distribution of Nissan Taiwan and Toyota Taiwan fan page.

	Nissan Taiwan		Toyota Taiwan	
Category	Number of messages	Percentage	Number of messages	Percentage
Promotional Activities	33	28.2%	70	39.8%
Product Recommendation	33	28.2%	56	31.8%
Knowledge Sharing	11	9.4%	17	9.7%
Scenery Sharing	13	11.1%	N/A	N/A
Others	27	23.1%	33	18.8%
Total	117	100%	176	100.0%

companies do not necessarily reflect consumer preference and needs. Other indicators need to be considered.

After reading the posted message, subscribers of the fan page could either click on "like" to indicate that they like the post, click on "comment" to leave a comment, or click on "share" to send the post to friends. Clicking the "like", "comment", "share" button reflects response to the posted message, and thus customers' preference and needs. In this study, the counts of "like", "comment" and "share" for each message are recorded and summarized, and the results are presented in Table 4.

According to Table 4, messages in the promotional activities category got an average of 179 "like" responses per message, indicating that the messages in this category were by far the most popular among consumers. Each message in this category also got an average of 10.5 "comment" and 9.7 "share" responses. Messages in the product recommendation category also got high counts of "like" responses (an average of 166 per message), and 13.6 counts of "share" responses.

In summary, the highest number of messages fell into the category of promotional activities. At the same time, these were the posts that the consumers were most interested in and

Table 4. Counts of "like", "comment", "share" for each message category (Nissan and Toyota data combined).

	Number of messages	Counts of response	"Like"	"Comment"	"Share"
Promotional Activities	103	Total counts	18,485	1,072	1,004
		Average per message	179	10.5	9.7
Product Recommendation	89	Total counts	14,808	676	1,180
		Average per message	166	7.6	13.6
Knowledge Sharing	28	Total counts	3,889	165	256
		Average per message	139	5.9	9.1
Scenery Sharing	13	Total counts	554	24	14
		Average per message	43	1.8	1.1
Others	60	Total counts	6,847	290	291
		Average per message	114	4.6	4.9
Total	293	Total counts	44,583	2,227	2,745
		Average per message	152	7.6	9.4

they were more willing to respond, discuss and share the information with others. Messages in the product recommendation category elicited the highest count of "share" responses.

Based on these results, we can see that through social networking site such as Facebook, companies can easily and speedily spread information on promotional activities and new products to consumers at a low cost.

5 CONCLUSION

New technology makes new business model. Although current business model has started to promote with the Internet instead of traditional television and newspaper advertising, interactive communication is still lacking. With the increasing popularity of social networking sites and the ease of such sites to connect people and spread information, a lot of companies have begun to set up Facebook Pages to establish customer relationship.

The study took a look at how two leading companies in the automobile industry in Taiwan (Toyota and Nissan) utilized the social networking site Facebook to reach out to their customers and how their customers responded. Based on the findings, the following suggestions are made regarding the management of social networking site.

1. Because of these sites, companies can share benefit and grow with consumers. The most popular items on Facebook are promotional activities and product recommendation.
2. Based on the idea of sharing, companies can choose sweepstakes and promotions as the core strategy on Facebook.
3. If any customer joins the fan page, the company can arrange activities to interact with the customers and retain them for creating more new values.

FUTURE RESEARCH

1. The study focuses on only one social networking site (Facebook). We will study the impact of other new social networks such as Line and WeChat.
2. The work chooses the automobile industry as the study case. We will try to investigate other industries.
3. We hope to compare customer's behavior of social networking sites between Taiwan and other countries.

ACKNOWLEDGEMENTS

The authors would like to thank the National Science Council of the Republic of China, Taiwan for financially supporting this research under Contract No. NSC 101-2221-E-324-021.

REFERENCES

[1] R. Buyya, C. Shin, S. Venugopal, J. Broberg and I. Brandic, "Cloud computing and emerging IT platforms: Vision, hype, and reality for delivering computing as the 5th utility," *Future Generation Computer Systems*, vol. 25, pp. 599–616 (2009).

[2] P. Mell and T. Grance, "The NIST Definition of Cloud Computing," *National Institute of Standards and Technology, Information Technology Laboratory*, Version 15, 10-7-09 (2009).

[3] IDC Frank Gens, IDC Analyze the Future, http://www.idc.com/research/viewfactsheet.jsp?containerId=IDC_P20179&sectionId=null&elementId=null&pageType=SYNOPSIS (2010).

[4] S. Subashini, V. Kavitha," A survey on security issues in service delivery models of cloud computing", Journal of Network and Computer Applications, 34, 1–11 (2011).

[5] Sean Marston, Zhi Li, Subhajyoti Bandyopadhyay, Juheng Zhang, Anand Ghalsasi, "Cloud computing—The business perspective", Decision Support Systems 51, 176–189 (2011).

[6] Isobel Claire Gormley, Thomas Brendan Murphy, "A mixture of experts latent position cluster model for social network data", Statistical Methodology, 7, 385–405 (2010).

[7] Saeedeh Shekarpour, S.D. Katebi1, "Modeling and evaluation of trust with an extension in semantic web", Web Semantics: Science, Services and Agents on the World Wide Web, 8, 26–36 (2010).

[8] Hsiang-Lin Cheng, "Seeking knowledge or gaining legitimacy? Role of social networks on new practice adoption by OEM suppliers", Journal of Business Research, 63, 824–831 (2010).

[9] Yusoon Kim, Thomas Y. Choi, Tingting Yan, Kevin Dooley, "Structural investigation of supply networks: A social network analysis approach", Journal of Operations Management, 29, 194–211 (2011).

[10] Toyota Taiwan(https://www.facebook.com/TOYOTA.Taiwan).

[11] Nissan Taiwan(https://www.facebook.com/nissan.tw).

New Ergonomics Perspective – Yamamoto (Ed.)
© *2015 Taylor & Francis Group, London, ISBN 978-1-138-02751-0*

Target distance and exposure time of glance to protect multiple choice examination dishonesty

Phairoat Ladavichitkul & Haruetai Lohasiriwat
Department of Industrial Engineering, Chulalongkorn University, Bangkok, Thailand

ABSTRACT: A major problem in academic is the dishonesty especially on examination. In a huge class, a large number of students is a major problem not only in the lecture period but also in the examination period. Multiple choice examination is a solution which can manage the test result in time. However, the multiple choice examination is easy to be dishonest by glancing. In order to protect glancing among the examinees, the examiner has to prepare the room space and number of staff for the examination. The room space directly effects the space or distance between the examination seat and the number of staff effects on the time of monitoring at examinee activities. The dishonesty problems always occur if there were not sufficient room space and number of staff. Therefore, the room space and number of staff requirement are the important information for managing the examination. The main objective of examination management is to protect the dishonest. This can be interpreted as that the glancing behavior have to be controlled or the glancing results have to get in very high of copying error.

The concept of Schmidt's law was applied to study the capacity of glancing at the multiple choice sheet. There were two independent factors such as 1. the distance between the target sheet and the examinee sheet and 2. the glance exposure time which became the index of difficulty of glancing task. The error number of answering is the dependent factor. The relationship between the error and the index of difficulty were examined. Thirteen subjects were recruited from the university students to do a glancing task in the simulated examination seat. Seven levels of the distance factors (varied from 100 cm to 220 cm) and three levels of the exposure time (1 s, 2 s and 4 s) were set as the research conditions (7 distance levels × 3 exposure time) with twice replication. In each replication, each subject had to glance and copy thirty questions on a random target sheets.

The research results showed the variable relationship as the following equation with $R^2 = 0.918\%$ Error = −21.8 + 48.0 (Distance − 0.09 × Time).

To use this equation, the examiner has to set the expected error rate and then prepare or trade-off between the examination space and the number of monitoring staff. On the other hand, this equation can be used to predict or evaluate the examination management.

Keywords: Speed and Accuracy Trade-off; Schmidt's Law; Protect the Examination Dishonesty; Glance

1 INTRODUCTION

Objectives of examinations may vary but they all have mutual parts to evaluate the examinee performance. Some may used as a recruitment tool (e.g., students to academic institution, employees to work organization) while others may used as evaluation tool to test for ability of achievement (e.g., integral parts of academic system). For high-stakes testing program, cheating could result in serious ramifications as it could mean under-qualified individuals receiving certification to practice.

For education institutes, academic dishonesty policy and penalty for violation are usually and clearly defined to ensure major foundation of all academic work, intellectual integrity and credibility. In paper-based examination, two major types of examinations are found; multiple choice and write-up. For both types, dishonesty has emerged in a variety of methods including bringing unauthorized materials into the examination room, text messaging or giving signage among examinees, or even the use of impersonator, and so on. Among these methods, simply looking at neighbour examinee's answer sheet and copying their answers are the most traditional method and still be used at present. This type of dishonesty relies mostly on the examination proctor conscientious which obviously cannot be perfect. To reduce successfulness of such technique, the use of video recording to the entire examination room has proposed in many academic institutes. At the same time, there are thrives to utilize statistical method to detect any unusual amount of similarity between answers of suspect examinee and the neighbours [1][2]. Not only checking multiple choice answers, nowadays write-up essay also checked against plagiarism using computer program such as Turn-It-In. Going beyond catching fraud once commit, there is also attempt to predict if one has more possibility to violate the honour code than others using individual difference characteristics (e.g., subclinical psychopathy and poor verbal ability [3]). As one may expect, the more variety of cheating techniques, the more solutions educators have come up with to tackle this long persistent problem. Hopefully, the techniques could be some objective measures and able to avoid "I said-You said" argument. Among all the solutions, the most basic method to reduce probability of copying answer from neighbour examinees is alternated or scrambled test forms among examinees seating locations accompany with space-out seat arrangement. In this study, we have objectively investigated seat arrangement factors which influencing cheating performance on multiple choice examination. More detail will be discussed in later sections.

2   VISUAL PERFORMANCE FACTORS

In order to successfully copy neighbour's answer, the observer has to first locate desired target among all other distractions (e.g., locate answer of a question in a range of marked answers) via visual orienting. This subtask could be performed by manipulating head and/ or eye movement to the approximate area of selected question. This step is most likely using top-down process. Then, s/he will focus visual attention to the area and performance of copying task will largely depend on spatial resolving capacity of his/her visual system. In order to make decision on whether the target is located in the column of choice A, B, C, D, or E, the observer has to discriminate the answer's location by comparing target answer with the columns in adjacent rows. This particular task thus falls in to target localization acuity subtype. Any factors evidenced to affect acuity performance are thus hypothesized to have influenced on the copying task as well.

First, discontinuity or ability to detect fine detail of location displacement between the two rows is specified in terms of its angular size or angle of displacement. As shown by fig. 1. the farther away, the smaller the angle of displacement and therefore, lower performance on acuity test is expected. This fact supports possible negative relationship between distance factor measured from observer eye to the target and the copying task performance.

Then, level of background luminance has repeatedly shown to affect visual acuity. For both scotopic and photopic functions, acuity performance is increased according to higher luminance level [4]. Although lighting itself cannot produce work output, highly visual

Figure 1.   Angle of displacement; an effecting factor on visual acuity.

326

related task like copying answer sheet is undoubtedly expected to influence greatly by such factor. However, under real examination setting, it is impossible to reduce illumination level to be lower than level specified by lighting standards as the examinee also has to be able to read and write with comfort and without distraction. This is major reason for not varied illumination factor in our current study. Having constant level of illumination set up, state of eye adaptation can also be neglected in our study. We assumed that our participants' eyes have adapted to the same level as the testing condition and thus should yield highest level of visual acuity possible.

Besides the two major adjustable factors (i.e., distance and lighting arrangements), there are also factors influencing visual acuity but rather depend on the observer's posture and movement. Hence, they are uncontrollable by the examiner who setting up the test room. These factors include area of retina stimulated and time exposure to target. Because densely packed cones at the fovea area, visual acuity is the greatest at the center of fixation (looking directly in front of the observer's eyes). Poorer acuity is found in the farther distributed area measured from the center point. Our present study has no intention to analyse possible effects on this factor as under real setting, the cheater usually intentionally move his/her head and direct toward neighbour answer sheet to allow possibility for fovea vision. This action implies that it is necessary to have direct vision for successful copy (otherwise, most cheater will able to copy without or with only slight movement of his/her head using peripheral vision). Therefore, the factor of exposure time gets into our attention more as it related directly to how we should manage proctor frequency of watching each examinee. Though there is no simple acuity-exposure time relationship for the resolution of the target, the acuity is usually accepted as proportional to the exposure time [5]. Testing on face recognition task, exposure time to targets and non-targets (i.e., target face among non-target people) is found to be a major factor on the task performance [6]. Normally, the longer time provided for observer to continually looking at target, the better acuity or visual related task performance are found until performance levels off. Finally, although we not included any factors regarding target characteristics which known to affect acuity performance (e.g., contrast sensitivity, size, color) but rather controlled such variable to stay constant. We will briefly discuss at the last section regarding future studies.

3 EXPERIMENTAL VARIABLES

As mentioned earlier, performance of the copying task depends largely on how far the target is from the observer eyes. The first independent variable in this study is horizontal distance between the center of observer's answer sheet and the center of target answer sheet rather than actual distance from observer's eyes to the target point (Fig. 2). Underlying reason is that our ultimate goal from the study is to set criteria for the actual seat arrangement which means direct horizontal distance will be more useful than actual viewing distance.

To identify level of this independent measure, a pilot study testing with five industrial engineering students was set up. These volunteers are asked to copy answers from a multiple

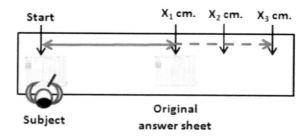

Figure 2.   First independent variable; distance between observer's and target's answer sheets.

choice answer sheet. The answer sheet has five multiple choices in each question (A to E) and we randomly marked down the total of thirty questions in the target sheet (Fig. 3.) Size of answer sheet is 21.6 x 28.0 cm.

More detail on the pilot study is described in [7]. In shorts, we were testing on ten distance levels to the observer's right side (i.e., 80, 100, 120, 140, 160, 180, 200, 220, 240, and 260 cm). The 260 cm was set as maximum distance because [8] has reported approximately 240 cm to be the farthest distance required for legibility in recognizing ¼ inch tall capital letter (i.e., A, B, C, F, T). However, as our study is focusing on multiple choices rather than letter recognition, we assumed possible longer maximum range and made decision on expanding test condition to 260 cm. Then, target exposure time was set as the second independent variable. To control exposure time, we required participants to perform specific head movement and controlled their speed by a metronome. Metronome used is a rhythmic sound generated by a computer program. The first "BEEP" was given cue to turning head and orient toward the target answer sheet. Later on, the second "BEEP" was given cue to rotate head back and write down the copied answer to the relevant question order. Then, the process is repeated to the next question order. The metronome is programmed so that the interval of exposure time to target can be manipulated into various levels (i.e., 1 s, 2 s, 4 s, and 6 s) whereas the marking answer time is set constant at one second. Participant was required to mark down answer without hesitation in order to allow the major use of short-term memory rather than decision making process.

In terms of dependent measure, percent of incorrect copied answer calculated from the total of thirty test questions was used. Lower performance or higher percent of incorrect answer were expected to relate with shorter distance and longer exposure time. Note that other major factors influencing visual task performance mentioned in previous section are mostly controlled. These factors include 300 lux illumination measured on the task table. This figure was set following the normal light intensity standard [9] using fluorescent lamp (warm white color). Additionally, contrast of target against its background is approximately equal to all tested target answer sheet by using 2B pencil for marking answers in all sheets. Using the same answer sheet format for all condition (but varied answers randomly) ensure that target size is kept constant.

Figure 3.   Answer sheet used in the study.

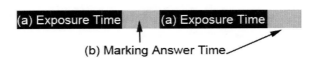

Figure 4.   Tasks performed by participant (a) Exposure time to target varied from 1 s-6 s (b) Marking answer time set at 1 s.

4 PILOT STUDY RESULTS

Our hypothesis was confirmed in this pilot study (Fig. 5). Using multiple comparisons, there were significant difference in the average of incorrect answer between different exposure time except those between the 4 and 6 second conditions (confidence level at 0.710). This could infer that participants required the maximum of 4 seconds to successfully search and locate to-be-copied answer from the target answer sheet. The longer exposure time beyond 4 seconds to the answer sheet will not increase copying performance whereas the shorter time would reduce performance. Note that the smallest time exposure is set up at 1 second since this is the minimum period of time participant reported able to perform head rotating and locate visual target. Likewise, our hypothesis regarding target distance was accepted.

Result shows positive correlation between measured distance and percent of incorrect answer (Fig. 5.) Note that perfect performance was found with 4 second exposure time testing at 80 cm distance while the shorter time period slightly increase incorrect answers. At the other end of 260 cm condition, all exposure time levels reach almost 80% incorrect which considered equal to probability of correct guessing by chance (i.e., get correct 5–6 correct answers from 30 questions).

5 METHODOLOGY

Process similar to the pilot session was repeated with the other 13 participants. However, time exposure to target answer sheet was reduced to only three levels (i.e., 1 s, 2 s, and 4 s) as we found no difference between 4 s and 6 s conditions earlier. Distance factor were reduced to 7 levels (i.e., 100, 120, 140, 160, 180, and 200 cm). Major objective of this second test is to come up with empirical equation to explain relationship between variables. After gaining such equation, another 16 participants were tested under same experimental setting and procedure to verify prediction accuracy of the equation. However, for verifying equation, each participant was given with only six testing conditions. At the same time, we ensured that all possible twenty one testing conditions were tested. All participants were volunteered to the study. Their age ranged from 18–27 years. All were students under the Industrial Engineering Department, Chulalongkorn University. Prior to the experimental session, all participants have to pass normal vision of 20/20 from Snellen visual acuity test protocol. All experiment sessions was conducted in Ergonomics Laboratory at the department. Each participant was required to repeat each testing condition (7 distance levels × 3 exposure time) twice.

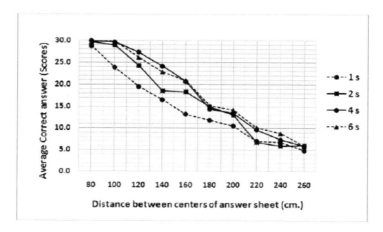

Figure 5.   Relationships between target distance/exposure time and percent of correct answer.

6 EXPERIMENTAL RESULTS

Data from our experiment suggest high linear relationship. Using multiple linear regression method, we can predict percentage of incorrect answers using equation 1 at R-squared 91.8%.

$$E = -21.8 + 48.0(D - 0.09T) \tag{1}$$

E is percentage of incorrect answer. D is distance measured between the centers of participant's and target answer sheets in meter ($1.00 \leq D \leq 2.20$). T is target exposure time in second ($1 \leq T \leq 4$). As seen by the equation, positive effect from distance factor and negative effect from exposure time were evidenced. However, effects from the looking distance found to be much more significant than the other factor, the looking time. Prediction performance of eq. 1 was later tested. The equation can correctly predict up to 89% using 95% prediction confidence interval criteria. Our equation seems to fail for extreme points.

7 DISCUSSION AND CONCLUSION

Setting seat to be apart from each other at least 220 cm is not usual in practice. For instance, at the Faculty of Engineering, Chulalongkorn University, we are setting around 1.30 m between each testing table. Using equation specified in this current study means cheater can reach around 68% correct copied answer with 4 second exposure time. Reduction to 63.72% correction still considered very high with 1 second exposure time. Even though, the authors believe copying task under real circumstances can be performed with lower performance than what predicted in our equation, some other possible factors that could reduce such probability should be studied in more detail. Major reason for expected lower performance than prediction is that part of target sheet will always be covered by the owner's body parts. Meaning that orienting vision direction toward target answer sheets will be much more difficult than simply turn his/her head to the right side and turn back before the examination proctor is able to catch. Eventually, all processes will add more time and result in more difficulty to perform copying task successfully. At the same time, having examination proctor walking around will create more stressful to dishonesty activity.

The authors encourage future studies regarding other possible factors that could reduce chance of copying success. Rather than focusing on seat arrangement alone, it could be interesting to find optimum size of marked area in which probability to successful copy is equal chance without enhancing seat space-out distance up to over 200 cm. At the same time, with computer technology has advanced and may able to correctly check answers with lower contrast, using HB pencil or any lighter color than 2B might be another way to reduce dishonesty performance. Finally, being hyperacuity related task, the authors believe that the number of choices on answer sheet could as well affect copying performance in similar manner as increasing choices effects reaction time task performance. It would be interesting to study possible combined effect from target exposure time and number of choices simultaneously. Incorporating with the other adjustable factors may reduce minimum requirement for between-seat distance and thus, more practical under real setting.

REFERENCES

[1] Wollack, J.A. Detecting Answer Copying on High-Stakes Exams. *National Conference of Bar Examiner*, **73, No. 2**; 2004, pp. 35–45.
[2] Wesolowsky, G.O. Detecting excessive similarity in answers on multiple choice exams. *Journal of Applied Statistics*, **27**; 2000, pp. 909–921.

[3] Lau, K.S.L., Nathanson, C., Williams, K.M., Westlake, B., and Paulhus, D.L. Investigating Academic Dishonesty with Concrete Measures. *Poster presented at the 17th annual meeting of the American Psychgological Society*; Download November 19, 2012 from http://neuron4.psych.ubc.ca/~dpaulhus/research/ED_PSYCH/CONFERENCES/APS.05.pdf.

[4] Johnson, C.A., and Casson, E.J. Effects of Luminance, Contrast, and Blur on Visual Acuity. *Optom Vis Sci*, **72(12)**; 1995, pp. 864–869.

[5] Waugh, S.J., and Levi, D.M. Visibility, timing and vernier acuity. *Vision Research*, **33(4)**; 1993, pp. 505–526.

[6] Laughery, K.R., Alexander, J.F., and Lane, A.B. Recognition of human faces: Effects of target exposure time, target position, pose position, and type of photograph. *Journal of Applied Psychology*, **Vol. 55(5)**; 1971, pp. 477–483.

[7] Wiwatwisawakorn, K. and Ladavichitkul, P. A Preliminary Study of Distance between Examination Seats for Preventing Cheat with Speed-Accuracy Tradeoff. *Proceedings of the International MultiConference of Engineers and Computer Scientists*, **vol. 11**; 2011.

[8] Pomales-Garcia, C., Carlo, H.J., Ramos-Ortiz, T.M., Figueroa-Santiago, I.M., and Garcia-Ortiz, S. Non-traditional Exam Seat Arrangements. *Computers and Industrial Engineering*, **57**; 2009, pp. 188–195.

[9] Ministerial Regulation No. 39 (B.E.2537), issued Parliament. *Act of control building B.E. 2522*; Download October 25, 2010 from http://203.155.220.239/yota/acrobat/yota39_37.pdf.

New Ergonomics Perspective – Yamamoto (Ed.)
© 2015 Taylor & Francis Group, London, ISBN 978-1-138-02751-0

Preliminary study of visual arc on visual performance base on Fitts' law concept

Nahapon Puttyangkura
Department of Industrial Engineering, Siam University, Bangkok, Thailand

Phairoat Ladavichitkul
Department of Industrial Engineering, Chulalongkorn University, Bangkok, Thailand

ABSTRACT: Visual perception is the most useful input sensory system. Most of work and living activities require visual information. The visual capacities were studied and presented in term of visual field and visual arc. Visual field explains the effects of incoming direction of visual information. Moreover visual arc explains the effects of the relationship between the object size and distance. Visual arc is used for designing the input character especially on computer display window. The improper size of character affects not only the outputs but also the working posture. The problem occurs if there are too many display information then the program designer has to reduce the character size. The working speed requirement including the small character size make the workers have to lean forward their body close to the monitor display. Long period working in this situation causes a chronic symptom on their lower back and neck.

Visual arc is calculated from the object size and the perception distance. Both factors would effect on the perception speed. Therefore, Visual arc is looked similar to the index of difficulty of Fitts' law. The proposes of this study were to study the effect of visual arc on visual perception speed and to find a visual performance from the linear relationship between visual speed and visual arc based on Fitts' Law concept. Test of the subjects 7 subjects aged 25–40 years who have experiment use of computers in daily life without any visual problems and experiment with programs that using the numeric characters 0–9 under the following on 7 conditions by fix for 2 factors are size of character and distance from head to LCD monitor.

The study is found relationships that were linear when the 1/MoA is increases, time response is increases together. From the relationship can create equation that consistent with Fitts' law's equation and trend in the same direction. It can be concluded for Visual Arc is important in working and living. The future studies other than principles of work station design then will include this study. The study of Visual arc and Visual performance would be useful to employer in order to assign the employee workload by considering the visual capacities especially in computer usage. Moreover, the results can be applied to design the warning sign or information board proper to the situation such as in the emergency or disaster situation.

Keywords: Visual Arc; Fitts' Law

1 INTRODUCTION

Visual perception is most of working and living activities. The capacities visual perception were studied and presented in term of visual field and visual arc. Visual field explains the limitation of incoming direction of visual information. Moreover visual arc defines as the relationship between the object size and distance from object. Visual arc is used for designing the input character especially on computer display window. The improper size of character effect is not only the outputs. The problem occurs when there are a lot of information display on the limited area, the program designer has to reduce the character size which effects to the user resulting to receive incomplete information or perceive more time.

The working speed requirement and the small character size make the workers have to forward their body close to the monitor display in order to increase their working performances. Long period working in this situation causes a chronic symptom on the lower back and the neck. Therefore, this study interested in the effect of the font size on display monitor and the distance from head to monitor on the working performance.

2   METHOD

2.1   Objectives of the study

Study the relationship of visual arc with response time for compare with the Fitts' Law.

2.2   Participants

There were 7 subjects aged 25–40 years who have experiment use of computers in daily life without any visual problems.

2.3   Apparatus

- The work station was at able and adjustable chair for general office.
- Acer LCD monitor, FT200HQL, 20"
- Keyboard, mouse,
- House programs used for display random characters on the screen that can change a size of characters.
- Head-Chin Rest for controlling the distance of the monitor and subject's head.
- Light at 600 Lux.

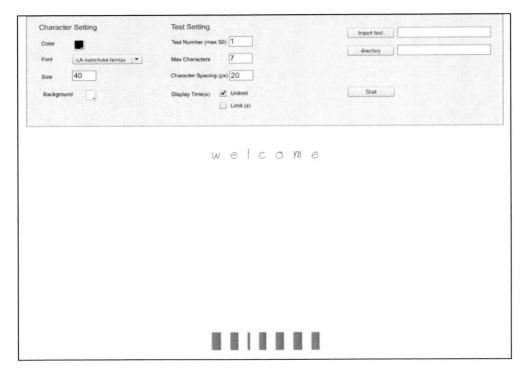

Figure 1.   Interface of house program.

Figure 2.   The experiment station.

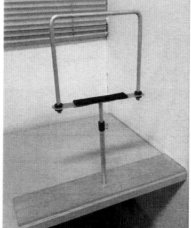

Figure 3.   The Head-Chin Rest.

Table 1.   All conditions in this experiment.

	MoA (Minute of Arc)	1/MoA
1) 2.3 mm Size, 1,600 mm Distance	0.08	12.5
2) 2.3 mm Size, 1,200 mm Distance	0.11	9.1
3) 2.3 mm Size, 800 mm Distance	0.16	6.3
4) 2.3 mm Size, 400 mm Distance	0.32	3.1
5) 1.5 mm Size, 400 mm Distance	0.21	4.8
6) 1.0 mm Size, 400 mm Distance	0.14	7.1
7) 0.5 mm Size, 400 mm Distance	0.07	14.3

2.4   Procedure

The first step is explaining an overall process for subjects and train subject for using the programs and answer buttons. Second step is experiment with programs that using the numeric characters 0–9 under the following on 7 conditions by fix for 2 factors are size of character and distance from head to LCD monitor. The all conditions show in Table.1 and MoA's equation that follow in (1)

Test for 2 replicates of each any conditions and 20 times each test in same period. Process for experiment is start from subjects click a start button on interface and program will show a random 1 number and touch the keyboard for answer when subject seen the number and decision.

The results were obtained will be plotting graph for consider the relationship occurred.

$$MoA = [2 \arctan (S/2D)]/60 \tag{1}$$

S = Object height.
D = Distance from eyes to object.

3   RESULT

Test of the subjects in range from 1/MoA (3.1–14.3. In Figure 4, the graphs show the relationship between response time with 1/MoA or 7 conditions of 7 subjects. The study is

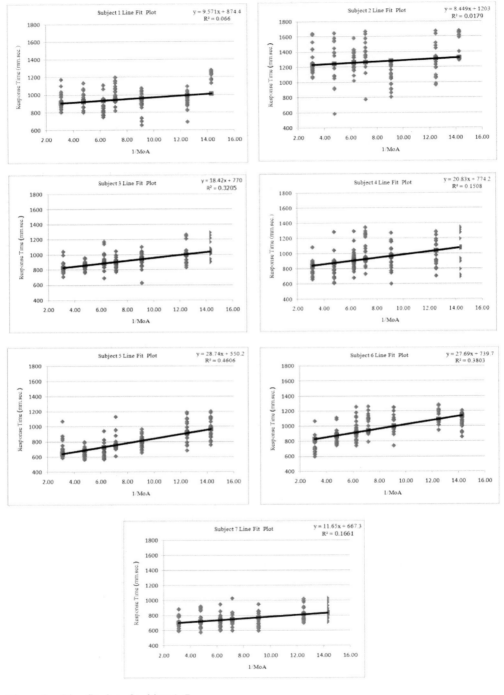

Figure 4. Line fit plot of subject 1–7.

found relationships that were linear when the 1/MoA is increases, time response is increases together. From the relationship can create equation in (2) that consistent with Fitts' law's equation in (3).

Table 2. Equation of all subjects.

	Equation	R^2
Subject 1	y = 9.571x + 874	0.0669
Subject 2	y = 8.449x + 1203	0.0179
Subject 3	y = 18.42x + 770	0.3205
Subject 4	y = 20.83x + 774	0.1508
Subject 5	y = 28.74x + 550	0.4606
Subject 6	y = 27.69x + 739	0.3803
Subject 7	y = 11.65x + 667	0.1661

$$MT = a + b(1/MoA) \qquad (2)$$

$$MT = a + b(ID) \qquad (3)$$

From table 2. It can be seen that the characteristics of all equations are similar and were linear equation that all of the y-axis are same direction.

4 CONCLUSION

The relationship between the MoA (Minute of Arc) and the response time from 7 subjects could be represented in linear regression. The regression line of all subjects had the similar pattern. Compare to Fitts' Law, it could be concluded that the Visual Arc effected on working time and could be become the index of difficulty for perception workload.

The study of Visual arc and Visual performance would be useful in order to assign the employee workload by considering the visual capacities especially in computer usage and paper reading. Moreover, the results can be applied to design the warning sign or information board in the proper situation such as the emergency or disaster situation.

ACKNOWLEDGEMENTS

I would like to thank Miss Punyisa Kuendee, Miss. Siwalee jetthumrong and Miss. Sutthapa Sutthisang for encouragement and support. The last thank for all subjects who joined in this experiment.

REFERENCES

[1] I. Scott Mac Kanzie, Fitts' Law as a Research and Design Tool in Human-Computer Interaction, Human-Computer Interaction, 1992, Volume 7, pp. 91–139.
[2] Kaiser, Peter K. "Calculation of Visual Angle". The Joy of Visual Perception: A Web Book. York University.
[3] Krongkarn Wiwatwisawakorn and Phairoat Ladavichitkul, A Preliminary Study of Distance Between Examination Seats for Preventing Cheat With Speed-Accuracy Tradeoff, Proceedings of the International MutilConference of Engineers and Computer Scientists 2011 VOL II, IMECS 2011, March 16–18, 2011 Hong Kong.

New Ergonomics Perspective – Yamamoto (Ed.)
© 2015 Taylor & Francis Group, London, ISBN 978-1-138-02751-0

Predicting brain metastasis from lung cancer by Bayesian Network

Kung-Jeng Wang
Department of Industrial Management, National Taiwan University of Science and Technology, Taipei, Taiwan, R.O.C.

Bunjira Makond
Department of Commerce and Management, Prince of Songkla University, Trang, Thailand

Kung-Min Wang
Department of Surgery, Shin-Kong Wu Ho-Su Memorial Hospital, Taipei, Taiwan, R.O.C.

ABSTRACT: Bayesian Network (BN) is a model graphically represented using bioinformatics variables to support informative medical decisions and observations using probabilistic reasoning. This study proposes a BN model to predict the occurrence of brain metastasis from lung cancer. A nationwide database of cancer patients in Taiwan is used in this study. Accuracy, sensitivity, and specificity are used to evaluate the performances of the proposed BN model. Experimental results show that the proposed BN performs well. The proposed model has advantages compared with the other approaches in interpreting how brain metastasis develops from lung cancer.

Keywords: Bayesian Network; brain metastasis; lung cancer

1 INTRODUCTION

Lung cancer is a leading cause of death worldwide and often spreads to the brain given that 65% of patients diagnosed with a primary tumor in their lungs will have brain metastases [1] while twenty to forty percent of cancer patients develop brain metastases during their illness [2–3]. Survival time decreases and quality of life deteriorates once lung cancer metastasizes to the brain [4]. Predicting the development of brain metastasis from lung cancer become necessary in the early detection of brain metastasis [5].

Although traditional statistical and machine learning model (e.g., [6–7]), such as LR and Support Vector Machine (SVM), are popularly used for cancer prediction, these models are not as promising as the Bayesian Network (BN) given that BN can use reasoning under uncertainty whereas both LR and SVM cannot. BN is a powerful tool for representing stochastic events and conducting prediction tasks. Oh et al., [8] stated that this tool can approximate complex multivariable probability distributions of heterogeneous variables as interpretable local probabilities to incorporate prior clinical and biological knowledge as well as to visualize and interpret the interactions among variables of interest for clinical use. BN can also be used as a classifier based on a learned network structure. Traditional statistical models cannot compute for posterior probabilities. BN is thus more effective than these traditional methods because it can represent the relations between variables. As a result, each node can compute for the posterior probability distribution, which is useful for decision-makers. In addition, BN can be applied in both linear and non-linear relation problems, including interaction problems such as a parent-child relation. By contrast, traditional statistical models have rigid assumptions that the variables are independent [9–10].

2 MATERIALS AND METHODS

In this study, six variables are used to construct the proposed BN model: (1) age; (2) gender; (3) region of residence, environment of patients; (4) location of lung cancer within human body; (5) treatment (primary treatment described in the medical database of patients); and (6) occurrence, which represents the development of brain metastasis from lung cancer (also refer to [11–12]. This occurrence is designated as "yes" if the second or latter diagnosis has brain metastasis; otherwise, it is designated as "no." Occurrence likewise functions as a response variable. Moreover, the possible values of these variables are presented in Table 1.

This study used the NHI database from 1996–2010 collected by the Bureau of NHI, Taiwan (BNHI). The two groups of files are registration and original claim data for reimbursement [13]. In this study, we retrieved data from the "Ambulatory care expenditures by visits file" (CD file), which is in the "Original claim data for reimbursement" file. We obtained legal records of more than 36,000 lung cancer patients from the database. As shown in Table 1, 35,605 patients were diagnosed with only lung cancer and only 438 patients developed brain metastasis.

The data set is highly imbalanced. Given the imbalanced data set, a classification model can be made ineffective by the instances in the majority class, which results in high accuracy for the majority class but poor accuracy for the minority class. To deal with this problem, we used both random under-sampling and random over-sampling method to adjust our data distribution.

To evaluate the prediction performance of brain metastasis, we use 80% of the data for the training set and 20% as a testing set. The performance of the model is determined from three indexes, namely, accuracy index, which calculates if the overall prediction is correct, as well as sensitivity and specificity indexes, which measure the positive and negative predictive performances, respectively.

Table 1. Data profile of lung cancer patients.

Characteristics		Number of patients	(%)
Gender	Female (F)	12,473	35
	Male (M)	23,570	65
Age	Less than 50 yr (<50)	4,396	12
	5–0–60 yr (50–60)	6,184	17
	60–70 yr (60–70)	9,340	26
	More than 70 yr (>70)	16,123	45
Region	Central branch (C)	6,687	19
(in Taiwan)	Eastern branch (E)	1,250	3
	Northern branch (N)	18,174	50
	Southern branch (S)	9,932	28
Site of lung	Trachea, bronchus and lung (162)	3,404	9
cancer	Trachea (1620)	1,344	4
	Bronchus (1622)	262	1
	Upper lobe, bronchus or lung (1623)	472	4
	Middle lobe, bronchus or lung (1624)	243	1
	Lower lobe, bronchus or lung (1625)	872	2
	Other parts of bronchus or lung (1628)	1,449	4
	Bronchus and lung, unspecified (1629)	26,997	75
Treatment	Radiotherapy (Ra)	2,045	6
	Chemotherapy (Ch)	1,201	3
	Drugs (Dr)	32,797	91
Occurrence	Yes (Y)	438	1
(brain metastasis)	No (N)	35,605	99

3 EXPERIMENTS

The graphical model with six variables is constructed and shown in Figure 1. The model represents causal relationships among variables with the arc indicating the direction of cause and effect by specifying the probabilities of variables. Gender, age, and region would affect site of lung cancer; the treatment is affected by gender, age, region and site of lung cancer; as well as, the occurrence of brain metastasis depends on site of lung cancer and treatment. The data are used to calculate the conditional probabilities which explain the changes of each variable affected by the evidences would provide the useful information for determining the possibility of brain metastasis.

Table 2 shows the resulting conditional probability for Site prediction as well as the findings from the changes in lung cancer site after the presentation of gender, age, and region. Examples are as follows. (1) Lung cancer in the trachea, bronchus and lung (162) mostly occurs in male patients aged of 50 to 60 and living in the east, which accounted for 27.87%. (2) Lung cancer in the trachea (1620) mostly occurs in male patients over 70 years old and living in the east, which accounted for 16.44%. (3) Lung cancer in the bronchus (1622) mostly occurs in female patients over 70 years old and living in the east, which accounted for 13.51%. (4) Lung cancer in the upper lobe, bronchus, and lung (1623) mostly occurs in male patients aged 50 to 60 and living in the south, which accounted for 19.23%. (5) Lung cancer in the middle lobe, bronchus, and lung (1624) mostly occurs in female patients aged of 50 to 60 and living in the center, which accounted for 10.42%. (6) Lung cancer in the lower lobe, bronchus and lung (1625) mostly occurs in male patients over 70 years old and living in the south, which accounted for 14%. (7) Lung cancer in other parts of the bronchus or lung (1628) mostly occurs in male patients aged of 61 to 70 and living in the center, which accounted for 24.35%. (8) Lung cancer in the bronchus and lung or an unspecified site (1629) mostly occurs in male patients aged of 50 to 60 and living in the east, which accounted for 24.35%.

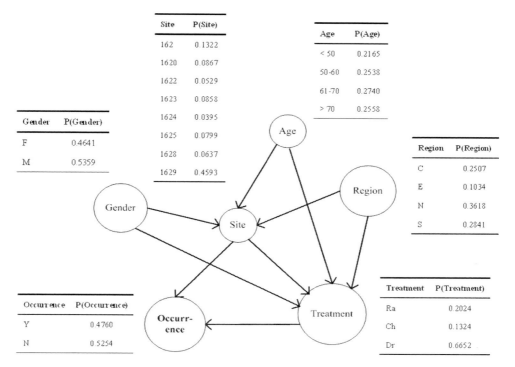

Figure 1.   The topology of brain metastasis occurrence from lung cancer with the prior probability of each node.

Table 2. Resulting conditional probability for Site prediction.

G	A	R	P(S|G,A,R) 162	1620	1622	1623	1624	1625	1628	1629
F	50	C	0.1221	0.0916	0.0992	0.0992	0.0153	0.0992	0.0687	0.4046
M	50	C	0.1791	0.0373	0.0373	0.0522	0.0597	0.0821	0.1269	0.4254
F	50–60	C	0.1736	0.0764	0.0278	0.0208	0.1042	0.0903	0.1875	0.3194
M	50–60	C	0.1067	0.1533	0.0600	0.0867	0.0600	0.1067	0.0667	0.3600
F	61–70	C	0.0348	0.1304	0.0609	0.0609	0.0087	0.1043	0.2435	0.3565
M	61–70	C	0.1968	0.0745	0.1011	0.0479	0.0106	0.1170	0.0851	0.3670
F	>70	C	0.2039	0.1456	0.0097	0.1068	0.0388	0.0388	0.1165	0.3398
M	>70	C	0.2034	0.1356	0.0565	0.1469	0.0226	0.1073	0.0960	0.2316
F	<50	E	0.2128	0.0851	0.0000	0.0000	0.0851	0.0000	0.1277	0.4894
M	<50	E	0.0962	0.0962	0.0000	0.1154	0.0000	0.1346	0.0385	0.5192
F	50–60	E	0.1111	0.0000	0.0000	0.1111	0.0556	0.0556	0.0278	0.6389
M	50–60	E	0.2787	0.0984	0.0000	0.0984	0.0984	0.0820	0.0164	0.3279
F	61–70	E	0.1231	0.0308	0.0769	0.1385	0.0000	0.0615	0.0000	0.5692
M	61–70	E	0.0635	0.1111	0.0794	0.1111	0.0794	0.0000	0.0159	0.5397
F	>70	E	0.1622	0.0405	0.1351	0.0405	0.0000	0.0270	0.0811	0.5135
M	>70	E	0.2603	0.1644	0.0000	0.0685	0.0411	0.0548	0.0548	0.3562
F	<50	N	0.1171	0.0829	0.0732	0.0537	0.0390	0.1024	0.0439	0.4878
M	<50	N	0.1898	0.1095	0.0292	0.1095	0.0511	0.0438	0.0292	0.4380
F	50–60	N	0.1587	0.1058	0.0212	0.0741	0.0212	0.0794	0.0159	0.5238
M	50–60	N	0.1167	0.0625	0.0750	0.0417	0.0042	0.0792	0.0583	0.5625
F	61–70	N	0.0920	0.1166	0.0307	0.0368	0.0245	0.0920	0.0123	0.5951
M	61–70	N	0.1722	0.0769	0.0183	0.0513	0.0733	0.0733	0.0293	0.5055
F	>70	N	0.0977	0.0837	0.0465	0.0698	0.0186	0.0837	0.0977	0.5023
M	>70	N	0.1681	0.0575	0.0265	0.0752	0.0398	0.0575	0.0619	0.5133
F	<50	S	0.0621	0.1034	0.0345	0.1793	0.0828	0.0828	0.0345	0.4207
M	<50	S	0.0741	0.0222	0.0296	0.1630	0.0444	0.0963	0.0222	0.5481
F	50–60	S	0.0680	0.0194	0.0874	0.1456	0.0243	0.0922	0.0680	0.4951
M	50–60	S	0.0692	0.1308	0.0231	0.1923	0.0692	0.0308	0.0692	0.4154
F	61–70	S	0.0682	0.1023	0.1080	0.0511	0.0398	0.0568	0.0284	0.5455
M	61–70	S	0.1122	0.1220	0.0683	0.1073	0.0244	0.0585	0.0439	0.4634
F	>70	S	0.1900	0.0800	0.0900	0.0900	0.0200	0.1400	0.0400	0.3500
M	>70	S	0.0964	0.0457	0.0711	0.0863	0.0609	0.0964	0.0457	0.4975

The resulting conditional probabilities of treatment when presenting gender, age, region, and site of lung cancer offer informative findings on Treatment prediction. For instance, patient who are male, aged of 50 to 60, living in central Taiwan, and have lung cancer in bronchus (1622), are highly likely to undergo radiotherapy. Female patients below 50 living in the south, and having cancer in the middle lobe, bronchus, and lung (1624) are very likely to undergo chemotherapy (83.33%). Likewise, the appearance of specific characteristics in patients results in the certainty of undergoing treatment with drugs. However, most patients with lung cancer in bronchus and lung, or an unspecified site (1629) rarely undergo drug treatment.

Moreover, when the proposed BN is used, we can compute for the posterior probabilities for any query/prediction given the appropriate evidence. BN allows the model to find the probability despite incomplete evidence. For example, brain metastasis is difficult to diagnose. Thus, physicians can use the proposed BN to predict the likelihood of the patient experiencing brain metastasis given particular information regarding the patient. If the patient is a 55-year-old man diagnosed with lung cancer in the trachea (1620) and living in the eastern part of Taiwan (without any information concerning treatment), the probability of the occurrence of brain metastasis can be express as $P(O \mid G = M, A = 50\text{–}60, R = E, S = 1620)$.

Based on the BN, we can compute the probability of the occurrence of brain metastasis as follow:

$$P(O = Y|G = M, A = 50 - 60, R = E, S = 1620) = \frac{P(O = Y, G = M, A = 50 - 60, R = E, S = 1620)}{P(G = M, A = 50 - 60, R = E, S = 1620)}$$

$$= \frac{\sum_T P(G=M)P(A=50-60)P(R=E)P(S=1620|G=M,A=50-60,R=E)P(T|G=M,A=50-60,R=E,S=1620)P(O=Y|S=1620,T)}{\sum_{T,O} P(G=M)P(A=50-60)P(R=E)P(S=1620|G=M,A=50-60,R=E)P(T|G=M,A=50-60,R=E,S=1620)P(O|S=1620,T)}$$

$$= \frac{P(G=M)P(A=50-60)P(R=E)P(S=1620|G=M,A=50-60,R=E)\sum_T P(T|G=M,A=50-60,R=E,S=1620)P(O=Y|S=1620,T)}{P(G=M)P(A=50-60)P(R=E)P(S=1620|G=M,A=50-60,R=E)\sum_T P(T|G=M,A=50-60,R=E,S=1620)\sum_O P(O|S=1620,T)}$$

$$= \frac{0.2715}{1} = 0.2715.$$

4 CONCLUSION

This study has proposed a BN to predict the occurrence of brain metastasis from lung cancer in patients. Data obtained from BNHI in Taiwan are used to evaluate the effectiveness of the proposed model. We concluded that the proposed BN can well predict the occurrence of brain metastasis from lung cancer. The experimental findings help physicians make more reliable decisions regarding lung cancer diagnosis and treatment.

ACKNOWLEDGEMENTS

The authors gratefully acknowledge the comments and suggestions of the editor and the anonymous referees. This work is partially supported by the National Science Council, the top-research-university project and the model-of-vocational-university project of Ministry of Education (Taiwan), and National Taiwan University of Science and Technology—Taipei Medical University Joint Research Program.

This study is based in part on data from the National Health Insurance Research Database provided by the Bureau of National Health Insurance, Department of Health and managed by National Health Research Institutes, Taiwan, R.O.C. The interpretation and conclusions contained herein do not represent those of Bureau of National Health Insurance, Department of Health or National Health Research Institutes.

REFERENCES

[1] Chi, A. & Komaki, R. (2010). Treatment of Brain Metastasis from Lung Cancer. Cancers, 2, 2100–2137. doi:10.3390/cancers2042100
[2] Pietzner, K., Oskay-Oezcelik, G., Khalfaoui, K., Boehmer, D., Lichtenegger, W., & Sehouli, J. (2009). Brain Metastases from Epithelial Ovarian Cancer: Overview and Optimal Management. Anticancer Research, 29, 2793–2798.
[3] Penel, N., Brichet, A., Prevost, B., Duhamel, A., Assaker, R., Dubois, F., & Lafitte, J.-J. (2001). Prognostic factors of synchronous brain metastases from lung cancer. Lung Cancer, 33, 143–154.
[4] Gavrilovic, I.T., & Posner, J.B. (2005). Brain metastases: epidemiology and pathophysiology. Journal of Neuro-Oncology, 75, 5–14. doi:10.1007/s11060–004–8093–6
[5] Graesslin, O. (2010). Nomogram to Predict Subsequent Brain Metastasis in Patients With Metastatic Breast Cancer. Journal of Clinical Oncology, 28 (12), 2032–2037.
[6] Hosmer, D.W., & Lemeshow, S. (2000). Applied logistic regression(2nd ed.). New York, New York, USA: A Wiley-Interscience Publication, John Wiley & Sons Inc.
[7] Badriyah, T., Briggs, J.S., & Prytherch, D.R. (2012). Decision trees for predicting risk of mortality using routinely collected data. International Journal of Social and Human Sciences, 6, 303–306.

[8] Oh, J.H., Craft, J., Lozi, R.A., Vaidya, M., Meng, Y., Deasy, J.O., Bradley, J. D, & Naqa, I.E. (2011). A Bayesian network approach for modeling local failure in lung cancer. Physics in Medicine and Biology, 56(6), 1635–1651.doi: 10.1088/0031–9155/56/6/008

[9] Kahn Jr, C.E., Roberts, L.M., Shaffer, K.A., & Haddawy, P. (1997). Construction of a Bayesian network for mammographic diagnosis of breast cancer. Comput Biol Med, 27, 19–29.

[10] Visscher, S., Lucas, P.J.F., Schurink, C.A.M., & Bonten, M.J.M. (2009). Modelling treatment effects in a clinical Bayesian network using Boolean threshold functions. Artificial Intelligence in Medicine, 46, 251–266.

[11] Medina, F.M., Barrera, R.R., Morales, J.F., Echegoyen, R.C., Chavarria J.G., & Rebora, F.T. (1996). Primary lung cancer in Mexico city: a report of 1019 cases. Lung Cancer, 14, 185–193.

[12] Hubbs, J.L., Boyd, J.A., Hollis, D., Chino, J.P., Saynak, M., & Kelsey, C.R. (2010). Factors Associated With the Development of Brain Metastases. Cancer, 5038–5046. doi: 10.1002/cncr.25254

[13] National Health Insurance Research Database, Taiwan, Bureau of National Health Insurance, Department of Health and managed by National Health Research Institutes. Available from URL: http://www.nhri.org.tw/nhird/en/index.htm [Accessed 2013 Jan.]

New Ergonomics Perspective – Yamamoto (Ed.)
© 2015 Taylor & Francis Group, London, ISBN 978-1-138-02751-0

Research on the body characteristics, figure types and regression model of elderly Taiwanese men

Chih-Hung Hsu
Institute of Lean Production Management, Hsiuping University of Science and Technology, Dali District, Taichung, Taiwan

Hui-Ming Kuo
Department of Logistics Management, Shu-Te University, Yen Chau, Kaohsiung County, Taiwan

ABSTRACT: Production and marketing standards of garment are important ergonomics standards for the garment value chain. However, little research has also been applied to standard sizing systems development for ergonomics standards. Accordingly, this study attempts to identify the body characteristics, figure types and regression model of elderly Taiwanese men using statistics as obtained from anthropometric data for sizing system development. Focusing on the elderly anthropometric data in Taiwan, this study was to develop production and marketing standards. Certain advantages may be observed when ergonomics standards are developed. The results of this study can provide a systematic approach of identifying the body characteristics, figure types and regression model of elderly Taiwanese men to improve the lean value chain-oriented production and marketing standards in the garment industry. This study provides a precise approach for identifying figure types for establishing standard elderly aged male size systems based on anthropometric data to facilitate apparel production. These new standards can help garment factories improve the fit of mass produced clothing, and their application in manufacturing can reduce production costs and increase competitiveness.

Keywords: body characteristics; figure types; regression model; elderly Taiwanese men

1 INTRODUCTION

Among textile manufacturing industries, garment manufacturing produces products with the highest added value [1]. In the garment industry, large-scale machine production has replaced manual production, which has greatly increased output and reduced manufacturing costs. Because certain standards and specifications must be followed in large-scale machine production, each country must have its own standard-sizing systems for manufacturers to follow. Furthermore, standard-sizing systems can correctly predict production numbers and proportion of sizes to be produced, resulting in accurate production planning and material control [2, 3].

Garment-sizing systems were originally based on those developed by tailors in the late 18th century. Before then, all garments were hand-made to order. Tailors measured the body dimensions of each customer, and then drew and cut patterns for each garment, which yielded the most satisfactory fitting possible, but was only suitable for individuals whose body dimensions had been measured. After many original patterns had been accumulated, the tailors discovered correlations between bodily dimensions, regardless of the individual differences. Tailors gradually developed these patterns into a system of garment storage, which could be used to make clothes for people with similar figures [4].

In 1959, Emanuel determined a set of procedures for formulating standard sizes for all figure types [5]. According to this system, people of all figure types were first classified into one of four bodyweight groups. These bodyweight groups were further subdivided into tall and short people. People were, thus, divided into eight categories based on similar heights and weights. The sizing systems in other countries were also similar, with classifications based on two or three sizing variables [6]. The sizing variables for female garments are height, bust girth and hip girth; the sizing variables most commonly used for male garments are typically height, chest girth and waist girth [7, 8].

In 1998, McCulloch, et al. proposed criteria by which sizing systems could be evaluated [9]. They stated that the sizing systems should ideally.

- Cover the greatest number of people.
- Cover them with the fewest number of sizes.

At times these criteria conflict with each other. At other times one takes priority of another, depending on the type of garment and the needs of the customer.

Data mining, a major step of Knowledge Discovery in Database (KDD), has been successfully applied in many fields, including finance [10], health insurance [11], biomedicine [12], manufacturing [13], and e-commerce [14]. However, research on establishing sizing systems using data mining is lacking.

One of the most important data mining methods is the decision tree technique, which can be used to classify data according to rules deduced from input variables, display the data in a tree-shaped form and explore the significance of influencing factors. Data can be divided into two types: discrete and continuous. Different decision tree algorithms suit different types of data. Of the many available algorithms, the most appropriate method for processing continuous data is the Classification and Regression Tree (CART) [15, 16], which this study uses, since anthropometric data is continuous. CART puts pre-processed data into the tree root, and then, through a series of classifying processes, it determines the best separating points, which it uses to form a tree-shaped structure. Data is split according to the purity or impurity of the child nodes by classification rules. These rules incorporate the valuable information gained from mining the data and they can be used to classify these data [17].

The regression analysis is also the most important data mining methods. In statistics, regression analysis is a statistical process for estimating the relationships among variables. It includes many techniques for modeling and analyzing several variables, when the focus is on the relationship between a dependent variable and one or more independent variables. More specifically, regression analysis helps one understand how the typical value of the dependent variable (or 'criterion variable') changes when any one of the independent variables is varied, while the other independent variables are held fixed. Most commonly, regression analysis estimates the conditional expectation of the dependent variable given the independent variables—that is, the average value of the dependent variable when the independent variables are fixed. Less commonly, the focus is on a quantile, or other location parameter of the conditional distribution of the dependent variable given the independent variables. In all cases, the estimation target is a function of the independent variables called the regression function. In regression analysis, it is also of interest to characterize the variation of the dependent variable around the regression function which can be described by a probability distribution.

2 METHODS

2.1 *Defining the subject for data mining*

Because of the urgent need for accurate sizing systems, an anthropometric database was created based on static anthropometric variables measured in Taiwan. Because standard-sizing systems for elderly Taiwanese men's pants were outdated and incomplete, this study uses the CART and regression analysis data mining technique to explore and analyze the anthropometric data to identify systematic patterns in body dimensions. Based on these body dimension patterns, the bodies of elderly Taiwanese men can be classified into representative figure types and standard-sizing systems can be established. This work will be beneficial to the production in Taiwan.

2.2 *Data preparation and analysis for data mining*

Before the data are mined, they must be examined and purified. In this study, all missing or abnormal data were omitted. Because all body dimensions were measured in millimeters with decimals, they were converted into integers in centimeters, allowing comparison and calculation with commonly used international garment-sizing units. Additionally, two anthropometric variables, body weight and height, were converted into a new variable, the Body Mass Index (BMI). The World Health Organization defines BMI as weight divided by height squared. It is used in medical science as a reference for judging obesity-related diseases and whether a figure type is standard or not. In this study BMI served as the target variable of the decision tree and could be used to facilitate data mining.

Not all of the 265 anthropometric variables were useful for establishing the sizing systems of pants; hence, the body dimensions commonly used in garment manufacturing are first considered. Domain experts were consulted to determine 8 anthropometric variables that are strongly associated with garment production.

Using all 8 anthropometric variables to establish sizing systems would be very difficult. To simplify, we attempted to extract the most important factors. Using Kaiser's eigenvalue criterion, we selected two factors with eigenvalues over one [19]. Factor loadings were calculated to determine the coefficients of correlation between the two factors and the anthropometric variables. Consequently, anthropometric variables with factor loadings of greater than 0.7 were found to be clustered in Factors 1 and 2. Most of the variables that appeared in Factor 1 were girth-related variables and included waist girth, hip girth, thigh girth and weight. The variables appeared in Factor 2 were height-related variables and included outside leg length, crotch length and height. Thus, Factor 1 was called the girth factor, and Factor 2 was called the height factor.

The results of the factor analysis presented in Table reveals that, besides weight, the top three anthropometric variables most closely correlated with the girth factor were waist girth, hip girth and thigh girth. Although hip girth was the anthropometric variable most closely correlated with the girth factor, waist girth is the most important variable in establishing sizing systems of pants in garment making [8, 20]. Therefore, waist girth was selected to represent the girth factor. Because the anthropometric variable that correlated most closely with the height factor was outside leg length, it was selected to represent the height factor.

2.3 *CART analysis and regression analysis*

Once waist girth and outside leg length were selected by factor analysis to be the most important sizing variables, the CART decision tree technique was used to mine data. This study takes BMI as the target variable, with waist girth and outside leg length being predictors used to classify the target variable. The following stopping rules were set as follow.

- The greatest depth of the tree extends to the fourth level beneath the root node.
- The minimum number of samples in the parent node is 100, and the minimum number of samples in the child node is 50.

Figure 1 plots a regression and distribution graph of waist girth as the X-axis against hip girth on the Y-axis to show the distribution of all of figure types. Hip girth is also an important variable for sizing male pants in many countries and it significantly correlates with the girth factor. This study identifies the figure type of 118 people with smaller waist girth and hip girth as Y, the figure type of 299, the majority, as A, the figure type of 88 soldiers with larger waist girth and hip girth as C, and the figure type of the remaining 208 samples as B.

A line graph was plotted to yield better insight into the differences among the four CART-classified figure types. As shown in Figure 2, the four figure types by marked differences in waist girth, hip girth and thigh girth. ANOVA results and Duncan's multiple range tests further confirmed the existence of the different girth among the four figure types.

2.4 *Establishing the sizing systems*

Waist girth, hip girth and outside leg length are the most commonly used variables for sizing male pants. This section includes relevant scatter plots of waist girth on the X-axis against hip girth on the Y-axis at intervals of 4 cm. The waist girths were classified as follows.

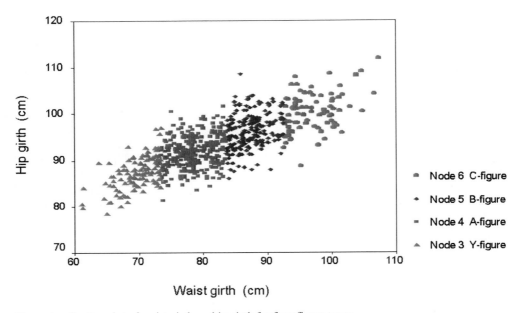

Figure 1.   Scatter plot of waist girth vs. hip girth for four figure types.

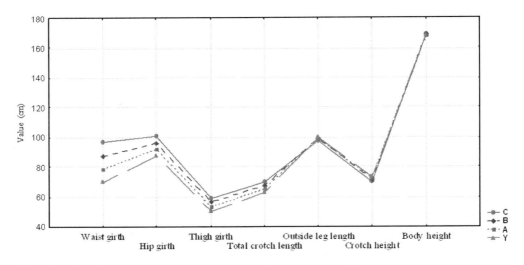

Figure 2.   Four figure types and corresponding anthropometric variables.

- Y figure, 68 cm and 72 cm;
- A figure, 76 cm, 80 cm and 84 cm;
- B figure, 88 cm and 92 cm;
- C figure, 96 cm, 100 cm, and 104 cm.

Using the methods of Emanuel [5], the sizing systems for the four figure types were established.

Figure 3 shows a scatter plot for the Y figure of hip girth on the Y-axis against waist girth on the X-axis. The waist girths ranged from 61 cm to 74 cm. Since most countries use 4 cm as the interval of waist girth [21], the 68 cm and 72 cm waist girths were used for the Y figure. For the 74 cm waist girth, there was a small overlap between the A figure and the Y figure. It resulted from the CART rules, which divided the Y figure and the A figure by 73.5 cm. The small overlap is allowable, because it can offer greater size diversity for elderly Taiwanese men when they choose clothes.

Figure 3 shows a scatter plot of 118 samples of the Y figure. The scatter plots of the other types of figures are similar to that of the Y figure type. Table shows complete pants sizing systems for all four-figure types. Samples less than 0.3% were deleted in each size group. Out

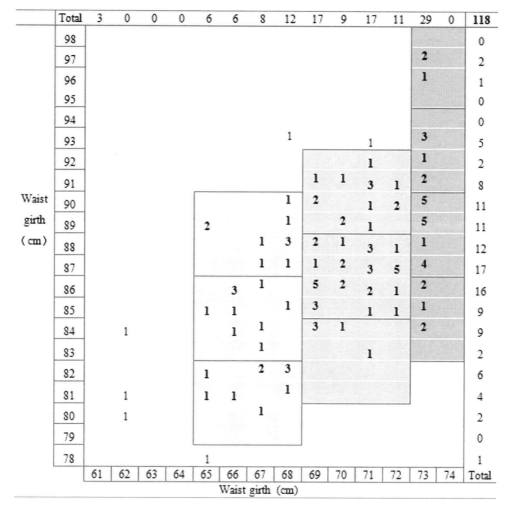

Waist girth (cm)	Total	61	62	63	64	65	66	67	68	69	70	71	72	73	74	Total
		3	0	0	0	6	6	8	12	17	9	17	11	29	0	118
98																0
97														2		2
96														1		1
95																0
94																0
93									1			1		3		5
92												1		1		2
91										1	1	3	1	2		8
90								1		2		1	2	5		11
89						2		1		2		1		5		11
88							1	3		2	1	3	1	1		12
87							1	1		1	2	3	5	4		17
86							3	1		5	2	2	1	2		16
85						1	1		1	3		1	1	1		9
84		1						1	1	3	1			2		9
83									1			1				2
82						1		2	3							6
81			1			1	1		1							4
80			1					1								2
79																0
78						1										1

Figure 3. Scatter plot for Y-figure.

349

of the 713 samples, only 45 samples were excluded. Therefore, the coverage of the proposed "waist girth and hip girth" sizing systems was 93.69%.

3 DISCUSSION AND CONCLUSION

The four figure types were obtained by applying the classification rules based on the data mining technique. The newly developed sizing systems exhibit the following three characteristics.

1. High coverage rates and few sizing groups
2. Regular patterns and rules
3. Manufacturing reference points to facilitate production

This study applies data mining to develop systems for sizing pants of elderly Taiwanese men in Taiwan. The obtained systems provide the following advantages.

- The total coverage of the sizing systems is 91.16%, which is relatively high.
- The sizing systems based on the decision tree show regular patterns and rules.
- The total number of the size groups comes to only 47, much fewer than the number of group in the previous sizing systems. It has outlined the percentages of people for every figure type in each size group, to provide manufacturers with detailed reference points for garment production.

The obtained systems for sizing elderly Taiwanese men' pants can be used to generate a realistic production plan. Unnecessary inventory costs due to sizing mismatches between number of elderly Taiwanese men and produced pants can be minimized.

ACKNOWLEDGEMENTS

This study was partial supported by a grant from the Ministry of Science and Technology, Taiwan, and Project No. NSC 102–2221-E-164 -010 -

REFERENCES

[1] C.F. Chang, The model analysis of female body size measurement from 18 to 22, Journal of Hwa Gang Textile, 6(1), pp. 86–94, 1999.
[2] K.M. Hsu and S.H. Jing, The chances of Taiwan apparel industry, Journal of the China Textile Institute, 9(2), pp. 1–6, 1999.
[3] Y.M. Tung and S.S. Soong, The demand side analysis for Taiwan domestic apparel market, Journal of the China Textile Institute, 4(5), pp. 375–380, 1994.
[4] L.D. Burns and N.O. Bryant, The Business of Fashion: Designing, Marketing, and Manufacturing, Fairchild, U.S.A., 2000.
[5] Emanuel, A height-weight sizing system for fight clothing, WADCTR, Aero Med lab, 56–365, 1959.
[6] G. Cooklin, Pattern Grading for Women's Clothes, Blackwell, Oxford, 1992.
[7] S.P. Ashdown, An investigation of the structure of sizing systems, International Journal of Clothing Science and Technology, 10(5), pp. 324–341, 1998.
[8] C.Y. Jongsuk and C.R. Jasper, Garment-sizing systems: an international comparison, International Journal of Clothing Science and Technology, 5(5), pp. 28–37, 1993.
[9] C.E. McCulloch, B. Paal and S.A. Ashdown, An optimal approach to apparel sizing, Journal of the Operational Research Society, 49(5), pp. 492–499, 1998.
[10] R. Gerritsen, Assessing loan risks: a data-mining case study, IT Professional, 1(6), pp. 16–21, 1999.
[11] Y.M. Chas, S.H. Ho and K.W. Cho, et al., Data mining approach to policy analysis in a health insurance domain, International Journal of Medical Informatics, 62, pp. 103–111, 2001.
[12] M. Maddour and M. Elloumi, A data mining approach based on machine learning techniques to classify biological sequences, Knowledge-Based Systems, 15 (4), pp. 217–223, 2002.

350

[13] C.X. Feng and X. Wang, Development of empirical models for surface roughness prediction in finish turning, International Journal of Advanced Manufacturing Technology, 20(5), pp. 348–356, 2002.

[14] W.T. Ying and S.P. Lee, Data mining using classification techniques in query processing strategies, Computer Systems and Applications, ACS/IEEE International Conference, pp. 200–202, 2002.

[15] C. Apte and S.Weiss, Data mining with decision trees and decision rules, Future Generation Computer System, 13, pp. 197–210, 1997.

[16] L. Breiman, J.H. Friedman, and R.A. Olshen, et al., Classification and Regression Tree, Chapman & Hall, Florida, 1998.

[17] M. Berry and G. Linoff, Data Mining Techniques: for Marketing, Sales, and Customer Support, Wiley, New York, 1997.

[18] Ministry of National Defense, Defense Report for Republic of China: Recruit Procedures, Taiwan, 2002.

[19] S.Y. Chen, Multivariate Analysis, Hwa-Tai, Taipei, 2000.

[20] G. Cooklin, Pattern Grading for Men Clothes, Blackwell, Oxford, 1992.

[21] J.M. Winks, Clothing Sizes: International Standardization, Redwood, U.K., 1997.

New Ergonomics Perspective – Yamamoto (Ed.)
© 2015 Taylor & Francis Group, London, ISBN 978-1-138-02751-0

Anthropometric data of Malaysian workers

Siti Nurani Hassan

Consultation, Research and Development Department, National Institute of Occupational Safety and Health (NIOSH), Selangor, Malaysia
Department of Mechanical and Manufacturing Engineering, Faculty of Engineering, Universiti Putra Malaysia, Malaysia

Rosnah Mohd Yusuff

Department of Mechanical and Manufacturing Engineering, Faculty of Engineering, Universiti Putra Malaysia, Malaysia

Raemy Md Zein

Consultation, Research and Development Department, National Institute of Occupational Safety and Health (NIOSH), Selangor, Malaysia

Mohd Rizal Hussain

Institute Gerantology, Universiti Putra Malaysia, Serdang, Malaysia

Hari Krishnan Tamil Selvan

Consultation, Research and Development Department, National Institute of Occupational Safety and Health (NIOSH), Selangor, Malaysia

ABSTRACT: Studies on anthropometry in Malaysia have been conducted by many researchers but mostly focused on specific groups of people. The objective of this study to develop anthropometric database for Malaysian workers. In this study, 23 static anthropometric dimensions of 1134 Malaysian workers comprising of 863 males and 261 females from 10 industrial sectors classified under the Law of Malaysia Occupational Safety and Health Act 514. The measurements taken followed the guidelines recommended by MS ISO 7250–1:2008, Basic Human Body Measurement for Technological Design. Anthropometry tool and an anthropometric grid was designed and developed to facilitate the measurement. Results showed that the mean statute height and standard deviation for males and females were 169.57+/–7.57 cm and 156.83+/–5.97 cm respectively. In addition, there were significant differences found between male and female for all measurements except for sitting thigh clearance. The anthropometry data collected can provide some guidelines in designing a safer and healthier workplace for Malaysian workers. Future work should be emphasizing in other states and to have a representative for each industry in Malaysia.

Keywords: Ergonomics; anthropometrics; malaysian workers

1 INTRODUCTION

The word "anthropometry" is derived from the Greek words "anthropos" (man) and "metron" (measure) and means measurement of the human body (Bridger, 2009). Anthropometry focuses on the measurement of bodily features such as body shape and body composition ("static anthropometry"), the body's motion and strength capabilities and use of space ("dynamic anthropometry"). Anthropometric measurements are used widely in a variety of scientific and technical fields. Within the field of ergonomics, the application of anthropometric measurements is primarily associated with different aspects of design for human use.

An important science in design of machines and workstation involve anthropometric data in order to fit machines and workstation with human body (Hanson et al., 2009). Karmegam et al., (2011) reported that there is lack of anthropometric data involving major ethnic group in Malaysia. Various studies on anthropometry in Malaysia were conducted by Deros et al., (2009), Rosnah and Wong (1996), Karmegam et al., (2011) and Ngeow and Aljunid (2009) focusing on young adults. Anthropometric for older Malaysian were also reported by Sharifah Norazizan et al., (2006), Rosnah et al., (2006) and Suriah et al., (1998).

However, there is lack of documentation of Malaysian anthropometry especially among the workers. Since there is no anthropometry database that can be used as reference to design Malaysian workers workplace, they are exposed to musculoskeletal disorders (MSDs). According to the Social Security Organization (SOCSO), Malaysia, there were only 14 cases of MSDs in 2006, 268 cases in 2011 and increase to 449 in year 2012 (Thomas, 2013).

Therefore, there is need to develop anthropometric database for Malaysian workers. This study is the first of its kind to establish a database to be used as reference value in designing workplace for Malaysian workers.

2 METHODOLOGY

Twenty three static anthropometric dimensions of 1134 Malaysian workers comprising of 863 males and 261 females from 10 industrial sectors were taken. Participants were selected from those attending training at National Institute of Occupational Safety and Health (NIOSH), Bandar Baru Bangi, Selangor, Malaysia from August 2012 until May 2013. The participants came from various parts of Malaysia and their signed consent was obtained prior to the measurement.

A customized anthropometric measurement form with body landmark figures was used as a guide. The measurements were conducted as recommended by MS ISO 7250–1:2008, Basic Human Body Measurement for Technological Design. An anthropometric grid, measuring 213.36 cm × 274.32 cm and 121.92 cm × 167.64 cm for standing and sitting respectively was designed and developed to facilitate the measurement of 19 anthropometric dimensions. The anthropometric grid was fixed to the wall. Four sitting measurements (shoulder-elbow length, buttock popliteal length, thigh clearance and knee height) were measured using the anthropometer for better accuracy. The grid was copyrighted as shown in Figures 1 and 2.

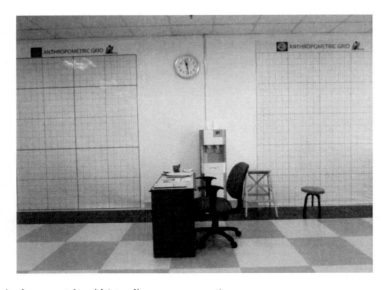

Figure 1.   Anthropometric grid (standing measurement).

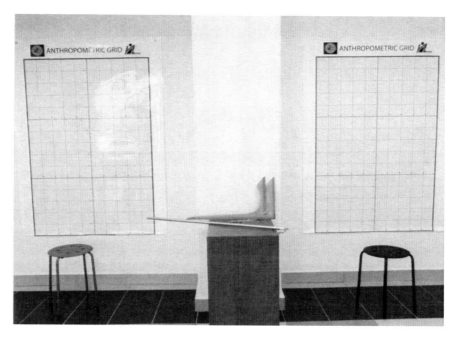

Figure 2.   Anthropometric grid (sitting measurement).

Measurements were conducted by four trained facilitators consisting of two males and two females. Due to ethical and cultural consideration, the male respondents were measured by male facilitators and female respondents by female facilitators. All respondents were lightly clothed for the measurement. The data was analyzed using the **IBM SPSS** Statistics version 21. Level of significance used for the data was set at $p < 0.05$ (two-tailed). Descriptive analysis was performed to identify frequency, percentage, Mean, Standard Deviation and Percentile Value of the respondent's demographic background and anthropometric dimensions. Anthropometric dimensions taken are as shown in Table 1, Figure 3 and Figure 4.

Table 1.   List of anthropometric dimensions.

No.	Body dimensions	No.	Body dimensions
	Standing Posture		*Sitting Posture*
1.	Vertical grip reach	13.	Elbow span
2.	Stature	14.	Span
3.	Eye height	15.	Sitting height (erect)
4.	Shoulder height	16.	Eye height, sitting
5.	Armpit height	17.	Shoulder height, sitting
6.	Elbow height	18.	Elbow height, sitting
7.	Hip height	19.	Elbow-Fingertip length
8.	Knuckle height	20.	Shoulder-elbow length
9.	Fingertip height	21.	Buttock popliteal length
10.	Tibial height	22.	Thigh clearance
11.	Biacromial breadth	23.	Knee height
12.	Bideltoid breadth		

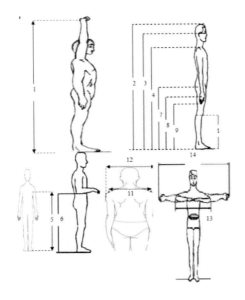

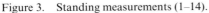

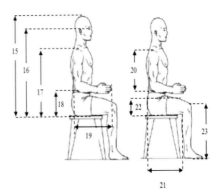

Figure 3.   Standing measurements (1–14).   Figure 4.   Sitting measurements (15–23).

3   RESULTS AND DISCUSSION

The demographic background of respondents is shown in Table 2. Majority the respondents were 30 years and below (54.2%) with the mean age 32.56 (SD = 9.49), male (76.1%), Malays (81.7%), married (56.0%), technician and associates professionals (36.9%) and public services and statutory authorities (33.4%). The results of the mean and standard deviation of the anthropometry data are shown in Table 3. In all 22 measurements, significant differences were found between male and female in term of standing vertical grip strength (t = 22.88, p < 0.01), stature (t = 25.34, p < 0.01), standing eye height (t = 24.99, p < 0.01), standing shoulder height (t = 25.02, p < 0.01), standing armpit height (t = 25.46, p < 0.01), standing elbow height (t = 21.96, p < 0.01), standing hip height (t = 5.08, p < 0.01), standing knuckle height (t = 15.27, p < 0.01), standing fingertip height (t = 14.95, p < 0.01), standing tibial height (t = 13.58, p < 0.01), standing biacromial breadth right (t = 9.19, p < 0.01), standing biacromial breadth left (t = 10.58, p < 0.01), standing bideltoid breadth right (t = 9.47, p < 0.01), standing bideltoid breadth left (t = 11.56, p < 0.01), standing elbow span right (t = 16.10, p < 0.01), standing elbow span left (t = 17.83, p < 0.01), standing span right (t = 22.71, p < 0.01), standing span left (t = 25.16, p < 0.01), sitting height (t = 13.00, p < 0.01), sitting eye height (t = 11.04, p < 0.01), sitting shoulder height (t = 9.60, p < 0.01), sitting elbow height (t = −3.00, p < 0.01), sitting elbow-fingertip length (t = 23.55, p < 0.01), sitting shoulder-elbow length (t = 18.54, p < 0.01), sitting buttock-popliteal length (t = 8.51, p < 0.01) and sitting knee height (t = 19.98, p < 0.01). However, only sitting thigh clearance was found no differences (t = 1.10, p = 0.273) between male and female workers in this study. These finding were similar to studies by Rosnah, Rizal & Sharifah Norazizan (2009), Haitao et al., (2007) and Onuoha, Idike & Oduma (2012) where anthropometric dimensions are different between gender.

The lower percentile value (5th) should be used when designing reachability (example vertical grip reach) and 95th percentile value for clearance (example stature, thigh clearance) as shown in Tables 4 and 5. This anthropometry data collected can provide some guidelines in designing a safer and healthier workplace for Malaysian workers.

Table 2. Demographic background of respondents.

Variable	Number (N)	Percentage (%)	Mean	Std. Deviation
Age			32.56	9.49
30 years and below	613	54.2		
31–44 years old	363	32.0		
45 years and above	156	13.8		
Gender				
Male	863	76.1		
Female	271	23.9		
Ethnicity				
Malay	927	81.7		
Chinese	71	6.3		
Indian	105	9.3		
Others	31	2.7		
Marital Status				
Single	495	43.6		
Married	635	56.0		
Divorced	4	0.4		
Occupational Category				
Managers	86	7.8		
Professionals	340	30.7		
Technician & Associate Professionals	408	36.9		
Clerical Support Workers	96	8.7		
Service and Sale Workers	22	2.0		
Craft & Related Trades Workers	69	6.2		
Plant & Machine: Operators and Assembler	85	7.7		
Industry Category				
Manufacturing	216	19.1		
Mining and Quarrying	164	14.5		
Construction	190	16.8		
Agriculture, Forestry and Fishery	1	0.1		
Utilities	23	2.0		
Transportation, Storage and Communication	98	8.7		
Wholesale and Retail Trades	4	0.4		
Finance, Insurance, Real Estate and Business Services	54	4.8		
Public Services and Statutory Authorities	378	33.4		

Table 3. Mean and standard deviation for malaysians male and female workers.

Anthropometry dimension (cm)	Male		Female	
	Mean	SD	Mean	SD
Standing Posture				
Vertical Grip Reach	200.85	10.71	184.27	9.38
Stature	169.57	7.57	156.83	5.97
Eye Height	157.02	7.25	144.50	7.02
Shoulder Height	140.73	6.19	130.02	6.03
Armpit Height	125.31	5.84	115.81	5.20
Elbow Height	104.49	6.30	96.86	4.50
Knuckle Height	71.16	4.34	66.49	4.56
Fingertip Height	61.92	3.92	58.21	3.45

(*continued*)

Table 3.　Continued.

Anthropometry dimension (cm)	Male		Female	
	Mean	SD	Mean	SD
Tibial Height	44.47	3.11	41.37	3.33
Biacromial Breadth Right	16.50	2.51	14.90	2.49
Biacromial Breadth Left	16.20	2.42	14.40	2.51
Bideltoid Breadth Right	23.64	2.74	21.80	2.94
Bideltoid Breadth Left	23.51	2.75	20.98	3.25
Elbow Span Right	44.20	2.96	40.91	2.89
Elbow Span Left	43.94	3.22	40.01	3.00
Span Right	85.44	4.39	78.47	4.47
Span Left	85.43	4.55	77.63	4.10
Sitting Posture				
Height	85.59	4.48	81.69	3.70
Eye Height	73.02	4.91	69.35	4.30
Shoulder Height	57.75	3.90	55.28	3.62
Elbow Height	21.73	3.95	22.53	3.48
Elbow Fingertip Length	46.61	2.47	42.28	3.13
Shoulder Elbow Length	35.66	2.43	32.47	2.59
Buttock Popliteal Length	47.81	2.97	45.73	3.66
Thigh Clearance	15.48	2.46	15.28	3.00
Knee Height	50.91	3.24	46.30	3.56

Table 4.　Percentile values (P) of anthropometry dimensions of malaysian male workers (n = 863).

Anthropometry dimension (cm)	P5	P25	P50	P75	P95
Standing Measurement					
Vertical Grip Reach	186.00	194.60	200.40	206.70	216.78
Stature	159.00	165.10	169.20	173.60	180.48
Eye Height	145.74	152.70	157.00	161.50	168.50
Shoulder Height	131.40	136.80	140.20	144.60	150.58
Armpit Height	116.30	121.30	125.00	129.00	135.30
Elbow Height	96.40	101.00	104.20	107.50	112.98
Hip Height	77.10	82.20	85.10	89.00	94.90
Knuckle Height	64.30	68.10	71.00	73.80	78.50
Fingertip Height	55.42	59.30	61.80	64.50	68.30
Tibial Height	39.40	42.50	44.50	46.30	49.08
Biacromial Breadth Right	12.32	14.80	16.50	18.10	20.80
Biacromial Breadth Left	12.50	14.60	16.00	17.80	20.00
Bideltoid Breadth Right	19.22	22.00	23.70	25.30	27.98
Bideltoid Breadth Left	19.72	21.90	23.30	25.00	27.78
Elbow Span Right	39.02	42.20	44.30	46.10	49.00
Elbow Span Left	38.32	42.00	44.00	46.20	48.88
Span Right	78.50	82.60	85.30	88.20	93.08
Span Left	77.82	82.50	85.20	88.50	92.78
Sitting Measurement					
Height	77.64	82.90	85.70	88.60	92.88
Eye Height	65.40	70.20	73.10	76.20	80.28
Shoulder Height	51.02	55.30	58.00	60.20	63.70
Elbow Height	15.50	19.20	21.90	24.10	27.70
Elbow Fingertip Length	43.00	45.00	46.60	48.00	50.48
Shoulder Elbow Length	31.80	34.30	35.60	37.00	39.20
Buttock Popliteal Length	43.30	45.80	47.80	49.90	52.60
Thigh Clearance	11.90	13.90	15.20	16.80	19.50
Knee Height	46.50	49.00	50.80	52.70	56.00

Table 5. Percentile values (P) of anthropometry dimensions of malaysian female workers (n = 271).

Anthropometry dimension (cm)	P5	P25	P50	P75	P95
Standing Measurement					
Vertical Grip Reach	170.84	178.80	183.00	190.00	196.90
Stature	148.30	152.70	156.20	160.70	166.52
Eye Height	134.50	140.00	144.30	149.00	154.64
Shoulder Height	122.00	126.10	129.60	133.80	138.70
Armpit Height	108.00	112.00	115.70	118.90	124.50
Elbow Height	90.24	94.00	96.50	99.30	104.80
Hip Height	71.92	79.50	83.70	88.00	92.64
Knuckle Height	60.46	63.80	66.40	69.00	72.34
Fingertip Height	52.92	55.90	58.10	60.40	63.44
Tibial Height	36.82	39.00	41.00	43.50	47.34
Biacromial Breadth Right	11.20	13.40	14.80	16.40	18.50
Biacromial Breadth Left	9.86	13.00	14.50	16.00	18.60
Bideltoid Breadth Right	17.70	20.00	21.60	23.40	26.32
Bideltoid Breadth Left	16.12	19.00	20.80	22.70	25.68
Elbow Span Right	36.12	39.00	40.80	43.00	45.84
Elbow Span Left	35.56	37.80	39.80	42.00	45.14
Span Right	71.50	75.80	78.30	81.50	85.16
Span Left	71.00	75.00	77.30	80.00	85.00
Sitting Measurement					
Height	75.66	79.50	82.00	84.00	87.60
Eye Height	62.10	66.70	69.60	72.30	76.00
Shoulder Height	49.76	53.40	55.10	57.00	61.36
Elbow Height	17.00	20.00	22.90	24.80	28.34
Elbow Fingertip Length	38.80	40.90	42.40	43.90	46.62
Shoulder Elbow Length	29.36	30.90	32.30	33.70	35.50
Buttock Popliteal Length	40.34	43.50	45.60	48.30	51.08
Thigh Clearance	11.00	13.50	15.00	16.80	19.28
Knee Height	42.00	44.80	46.50	48.20	50.44

4 APPLICATION OF ANTHROPOMETRY IN DESIGNING WORKPLACE

In workplace, it is a policy of employers to provide a safe and healthy work environment for all its employees and protect others who may be affected by its activities. This consideration should be taking into account when designing workplace. Improperly designed, it can cause the aches and symptoms around the shoulder, neck, nape and waist, and to the problems with the muscle and skeleton systems (Kalınkara et al., 2001).Workplace design is one of the major areas in which human factor professionals can help improve the fit between human, machines and environments. Therefore, if the anthropometry of working population is considered, a better design of a repetitive work task could be achieved.

For example, in designing seated work, the height of the seat and desk is critical dimensions for comfort. The seat height should be no higher than the popliteal height of female 5th percentile for both feet can be rested firmly on the floor. For desk height, it should be at sitting elbow height of female 5th percentile. However, to suit majority of workers population, it is highly recommended to design an adjustable chair and worktop (Rosnah, Anwarul & Muthamil, 1994).

5 CONCLUSION

With the anthropometry data, the Malaysian workers population can be better described. Design solution through anthropometric consideration will make it easier for work to be carried out and improve the performance of workers. The use of appropriate anthropometric measurements in designing will also prevent the occurrence of MSDs at the workplace.

ACKNOWLEDGEMENTS

The authors thank NIOSH Malaysia for funding, and also each Malaysian worker who participated in the study. We also would like to thank Amir Hamzah Harun, Mohd Abd Muiz Che Abdul Aziz, Nur Alyani Fahmi Salihen, Siti Hajar Mohd Yusuff who co-operated in the study and sampling session.

REFERENCES

[1] Bridger, R.S. 2009. *Introduction to ergonomics*. 3rd edition. (Taylor & Francis Group New York, USA).

[2] Deros, B.M., D. Mohamad, A.R. Ismail, O.W. Soon and K.C. Lee. 2009. "Recommended chair and work surfaces dimensions of VDT tasks for Malaysian citizens." *Eur. J. Sci. Res.* 34: 156–167.

[3] Haitao, H., Zhizhong, L., Jingbin, Y., Xiaofang, W., Hui, X., Jiyang, D. & Li, Z. 2007. "Anthropometric Measurement of the Chinese Elderly Living in the Beijing." *Int. Journal of Industrial Ergonomics* 37(4): 303–311.

[4] Kalinkara, V. et al. 2011. "Anthropometry measurements related to the workplace design for female workers employed in the textiles sector in Denizli, Turkey". *Eurasian J. Anthropol,* 2(2): 102–111.

[5] Karmegam, K., Sapuan, S.M., Ismail, M.Y., Ismail, N., Shamsul Bahri, M.T., Shuib, S., Mohana G.K., Seetha, P., TamilMoli, P. & Hanapi, M.J. 2011 "Anthropometric study among adults of different ethnicity in Malaysia." *Int. Journal of the Physical Sciences* 6(4): 777–788.

[6] M.Y. Rosnah, S.A.R. Sharifah Norazizan, S.H. Nurazrul, H. Tengku Aizan, H.H. Ahmad, M.S. Aini, G.S.C. Lina, W.C. Lo & H. Mohd Rizal. 2006. "Comparison of Elderly Anthropometry Dimensions amongst Various Population." *Asia-Pacific Journal of Public Health* 18 (supp): 20–25.

[7] Ngeow, W.C. & Aljunid, S.T. 2009. "Craniofacial anthropometric norms of Malays." *Singapore Medical Journal* 50: 525–528.

[8] Onuoha, S.N., Idike, F.I. & Oduma, O. 2012. "Anthropometry of South Eastern Nigeria Agricultural Workers." *Int. Journal of Engineering and Technology* 2(6): 1089–1095.

[9] Rosnah Mohd. Yusuff and O.O.Wong. 1996. "Ergonomically designed chair for Malaysian VDT users." *ASEAN Journal of Science and Technology Development* 13(2): 49–65.

[10] Rosnah, M.Y., Mohd Rizal, H. & Sharifah Norazizan, S.A.R. 2009. "Anthropometry dimensions of older Malaysians: comparison of age, gender and ethnicity." *Asian Social Science* 5(6): 133–140.

[11] Rosnah, M.Y., Anwarul, M. & Muthamil, C. 1994. "Anthropometry dimensions and preferred working surface heights for the electronic operators workbench." *Pertanika J. Sci & Technol* 2(1): 65–71.

[12] S.A.R. Sharifah Norazizan, H. Tengku Aizan, M.Y. Rosnah, H. Ahmad Hariza, M.S. Aini, H. Mohd Rizal & G.S.C. Lina. 2006. "Anthropometric Data of Older Malaysians." *Asia-Pacific Journal of Public Health* 18 (supp): 35–41.

[13] Suriah, A.R., Zalifah, M.K., Zainorni, M.J., Shafawi, S., Mimie Suraya, S., Zalina, N. and Wan Zainuddin, W.A. 1998. "Anthropometric measurement of the elderly." *Mal J Nutr* 4: 55–63.

[14] Thomas, S. 2013. "Musculoskeletal Diseases Reporting". Paper presented at Seminar Ergonomics for Occupational Diseases and Injury. June 26. Johor, Malaysia.

New Ergonomics Perspective – Yamamoto (Ed.)
© 2015 Taylor & Francis Group, London, ISBN 978-1-138-02751-0

Analysis of virtual makeup sensory reactions between young men and women

Takeshi Sato
Jissen Women's University, Hino, Tokyo, Japan

Hirohi Yasuoka
Tokyo University of Information Science, Wakaba-ku, Chiba, Japan

Shigeka Shioji
Jissen Women's University, Hino, Tokyo, Japan

Macky Kato
Waseda University, Mikajima Tokorozawa, Saitama, Japan

ABSTRACT: It was well known to apply makeup to get well daily life for especially dementia person. Then it was very important to makeup, enjoying paint own face beautifully for all women. Recent years there were many opportunities to makeup conditions during recruiting occupational period in Japanese university students. They had a training session about makeup and arrangement of hairstyle to best suited to getting jobs. That's the way that the purpose of this study was to investigate in difference between a man and a woman about virtual image of recruited makeup and hairstyle.

There were over 82% student from preliminary survey ($n = 145$), to learn important events markup for recruiting. There were 22 students (men: 6, female: 16) participate in this experiments to evaluate the 16 image of virtual markup patterns. It was produced the virtual markup images from jKiwi (GPL, OpenSource, ver0.9.5) to according to result of questionnaire markup materials (eye shadow: 2 colors, cheek color: 2 colors, lip liner: 2 color, and basemake: 1 type). It was applied sensory reactions in both pair comparison method and normalized standings methods. Additionally we tried to data acquisition of point of gaze compare two different images side-by-side by using EMR-9 (Eye Mark Recorder, Nac Image Technology, Japan).

There was the most popular image of pair comparison method that not too long hairstyle with elastic viewable ear, eye shadow with light brown color than pink, cheek with orange than pink, lip liner with pink than orange. However, it was finding of normalized standings methods that the most popular image was to markup cheek with pink than orange in a confined change. It was observed the different of gazing pattern between men and female that it was more long period of residence-time distribution at mouth in men.

Consequently, it was showed that it was mood shift the ill-loved onto heavy using of orange color makeup. It was indicated that perceive a need to worry about what the other person likes in acceptance wide variety for recruiting makeup.

Keywords: sensory evaluation; point of gaze; makeup

1 INTRODUCTION

It is a needful skill of cosmetic makeup for job hunting recruitment as an activity in Japanese university student. Almost university carrier develop of employment bureau was to hold sometimes the lecture of makeup own face to best suited to job-hunting activities, especially women's student.

There was one of the communication tools for imagination of personality at a glance. Moreover many young women were taken following fashion of the day to their own opportunities for self-realization. Therefore there is a gap in perception between makeup owner and others. It was considered that the choosey of just looking behaviour on face was to depend on gender or aged. The purpose of this investigate was to evaluate sensory reactions the pair-test and rank test using simulation makeup face pictures in men and women, and also point of glance as gazing pattern.

2 METHODS

Subjects: There were 22 subjects (man: 6, female: 16) participated in this study. They were in college Age: 19–24 old, There were no medical history in vision and eyesight.

It was performed preliminary survey for 145 students, which colour of their makeup kit items. It was find out highest common factor of makeup kit conditions, one base make, two eye line colour, two lip colour, two face cheek colour and two hair style.

Experimental preparations: It were making the 16 pattern picture of simulation makeup (see Fig. 1) for recruitment fashion style by using one typical Japanese young female photo: short and long of hair style, brown and pink of eye shadow make, pink and orange of cheek, pink and orange of lip, on the same base make. It was simulated making pattern on using jKiwi(ver0.9.5) of virtual makeover and hairstyle application. All simulation makeup pictures were fine printed out as A3 size longitudinal board.

Experimental procedure: First it was performed paired test, "Which picture of makeup is suitability for going about getting a job as student?". It was showed pair pictures for random combination(see Fig. 2). Second it was tested the evaluation of the normalized-rank

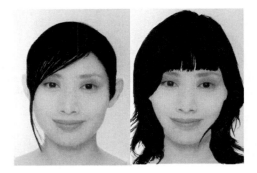

Figure 1. Example of virtual makeup pictures: two different hair styles on same makeup. There were 16 combination simulation pictures: two eye shadow, two face cheek, two lip colour and two hair styles.

Figure 2. Experimental setting in paired test for 16 pattern simulated makeup pictures.

362

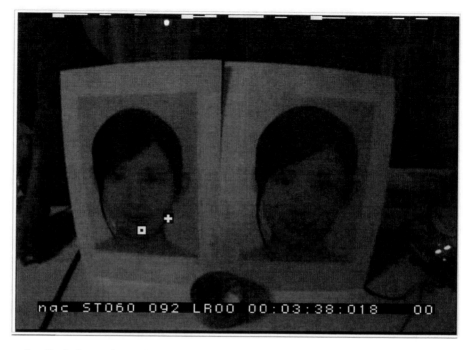

Figure 3. Typical experimental aspect point of gaze measurement in man two different pictures. Almost subject of men were to observe looking at around lip first.

approach. Third the subject was equipped point of gaze detection device (Eye mark recorder, NAC EMR-9, Japan) and compared two different pictures of simulation makeup pictures as a paired test in front of eyesight (see Fig. 3) within one minute, respectively. Moreover all subjects were tested rank order for all different pictures with just looking compared. This ranking test less than ten minute as soon as possible, line up each pictures on straight 3 m line by their evaluation values. After collecting of three experimental data, it was evaluated software to moving the point of gaze (EMR dFactoryVer.2.12b, NAC Image Technology Inc, Japan) and statistical software.

3 RESULTS

There was no significant different choosing the most suitable makeup by paired test 16 simulated make up pictures between man and women (Figs. 4 and 5). It was most popular make up as recruitment business activity that put hair style together with looking ears, natural light brawn eye shadow, lightly orange face cheek, pink colour lip. There was an acceptable response to select the choice between the two, paired tests that image of neat and clean pictures higher popularity. However the evaluation of the normalized-rank approach test was to choice the popular most of the same in paired test except for face pink cheek colour. Then one of the most unpopularity make up pattern was long hair with hide the ears, pink eye shadow, orange colour face cheek and orange colour lip. This resulet was decided to unsuited the job hunting activities in university student.

In gaze analysis experiment was performed to compare two different pictures within one minute. All subjects were not only gazing left or right picture, but also concerned left/right each other looking comparatively slow eye movement. It was observed cognition behaviour of seeing not only detection of spot of difference in makeup but also impression of seeing the picture. The results clearly revealed the association the visual environment considered as

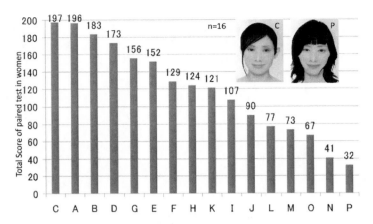

Figure 4. Result of total score of paired test in women between 16 pattern (A to K) of simulated makeup pictures. C: Truss hair with showing ear, brown eye make, orange face cheek, pink lip. D: Keep hair of her ear, pink eye make, orange cheek, orange lip.

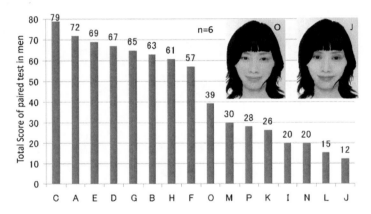

Figure 5. Result of total score of paired test in men between 16 pattern (A to K) of simulated makeup pictures. O: Keep hair of her ear, pink eye make, orange cheek, pink lip. J: Keep hair of her ear, brown eye make, pink cheek, orange lip.

an explanatory variable and the frequency of looking parts of makeup as an object variable. Especially residence time analysis of gaze movement, it was dominantly illustrated different gender behaviour for comparing the lips in men and eyes in women.

4 DISCUSSION

The major findings obtained from this study are summarised as follows: (1) participants were to choose the same picture for suite job hunting not dependent gender in paired test and the normalized-rank approach; (2) picture at the bottom was to differ between men and women in both paired test and the normalized-rank approach; (3) there were different gazing pattern in men and women. These results suggest that there was same cognitive image for job hunting activity in grooming and appearance. In addition to the gaze behaviour analysis, this study was not similar our previous study[1] in comparing images side-by-side, it had a few residence-time distribution at lips or eyes. However it was gender difference at the inadequacy makeup around the low order ranks simulation pictures. Thus, we conclude that decreased the gender effect in suitable job hunting was to have an acceptable sensory reactions in less fashionable makeup conditions.

ACKNOWLEDGEMENTS

We would like to offer our special thanks to Miss. Yuko Hara for helping this experiment.

REFERENCE

[1] Takeshi Sato, Macky Kato, Takayuki Watanabe, Hiroshi Yasuoka, Atsushi Sugahara, Analysis of human visual search performance in task of spot difference, IADIS International Conference Interfaces and Human Computer Interaction 2012; 331–334.

New Ergonomics Perspective – Yamamoto (Ed.)
© *2015 Taylor & Francis Group, London, ISBN 978-1-138-02751-0*

Body dimension measurements using a depth camera

Yueh-Ling Lin & Mao-Jiun Wang
Department of Industrial Engineering and Engineering Management, National Tsing Hua University, Hsinchu, Taiwan, R.O.C.

Ben Wang
Georgia Tech Manufacturing Institute, Georgia Institute of Technology, Atlanta, GA, USA

ABSTRACT: This paper proposes a virtual body measurement system by using a Kinect depth camera to extract human body features represented on the 3D body shape. The captured point clouds data are analyzed to extract feature curves and girth measurements of human body shape without pre-marking. The data are processed and aligned to obtain a smooth surface for 3D body shape construction. This method has been successfully tested on human subjects by obtaining the anthropometric data for fitting appropriately-sized clothes. Therefore, the newly developed method can achieve an efficient approach for anthropometric data collection and, in addition, the results can also be visualized through the virtual fitting of real people.

Keywords: Depth camera; Feature extraction; Body dimensions; Clothing making; Anthropometry

1 INTRODUCTION

For decades, the configuration of body shape and size were typically analyzed by both manual and image processing methods. Traditionally, the anthropometric data were collected by using the palpation technique to landmark the feature points on the human body. However, the manual measurement method is not only tedious and time consuming, it also contains human errors. In order to extract the features of a human body in an efficient way, current studies were based on expensive cameras or laser scanning systems, which captured rich information from the human body [1]. Presently, laser scanning systems have been used in several national anthropometric surveys, such as the Civilian American and European Surface Anthropometry Resource (CAESAR) Project for building the anthropometric database [2, 3]. Although laser scanning systems can help users of different body shapes fulfill the process of individual body measurement, the use of laser scanning systems are very costly and lack mobility. Additionally, it is only available to a small group of people, such as those who are subjects or experts in the laboratory through a dressing room. Because of the nature of the laser scanning systems, users must first scan their body in a specified dressing room set up to measure detailed body shapes. In order to simplify the entire process of anthropometric data collection, a low-cost system for body measurement is therefore needed.

Due to the varied body shapes of individuals, it is difficult to separate wide-ranging and diverse body shapes into appropriately-sized categories. As a result, there is a need for generating an efficient and effective way to measure the human body. In order to overcome the limitations of scanning in the dressing room with experimental gaps and clothes, McDonald [4] produced a light-structured 3D camera using his own web cam for the low cost toolkit of Kinect. An instantly popular depth-sensing camera, Kinect is an easy-to-use device containing both depth and color information. Compared with the conventional laser scanning system, Kinect is much cheaper and easier to use for general home-oriented applications. Although the complexity of human body shapes and varied clothing styles makes it hard to address both the suit and fit dilemmas. In addition, various body features such as neckline, princess line, and armhole are much too complex to be identified on the human body [5]. Thus, it is necessary to establish a system with body features that can be used for both anthropometric body feature recognition and clothing fashion applications to shorten the time of 3D cloth development to fit individual body sizes. Moreover, extracting the structural curves of human body shapes and integrating them to make the 3D virtual clothes, can help in providing a better fit performance to create well-fitted clothes.

This study proposes a low-cost approach to obtain body measurements. With the Kinect depth image, the captured point clouds data can be analyzed by using the feature extraction method. Then, the body features can be extracted on the concave and convex parts of the human body surface. For the extracted body feature points, the relative body measurements can be further collected by converting the point clouds data into coordinate values using the scaling relation method. Based on the obtained body measurements, the virtual clothes can be simulated in a proper size and fitted to the 3D body shape with realistic behavior. Furthermore, the visualization of the virtual try-on results can provide fitting recommendations for online shopping.

2 METHODS

2.1 Analyzing human depth data

Fig. 1 illustrates an overview of the proposed method. In this study, a Kinect depth camera was used to capture the 3D body shape. The human subject is asked to maintain a standard posture with their limbs straight and arms apart from the torso while the Kinect device acquires the depth views. With a standard camera calibration, two depth images were captured from the front and back views of a person to determine the 3D body shape. The depth images were formed by the point clouds data, including both depth and color information. The point clouds data from the front and back depth images acquire, respectively, both the front body shape of the person, as well as the back depth image. Consequently, the front and back point clouds data have a corresponding spatial relation to then generate an appropriate 3D body shape. Figs. 1(a) and (b) show the corresponding relationship between front and back point clouds data in forming the 3D body shape of a person. The point clouds data presenting the appearance of the human body are characterized by using a feature extraction algorithm to extract body features. For feature extraction, all points in the entire human body shape are separated into different layers to create a better representation of body shape curvature [6]. Since each sliced layer has the capability to clearly display the contour structure on each section, the slice keeps the original shape of the human body and shows the curve structure with contour points. The outermost contour points representing the complete shape of human body are characterized in the coordinate system. The feature extraction algorithm analyzes the coordinate values of the points to locate body feature. The detected feature points in Figs. 1(c) and (d) are further used for collecting the relevant feature dimensions.

2.2 Collecting body dimensions

This paper presents a method to measure the 3D body shape by obtaining measurements typically related to clothing sizes. First, the method extracts feature points from the 3D body shape. A feature extraction algorithm is applied to analyze the coordinate values of the point clouds data using a sequence theorem to locate the body features. The body features are then characterized by detecting the change between two connected sequences to define the turning points as the feature points. Based on this procedure, a series of feature points can be subsequently detected from the corresponding point clouds data. The correspondences between each feature point are measured by the Euclidean distances to define the sum of curvilinear distances as body measurements. Once the feature points are detected, the method can collect all of the relative body measurements as shown in Figs. 1(e) and (f) by converting the points data into coordinate values using the scaling relation approach.

For body measurements, feature dimensions such as the curvilinear dimensions and body girths can be obtained by calculating the arc lengths to find the body circumferences. The obtained circumferences include head, neck, shoulder, chest, waist, hip, hand, wrist, thigh, knee, and ankle girths. Also, the height and breadth dimensions can be obtained by

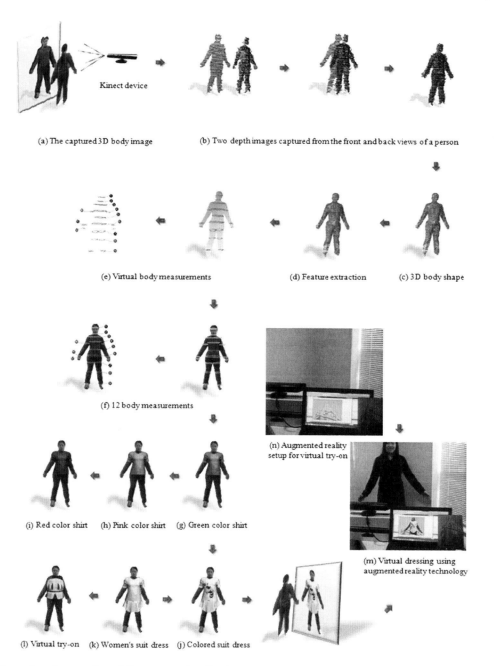

(a) The captured 3D body image

(b) Two depth images captured from the front and back views of a person

(e) Virtual body measurements

(d) Feature extraction

(c) 3D body shape

(f) 12 body measurements

(i) Red color shirt　(h) Pink color shirt　(g) Green color shirt

(n) Augmented reality setup for virtual try-on

(m) Virtual dressing using augmented reality technology

(l) Virtual try-on　(k) Women's suit dress　(j) Colored suit dress

Figure 1.　An overview of the proposed method.

connecting a line between two relevant feature points. Hence, personal body measurements can be measured conveniently for anthropometric data collection. The body measurements not only visualize human body features, but also help the individual to select a clothing size that fits better. The obtained measurements can be further used to track personal body shape and establish a database for clothing size recommendations. The recommended clothing size can be virtually dressed to the constructed 3D body shape with a virtual fitting. Thus, it can provide online apparel products with an easy solution for a virtual fit assistant.

2.3 *Applications*

Since the reconstruction of an individual body may not be animated interactively to perfectly match the current appearance of an individual, the process acquires an active character to display on-screen with a personal appearance. Instead of adapting a virtual character constructed by scanning the real person, the proposed method made it possible to create an interactive system with the real human visualized simultaneously by Augmented Reality (AR) technology. By improving the sensing quality for better reconstruction, Kinect Fusion technology is applied for Virtual Reality (VR) application [7]. The proposed method was programmed in Visual C#. With work applied on Kinect Fusion technology, a detailed model can be directly reconstructed in real time with a Kinect device. The proposed method can reconstruct a posing model for constant capture and make a segmentation of the captured person with a changeable background. Figs. 1(m) and (n) illustrate a virtual fitting that provides realistic dressing on a 3D image. With Kinect, the proposed method can produce a real-time human depth image and track the dynamic expression of the human body with fitting applications in virtual reality. Thus, the virtual try-on scenario can be tested directly to prove the available technology being achieved with the proposed method.

Additionally, the proposed virtual fitting system can be visualized with realistic cloth simulations associated with colors, fabric characteristics, and various clothing styles, along with segmenting the background into another scene at the same time. For clothing design, the outlines of the clothes patterns are edited as a set of 2D polygons using CAD software. The patterns are digitized and triangulated to enable the construction of the clothes. Based on the obtained anthropometric data, the clothing patterns are manufactured according to personal body size for virtual try-on. The proposed method provides a way to perform the clothing behavior on the 3D body shape interactively by using Kinect in augmented reality. As seen in Figs. 1(g) to (l), the clothes can be easily modified to produce the size scaling or color changing of the selected clothes. During the fitting process, virtual clothes can interact with a real human body simultaneously by displaying a realistic clothing appearance for virtual try-on realization. Therefore, the proposed method can help in decreasing tests of clothing samples, while representing a realistic look for the clothing end products.

3 CONCLUSION

The purpose of this study is to develop a body measurement system based on the 3D images of a person taken from both front and back views. The proposed method can go from Kinect data to body measurement utilizing only an inexpensive depth camera. This paper uses depth data for acquiring specific body measurements in practical clothing applications. The method has been tested on human subjects and the anthropometric data can be successfully obtained to provide recommendations for the clothing size that best fits the human body shape. The paper demonstrates application examples of a vision-based body measurement system, and describes a method for using a depth camera to measure a real human body. With a Kinect, the proposed method can generate a realistic replica of one individual for an interactive application of a virtual fitting system with a scenery background work-in-practice. Moreover, the work can also be used for personal styling design and proactive product applications.

REFERENCES

[1] Wang MJ, Wu WY, Lin KC, Yang SN, and Lu JM. Automated anthropometric data collection from three-dimensional digital human models, *International Journal of Advanced Manufacturing Technology*, 2007, vol. 32, pp. 109–115.

[2] Robinette KM, Daanen HA, and Paquet E. The CAESAR Project: a 3D Surface Anthropometry Survey, in *Proceedings of the 2nd International Conference on 3D Digital Imaging and Modeling*, Ottawa, 1999.

[3] Robinette KM and Daanen HA. Precision of the CAESAR Scan-extracted Measurements, *Applied Ergonomics*, 2006, vol. 37, pp. 259–265.

[4] Kean S, Hall J, and Perry P. Meet the Kinect: an Introduction to Programming Natural User Interfaces. *APress Publisher*, 2012.

[5] Huang H. Development of 2D Block Patterns from Fit Feature-aligned Flattenable 3D Garments, *PhD thesis*, Institute of Textiles and Clothing, The Hong Kong Polytechnic University, 2011.

[6] Dacorogna B. Introduction to the Calculus of Variations, published by Imperial College Press, distributed by Singapore: World Scientific, 2005, chapter 2, pp. 45–61.

[7] Izadi S, Newcombe R, Kim D, Hilliges O, Molyneaux D, Hodges S, Kohli P, Davison A, and Fitzgibbon A. *Kinect Fusion: Real-time Dynamic 3D Surface Reconstruction and Interaction*. ACM SIGGRAPH Publications, 2011.

New Ergonomics Perspective – Yamamoto (Ed.)
© 2015 Taylor & Francis Group, London, ISBN 978-1-138-02751-0

Human factors analysis of hydraulic leg press machine

Shu-Zon Lou
School of Occupational Therapy, Chung Shan Medical University, Taichung, Taiwan

Yu-Chi Chen
School of Sports Medicine, China Medical University, Taichung, Taiwan

Peng-Cheng Sung & Chen-Lung Lee
*Department of Industrial Engineering and Management, Chaoyang University
of Technology, Taiwan*

ABSTRACT: The aim of this study is to investigate effects of two types of leg press machines for young and elder people using motion analysis system. Ten young healthy men and women, and ten elder with an average age of are recruited. As many as 15 physically healthy male subjects were studied. The Vicon Motion System (Vicon 460, Oxford, UK) with six 120 Hz cameras and four uniaxial load cells installed on foot-plate (Transducer Techniques, MLP200) with 1080 Hz was used to measure relative joint positions and push force. The peak push force is significantly affected by types of leg press machine and push speeds, and the influence on young men and young women are higher than that on the elder. Result: There was not significant different in maximal flexion angles of knee and hip among two types of leg press machine, three types of push speed and three types of age group were not significant difference. There is significant difference of peak push forces for young men, women and the elder between Fast Group (FT) and Slow Group (SL) but not between Comfortable Group (CF) and SL group.

Keywords: hydraulic leg press machine; motion analysis

1 INTRODUCTION

The leg press machine is built for training the lower extremity by simulating the crouch or squats. Compared with these two movement, the leg press machine can be more efficiency in training muscle by adjustable loads control, but the personal trainer beside is needed for preventing the uncorrected posture or unsuitable external force that can damage the musculoskeletal system (Steinkamp et al. 1993, Bosco et al. 1999). Even though the resistant system progresses soon, the leg machine is designed for young healthy people. Netreba noted (Netreba et al. 2013) that responses of m. vastus lateralis to 8-week resistive training of various types at leg press machine in 30 male young subjects, the training-related increase of cross-sectional area. For the elder people, easy use, low force and high resistant force while occurring unsteadily, are main concerned. The hydraulic leg press machine seems to be suitable for these considerations (Ballor et al. 1987). However, there are many resources in patent of hydraulic leg machine (Davenport 1984, Shi 1992, Giannelli et al. 2003) but little in the research. The aim of this study is to investigate effects of two types of leg press machines on elder people using motion analysis system.

2 METHOD

2.1 Subject and experimental protocol

Ten young healthy men and women, and ten elder with an average age of are recruited with an average age of 24.2 (2.1), 23.6 (2.3) and 69.3 (3.2) year, an average height of 173.9 (2.72), 163.1 (5.42) and 160.4 (8.75) cm, and an average weight of 66.8 (7.39), 53.3 (5.5) and 56.8 (9.4) kg were studied. None had ever suffered from lower extremity injuries and disorders.

The Vicon Motion System (Vicon 460, Oxford, UK) with six 120 Hz cameras and four uniaxial load cells (Transducer Techniques, MLP200) installed on foot-plate with 1080 Hz was used to measure relative joint positions and push force.

A set of 19 reflective markers was place on selected anatomic landmarks bilaterally on the subject. The selected anatomic landmarks were intended to simulate the rigid body assumption for pelvic (sacrum, right and left anterior superior iliac spine), thigh (middle of the thigh, medial and lateral epicondyles of the knee), shank (middle of shank, medial and lateral malleolus of the ankle) and foot (heel, second metatarsal bone and medial and lateral malleolus of the ankle).

Each subject was asked to perform two different resistant types of leg machine, Weight Stack (WS) and Hydraulic (HD) (fig. 1), and three types of push speed, Comfortable (CF), Fast (FT) and Slow (SL) (as slow as possible, lasting 4 seconds). The resistant force was adjusted to the same at initial push while knee was in full extension.

2.2 Data reduction and analysis

Laboratory-developed kinematics and kinetics software were used to calculate the joint angles and push forces. A three-segment model, i.e. foot, shank and thigh, was employed in the analysis. Each segment was assumed to be a rigid body. Six CCD cameras were used to record 3-D position of the markers. The joint ankles were calculated using Euler's method with a y-x-z rotational sequence based on the attached markers (Haug 1989, Winter 1990). A strain gauge force plate was used to measure vertical push force. A Generalized Cross-Validation Spline Smoothing (GCVSPL) routine at a cut-off frequency of 6 Hz was used for data smoothing (Woltring 1986). Joint angles and push forces during leg press were calculated and then used for analysis.

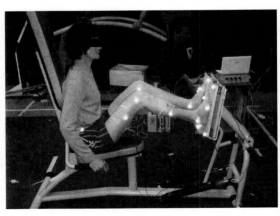
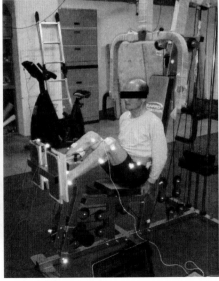

Figure 1. Two different resistant types of leg machine, Weight Stack (WS) and Hydraulic (HD).

The ANOVA with repeated measure was conducted to evaluate the differences between machine type, push speed and age group with p < 0.05 as statistical significance.

3 RESULTS

3.1 Joint angles

- There was not significant different in maximal flexion angles of knee and hip among two types of leg press machine, three types of push speed and three types of age group were not significant difference (Fig. 1 and Fig. 2). The peak angle of the hip and knee were approximate 100 and 90 degrees respectively.

3.2 Push forces

There is significant difference of peak push forces for young men, women and the elder between FT and SL but not between CF and SL (Fig. 3). The ratio of peak forces of FT to SL at hydraulic leg press machine for young men, women and the elder is 2.2, 2.4 and 1.5

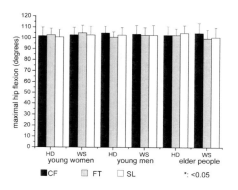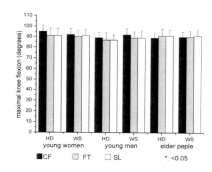

Figure 2. Mean and SD of maximal joint angles of Knee (left) and hip (right) at two types of leg press machine and three types of push speed.

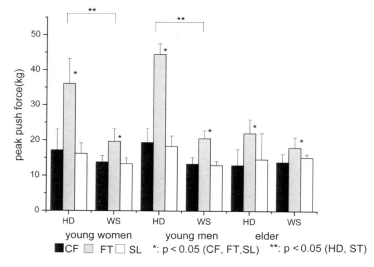

Figure 3. Mean and SD of peak push forces at two types of leg press machine and three types of push speed.

respectively, and that at weight stack leg press machine is 1.47, 1.59 and 1.18 (Fig. 3). There is significant difference peak push force between HD and WS for young men and women, but not for elderly people (Fig. 3).

4 DISCUSSIONS AND CONCLUSIONS

Even though these two leg machine provide the same joint movement of the lower extremity, the push forces are significantly different. The peak push force is significantly affected by types of leg press machine, push speeds and age group. The push force in HD is higher than WS, especially for FT group. For hydraulic system, it can provide large resistance at high speed. On the other hand it also decreased resistance at slow motion. This phenomenon is suit for elderly people. They need lower push force as they move slowly and they can also get larger force to avoid slipping or uncontrolled movement after they push fast. However, the hydraulic leg machine just can provide concentric muscle training not for eccentric muscle training. But if concentric muscle training is the issue for elderly people, the hydraulic leg machine is a good exerciser.

REFERENCES

[1] Steinkamp LA, Biomechanical considerations in patellofemoral joint rehabilitation. *The American journal of sports medicine* 1993;21: 438–444.
[2] Bosco C, Adaptive respsonses of human skeletal muscle to vibration exposure. *CLINICAL PHYSIOLOGY-OXFORD-* 1999;19: 183–187.
[3] Netreba A, Responses of knee extensor muscles to leg press training of various types in human. *Ross Fiziol Zh Im I M Sechenova* 2013;99(3): 406–416.
[4] Ballor DL, Metabolic responses during hydraulic resistance exercise. *Med Sci Sports Exerc* 1987;19: 363–367.
[5] Davenport DL, Hydraulic exercise device, Google Patents; 1984.
[6] Shi J, Hydraulic resistance type stationary rowing unit, Google Patents; 1992.
[7] Giannelli R, Leg press machine, US Patent 20,030,158,018. 2003.
[8] Haug EJ, Computer Aided Kinematics and Dynamics of Mechanical Systems Volume I: Basic Methods. Massachusetts, Allyn and Bacon; 1989.
[9] Woltring HJ, A FORTRAN package for generalized, cross-validatory spline smoothing and differentiation. *Advances in Engineering Software* 1986;8: 104–113.
[10] Winter DA, Biomechanics and motor control of human movement. New York, John Wiley and Sons; 1990.

New Ergonomics Perspective – Yamamoto (Ed.)
© 2015 Taylor & Francis Group, London, ISBN 978-1-138-02751-0

Study on vibration patterns using vibrating computer mouse

Hiroyasu Mitani & Daiji Kobayashi
Chitose Institute of Science and Technology, Chitose, Hokkaido, Japan

ABSTRACT: In this study, the appropriate design of vibration rhythm for qualitative and quantitative information was considered through experiments. In the experiments, the following two types of design method for the vibration patterns were evaluated from the viewpoint of easiness to identify the quantity: regular-type which was the patterns extended in accordance with the quantity or the level, and rhythmical-type which were made from some musical rhythms. In the experiments, we used the vibration mouse which was a custom computer mouse including a vibrating motor with eccentric mass for presenting the vibration patterns. The participants grasped the vibration mouse and felt the vibration patterns and answer the corresponded number to the vibration pattern presented after they learned the vibration pattern's meaning. The participants were twenty-five young individuals ranged from 20 to 22 years of age. As the results, we concluded that the regular-type was easier to understand than the rhythmical-type. However, the design method for quantitative vibration patterns such as the regular-type were not useful for implementing the vibration patterns in tactile interfaces because the vibration duration could be longer in accordance with its quantity presenting. Therefore, we should consider the alternative design method for quantitative vibration rhythms such as the rhythmical-type.

Keywords: vibration pattern; tactile information; haptic and tactile interaction

1 INTRODUCTION

Almost of mobile devices include an actuator oscillating the mobile device. The user of the mobile devices percepts the oscillation as vibration. There are some mobile devices presenting rhythmical vibration. Although the perception of vibration stimuli has studied from medical and physiological viewpoints [1–2], the appropriate design of vibration pattern corresponding to the information from the mobile device to the user has not been studied well. ISO 9241-910 [3], framework for tactile and haptic interaction, mentions that it is important to consider the age of potential users of tactile/haptic devices, since there is a considerable decline in haptic sensitivity with age. From this viewpoint, we have explored the way of designing rhythmical vibration which was named "vibration rhythm" from the view of memorability [4]. As the result, the characteristics of memorable vibration patterns for the young and elderly people were cleared to some extent. However, the use of the vibration rhythm has not considered well. As far as the results of our research about vibration rhythms, it seems to be useful to use the vibration rhythm for qualitative information from mobile devices. On the other hand, the validity of the vibration rhythms representing quantitative information has not been revealed. Thus, regardless of the vibration rhythms, the way of designing the vibration patterns corresponding to the quantitative information should be required to explore. Therefore, we tried to investigate the appropriate design for the vibration patterns presenting the quantitative information in this experimental study.

2 METHOD

2.1 *Apparatus*

In order to research the vibration patterns represented by tactile interfaces, we made and used a custom computer mouse. We called such mouse-type tactile interface "vibration mouse". The vibration motor in the vibration mouse rotated within a range of 0.3–0.7V and oscillate the vibration mouse. The power voltage for activating the vibration motor was controlled using a high-precision analog I/O terminal (CONTEC AIO-160802 AY-USB) and a personal computer (DELL XPS 420) running Windows 7 Professional Japanese edition. In other words, the voltage applied to the vibration motor was controlled by the I/O terminal with our custom software. The wave form of the amplitude of vibration on the top of the vibration mouse was not a sine curve but very rough. The resonant frequency of the amplitude of vibration was ranging from 74 to 116 Hz in accordance with the voltage applied to the vibration motor in the vibration mouse. In addition, the vibration mouse functioned also as the computer mouse with two buttons and a scroll wheel. The vibration velocity was 2.3 m/s which is the producible maximum vibration velocity by the vibration mouse.

2.2 *Designing vibration patterns for presenting quantitative information*

The vibration patterns was composed by vibration durations and gaps. Our previous research has determined these thresholds through experiments using the vibration mouse and then the minimum vibration duration included in the vibration pattern was determined 50 milliseconds and we estimated the threshold of the gap was 20 milliseconds in round figures. Based on these knowledge, we designed the experimental vibration patterns. Therefore, the minimum duration of vibration in the all vibration patterns was over 100 milliseconds.

The vibration patterns for presenting quantitative information we made were two types. One was a "regular-type" as shown in Fig. 2. This vibration pattern's duration were extended in accordance with the quantity and the other was "rhythmical-type." The rhythmical-type

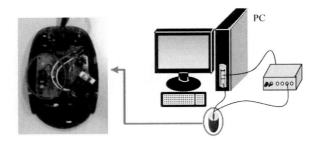

Figure 1.   Vibration mouse and the experimental components for controlling the vibration mouse.

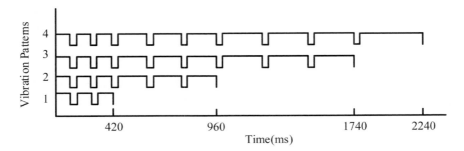

Figure 2.   Vibration patterns of regular-type.

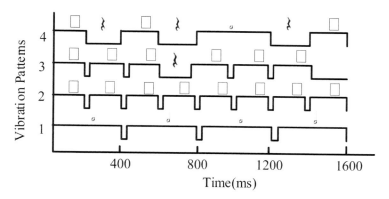

Figure 3. Vibration patterns of rhythmical-type.

was the aforementioned vibration rhythm; therefore the rhythmical-type was composed of second and forth musical notes and rests and the duration of fourth notes was 200 milliseconds as shown in Fig. 3.

2.3 Experimental design and procedure

For comparing the easiness to identify the number which correspond to the vibration pattern between the two types of vibration pattern, the experiment was within-subjects design. The measure for the easiness of identification was the number of correct answers by the participants. The experimental procedure was the followings: First, the participants intended to learn the number corresponding the represented vibration pattern which was regular-type. The participants grasped the vibration mouse and clicked the button on the mouse and then a vibration pattern was presented by the vibration mouse and the correspondent number was indicated on the PC's display. After the participants finished learning, they tried to answer the number corresponding to the randomly presented vibration pattern. This trial were repeated eight times. This set including eight trials were repeated three times. After finishing all sets for regular-type vibration patterns, the participants tried the next sets for rhythmical-type in the same way.

The participants were nineteen male and six female students ranged from 20 to 22 years of age (mean = 21.6, SD = 0.6).

3 RESULT

The number of collect answers by each type of vibration patterns were compared for each types of vibration pattern as shown in Fig. 4. The error bars as described in Fig. 4 indicates two-sided 95% confidence intervals. As the result, the number of collect answers for Regular-type was more than the case of Rhythmical-types significantly in the first set. However, the statistically-significant difference was not observed in the second and the third trials. Therefore, it seemed that Regular-type was more memorable than Rhythmical-types. On the other hand, the 5 of 10 participants who had skills of playing music answered that Rhythmical-types was easy to identify the means of the vibration pattern rather than Regular-type. In this regard, the easiness of identifying vibration patterns could be effected by the skills of music. Thus, it seemed that the phonological process effects on the cognitive process for processing information from vibration perception. These results suggested that Regular-types was appropriate for identifying such numbers from 1 to 4; however, the duration of the vibration patterns for quantitative information is longer in accordance with the quantity. Therefore, the alternative patterns such as Rhythmical-types should be designed.

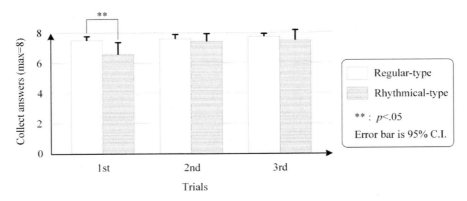

Figure 4.   Comparing the number of collect answers between the each cases of using Regular-type and Rhythmical-type.

4   CONCLUSION

In this study, we researched the way of designing the accessible vibration rhythms through experiments using the mouse-type tactile interface. We conclude the vibration patterns for quantitative information was easy to understand in case the duration of the patterns extended in accordance with the quantity or the level. However, the duration of vibration patterns for quantitative information is longer in accordance with the quantity or the level. Therefore, we must consider design method of vibration rhythm to convey the quantitative information.

REFERENCES

[1]   Wiles PG, Pearce SM, Rice PJS, Mitchell JMO. Vibration perception threshold: influence of age, height, sex, and smoking, and calculation of accurate centile values. Diabetic Medicine 1991; **8**:157–61.
[2]   Kaas AL, Stoeckel MC, Gorbrl R. The neural bases of haptic working memory. In: Grunwald M, editor. Human Haptic Perception—Basics and Applications, Basel: Birkhäuser; 2008, p. 113–29.
[3]   ISO 9241-910: Framework for Tactile and Haptic Interaction; 2011.
[4]   Daiji K. Study on perception of vibration rhythms. In: Sakae Y, editor. HIMI 2014 Part I, LNCS 8521, Springer Int. Pub. Switzerland; 2014, p. 208–16.

New Ergonomics Perspective – Yamamoto (Ed.)
© 2015 Taylor & Francis Group, London, ISBN 978-1-138-02751-0

Ergonomic floor limitation for a workplace nursery placed in outside buildings of workplaces

Sang-Hun Byun
Department of Safety Engineering, Graduate School in Chungbuk National University, Cheongju, Chungbuk, Korea

Jae-Hee Park
Department of Civil Safety Environment Engineering, Hankyong National University, Anseong, Kyungki, Korea

Hyeon-Kyo Lim
Department of Safety Engineering, Chungbuk National University, Cheongju, Chungbuk, Korea

ABSTRACT: In recent years, the number of workplace nurseries has been increased with the couples working together for their living so that mitigation of regulations on workplace nursery became a hot issue in Korea. Some workplace nurseries may be placed in buildings which were built for exclusive use whereas some may in ordinary buildings for general use. In the latter case, floor limitation and secure evacuation route is quite important for ensuring safety of infants in case of fires or disasters.

This study was carried out as a part of study on mitigation of regulation to depict problems related with safety and to develop a floor limitation from the viewpoint of ergonomists when a workplace nursery is placed in a high level building located in outdoor area of workplaces.

International guidelines and criteria on general building evacuation time, behavioral and response characteristics of infants, and workplace nursery placement standards were reviewed in order to comprehend the problems that may arise from placing a nursery on high level. According to the results, placement standards and floor limitations were different country by country, but in general, the floor limit that nursery could be placed were the first or the second floor height of the building.

Opinions of teachers who led evacuation practices of infants were collected through interviews. The result brought out that moving speed of a child under 6 years of age is said to be one layer or less per minute depending on the literature and interviews. Therefore, considering the speed of flame spread, around 5 minutes seemed to be a practical limit for evacuation.

Besides, debated on air quality blown from the ground level was appended from the view point of ergonomists, to which the infants may be exposed if they play on the top floor. As a consequence, it was concluded that the 5th floor would be a practical limitation to the floor for nursery placement.

Keywords: Workplace Nursery; Evacuation Time; Floor Limitation; Safety

1 INTRODUCTION

In June of 2013, The Korean Ministry of Health & Welfare declared counter-plans for revitalization of workplace nursery. The major contents were mainly 1) mitigation of regulation on the ratio of floor to the building where the nursery is located, 2) placement permission of workplace nursery on the 1st to 5th floor of buildings in general use, and 3) affirmative adjustment of upper limit for support fund.

When a workplace nursery is located in a building, sometimes it may be one for exclusive use, but sometimes not. In the latter cases, secure evacuation is one of the most important factors to be considered for the safety of toddlers so that usually floor limitation is set up for the requisite to nursery placements. However, in the counterplans declared, the intention to mitigate the floor limitation was included also.

Therefore, this study was carried out as a part of study on mitigation of regulation to depict the problems related with safety and to develop a floor limitation criterion from the viewpoint of ergonomists when a workplace nursery is placed in a high-storied building located in outside area of workplaces.

2 METHOD

To solve the problem confronted, The problems assigned to the project team were diverse and approaches were sophisticated though, only ergonomic approach will be introduced in this study.

The team members got together to share appropriate information and elicit practical solutions, and visited several workplace nurseries for confirmation of safety and ergonomic problems on one hand, and collection of teachers from the site on the other hand. Literal survey on domestic regulations and overseas references were conducted also. The whole project continued more than three months.

3 EVACUATION TIME

3.1 *General standards*

There have been a lot of researchers to study the parameters of evacuation and many of them were carried out for buildings in normal use as well as emergency conditions. In practical use, the Korean government apply "Six minute Rule" as a national standard for public transportation systems which means all the users could complete their evacuation in six minutes. It has the implication that most people including passengers should be get away from the site within 4 minutes, and get out the exit to be free from any smoke or toxic gas. Therefore, public infrastructure such as airports, subway stations, bus terminals, and so should be designed to fulfill the criterion of six minutes.

This time standard is based on movement speed shown in Table 1. Unfortunately, however, all the numeric figures were for an adult.

3.2 *Behavior characteristics of infants in emergency situations*

When a nursery is located on a high level of a building, a matter of great importance in the safety aspect is evacuation time of infants and toddlers. However, the number that studied safety of infants was few.

Referring to a study (Hong, 2008) which made a simulation to observe emergency behavior patterns of children, it would impossible for a child under age 9 to escape by oneself without assistance a nursing teacher. When faced an emergency situation, most infants tried to hide themselves in a corner of the room or cry with halting where they were. Major behavior characteristics of infants can be summarized as shown in Table 2.

Table 1. Movement speed of passengers.

Evacuation route	Movement speed
Horizontal route (platform, waiting room, passage)	60 m/min
Vertical route (stairs, stopped moving stairway)	15 m/min
On running moving stairway	36 m/min

3.3 *Movement speed of infants*

Movement speed of an infant can vary with its age, type and familiarity of stairway handle [1], and also be affected by clothing [2], [3]. According to the opinions of nursing teachers who led evacuation practices of toddlers, toddlers usually required 1 minute to 1 minute and a half to move one story in the dark. Average movement speed of a baby can be shown as Table 3 [4].

3.4 *Comparison of national standards on nursery placement*

National standards on nursery placement were compared with special reference to floor limitation in the aspect of safety and evacuation time. No consistent tendency was obtained, but most standards described 1 or 2 floors as a practical floor limitation in its height [5]–[11]. Compared results of standards on nursery height limitation can be shown as Table 4.

As a consequence, with considering the speed of flame spread and the speed of movement speed of infants, it might be concluded that the 5th floor would be a practical floor limitation to evacuation, and nursery placement.

Table 2. Behavioral response characteristics of a baby in an emergency situation.

Years of age	Behavior characteristics
Under 4	When faced an emergency situation, usually halt and cry
5~6	Intend to escape in early stages of a fire, soon give up and sit crouched in a corner with time
7~8	Eagerly try to escape, but only repeated trials due to insufficient skills
Over 9	Escape while groping with one's hands

Table 3. Average movement speed of a baby.

Years of age	Horizontal (m/sec)	Vertical (m/sec)	Reference
0–2	0.60	–	
3–6	0.84	Depends on the stairway type 0.13, 0.38, 0.58*	Larusdottir and Dederichs (2012)
3–4	0.77	–	
4–5	0.85	0.66	Kholshchevnikov et al. (2012)
5–7	0.86	0.73	

Table 4. Standards of nursery placement.

Country	Regulations on the floor limitation of nursery placement
Korea	General nursery—1st floor, workplace nursery—1~5 floor
United Kingdom	Ground floor, if not, bottom-most (Building Act) At highest, 2nd floor (in case of 20 ft or higher, special plan required
United States	In most states, 2nd floor at highest (Building Code)
Canada	No limitations in some states Ground floor for infant under 24 months of age, and at highest 2nd floor in most states (Building Code)
Hong Kong	12 m for infants under 24 years of age, 24 m for 3 years of age Ground floor in exceptional cases
Japan	No limitation

Until now, the Korean regulation allowed nursery playgrounds to be located in outdoor places in Korea. However, it is also be taken into consideration to mitigate nursery playgrounds, so that a playground may be located anywhere—outdoor place, indoor place, and alternative play field—freely.

It should not be allowed to place a playground in the roofs higher than the fifth floor from the viewpoint of evacuation time. Nevertheless, another problem may exists—the air the infants will face.

4.1   *Influence of wind speed*

Wind chill temperature of person is strongly affected by wind speed which varies with altitude. Wind speed can be estimated by a formula as follows [12];

$$U(z) = U(z_{ref}) \times \left(\frac{z}{z_{ref}} \right) \alpha \tag{1}$$

where α is power law exponent depending on the density of buildings in a certain area, and z is altitude of a certain point, and z_{ref} is altitude of measurement point, and $U(z)$ is wind speed at z, and $U(z_{ref})$ is wind speed at z_{ref}.

For instance, in an urban area, if wind speed is 4 m/sec at 1 meter altitude, it can change into more than 18 m/sec at 100 meter altitude place with power law exponent $\alpha = 0.33$. Wind speed 4 m/sec means light wind to shake leaves and twigs consistently whereas wind speed 18 m/sec means a fresh gale that breaks twigs, interferes normal walking, and makes high waves and clouds of spray on the sea. As a consequence, even a light wind on the ground may change into a strong wind on the roof of skyscrapers.

4.2   *Influence of pollutants*

If playground is located in the roofs of buildings, infants will be exposed to the air directly. The air in the urban area can be easily polluted by smog and pollutants.

The temperature distribution over time band can be diagrammed schematically as Fig. 1. During the daytime, the air above the ground in the urban area is heated up and moves upward. Thus, air pollutants also will be moved up. It means that, if infants play in the playground located in the roof of buildings, they will be exposed to air pollutants blown from ground level.

In recent years, the heights of buildings newly established in an urban area go higher and higher, so that there are many buildings higher than 30- or 40-stories. As a consequence, if

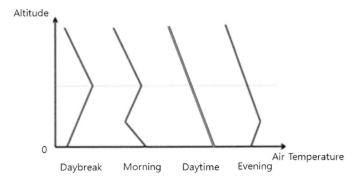

Figure 1.   The temperature distribution over time band.

the playground is located on the place higher than 20~30 stories, safety of infants will not be assured due to strong wind, and air pollutants.

5 CONCLUSION

Until now, the Korean regulation allowed nursery playgrounds to be located in outdoor places in Korea. However, it is also be taken into consideration to mitigate nursery playgrounds, so that a playground may be located anywhere.

This study was carried out to think over safety problems that can arise when a workplace nursery is allowed in high-story buildings, and to give a technical advice to regulation for floor limitation.

By reviewing evacuation time and effect of high altitude on play environment, and collecting opinions of nursing teachers who led evacuation practices of infants, it was concluded that five minutes and fifth floor would be practical limitation of a nursery. Yet, it would not be sufficient criteria for infant safety. When the number of infants is much greater than toddlers, it will take more time to make them evacuate altogether. Thus, more careful application will be required case by case, even with standard criteria.

REFERENCES

[1] Larusdottir, A.R. and Dederichs, A., Evacuation of children: Movement on Stairs and on Horizontal plane, *Fire technology,* 48(1): 43–53., 2012.
[2] Kholshchevnikov, V.V., Samoshin, D.A., Parfenenko, A.P., Study of children evacuation from preschool education institutions, *Fire and Materials, 36*(6–6): 349–366., 2012.
[3] Kholshchevnikov, V.V., Samoshin, D.A., Parfyenenko, A.P., Belosokhov, I.P., "Pre-school and school children building evacuation," *Proceedings of Fourth International Symposium on Human Behaviour in Fire 2009*, Robison College, Cambridge, UK., 2009.
[4] Taciuc, A. and Dederichs, S.A., *Determining Self-Preservation Capability in Pre-School Children*, Fire Protection Research Foundation, Denmark., 2013.
[5] Lee, Y.J., Hyun, S.H., Park, J.S., Lim, J.H., *Study on Improvement Plans of Workplace Nursery Placement Standards*, Ministry of Health & Welfare, Korea., 2010.
[6] CFPA, Eurpean Guideline *CFPA-E No.19*, CFPAEurope, 2009.
[7] Beach, J. and Friendly, M., *Childcare centre physical environment*, Childcare Resource and Reasearch Unit, Canada, 2005.
[8] US Department of Health and Human services, *Trends in childcare center licencing regulations and policies for 2011*, 2011.
[9] Ofsted, *Requirements for the Childcare Register: childcare providers on nondomestic or domestic premises*, 2013.
[10] Building Department, *Practice Note for Authorized Persons and Registered Structural Engineers, Licensing of Child Care Centres, Kindergartens and Restaurants*, 2009.
[11] Hong Kong Fire Service Department, *Fire Safety Requirements for Child Care Centre*.
[12] Han, S.E., Park, J.S., "A Study on the Wind-induced Pressure Characteristic for the Shell Structures using Computational Fluid Dynamics," *Journal of AIK*, Vol. 24, No. 4, pp. 51–58, 2008.

New Ergonomics Perspective – Yamamoto (Ed.)
© 2015 Taylor & Francis Group, London, ISBN 978-1-138-02751-0

Author index